Integrated Solid Waste Management:
A Lifecycle Inventory

Frontispiece Lifecycle inventory of an integrated waste management system.

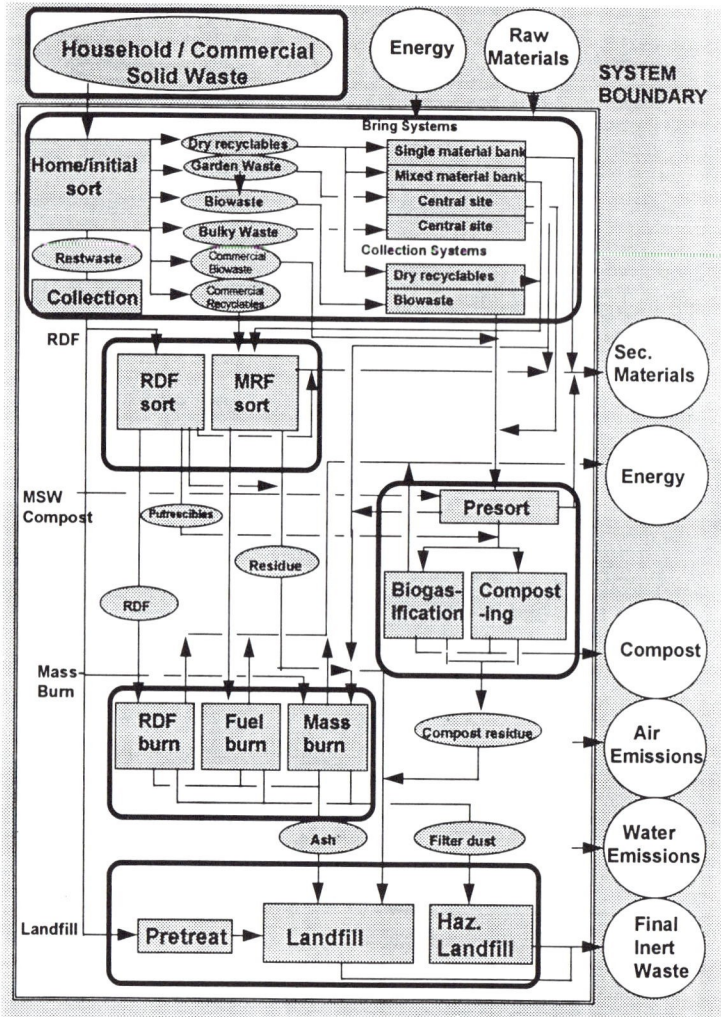

Integrated Solid Waste Management: A Lifecycle Inventory

Dr. P. White
Senior Environmental Scientist
Procter & Gamble Ltd. (UK)

Dr. M. Franke
Senior Environmental Scientist
Procter & Gamble GmbH (Germany)

and

P. Hindle
Director
Environmental Quality — Europe
N.V. Procter & Gamble (Belgium)

BLACKIE ACADEMIC & PROFESSIONAL
An Imprint of Chapman & Hall
London · Glasgow · Weinheim · New York · Tokyo · Melbourne · Madras

Published by
Blackie Academic & Professional, an imprint of Chapman & Hall,
Wester Cleddens Road, Bishopbriggs, Glasgow G64 2NZ

Chapman & Hall, 2-6 Boundary Row, London SE1 8HN, UK

Blackie Academic & Professional, Wester Cleddens Road, Bishopbriggs, Glasgow G64 2NZ, UK

Chapman & Hall GmbH, Pappelallee 3, 69469 Weinheim, Germany

Chapman & Hall USA, 115 Fifth Avenue, Fourth Floor, New York, NY 10003, USA

Chapman & Hall Japan, ITP-Japan, Kyowa Building, 3F, 2-2-1 Hirakawacho, Chiyoda-ku, Tokyo 102, Japan

DA Book (Aust.) Pty Ltd, 648 Whitehorse Road, Mitcham 3132, Victoria, Australia

Chapman & Hall India, R. Seshadri, 32 Second Main Road, CIT East, Madras 600 035, India

First edition 1995
Reprinted 1995, 1996

© 1995 Chapman & Hall

Typeset in 10/12pt Times by Colset Pte Ltd, Singapore
Printed in Great Britain by T.J. Press (Padstow) Ltd., Padstow, Cornwall

ISBN 0 7514 0046 7

A catalogue record for this book is available from the British Library

Library of Congress Catalog Card Number: 94-72266

∞ Printed on permanent acid-free text paper, manufactured in accordance with ANSI/NISO Z39.48-1992 (Permanence of Paper)

Preface

Life is often considered to be a journey. The lifecycle of waste can similarly be considered to be a journey from the cradle (when an item becomes valueless and, usually, is placed in the dustbin) to the grave (when value is restored by creating usable material or energy; or the waste is transformed into emissions to water or air, or into inert material placed in a landfill). This preface provides a route map for the journey the reader of this book will undertake.

Who?

Who are the intended readers of this book?

Waste managers (whether in public service or private companies) will find a holistic approach for improving the environmental quality and the economic cost of managing waste. The book contains general principles based on cutting edge experience being developed across Europe. Detailed data and a computer model will enable operations managers to develop data-based improvements to their systems.

Producers of waste will be better able to understand how their actions can influence the operation of environmentally improved waste management systems.

Designers of products and packages will be better able to understand how their design criteria can improve the compatibility of their product or package with developing, environmentally improved waste management systems.

Waste data specialists (whether in laboratories, consultancies or environmental managers of waste facilities) will see how the scope, quantity and quality of their data can be improved to help their colleagues design more effective waste management systems.

Policy makers will gain an overview of the considerations that must be given to the development of environmentally improved waste management systems that are considered affordable by voters and businesses.

Regulators will see the impact of existing regulations on the development of waste management systems and the future challenges they face to help the rapid development of improved systems.

Politicians (whether European, national or local) will see how specialists in many areas are combining their expertise to seek better ways of handling society's waste. They will find data and management approaches that they can use and support as they seek to provide direction to the societal debate on the emotive issue of solid waste.

Lifecycle inventory specialists will see how their emerging science can be used for practical purposes to help the development of environmental improvements in the area of waste management.

Total quality specialists will see how the principles of this management approach can underpin a systematic approach to environmental improvements.

Environmentalists (whether or not in environmental organisations) will see how the application of science, financial management and total quality systems management can be combined in the search for solid waste management systems that do not cost the earth.

Concerned citizens will gain an overview of the efforts being made to improve waste management systems in order to reduce environmental impacts in ways that are considered to be affordable. They will see a recognition that science and management combined do not have all of the answers. There is a role, within the democratic process, for the concerned citizen to influence developments and to ensure that reasoned decisions are made.

Why?

Why has this book been written? Why should anyone undertake the journey it describes? Why has a company that manufactures and markets a broad range of laundry, cleaning, paper, beauty care, health care, food and beverage products in more than 140 countries around the world encouraged three of its employees to research and publish this book?

The book has been written because there is a clear need to pull together the existing data on the environmental effects of solid waste management techniques and the associated costs. In doing this the authors recognised that there was a need for a systematic approach in dealing with the data and using it to develop improved solid waste management systems. From our work on lifecycle inventories (and especially on our computer spreadsheet programme for performing lifecycle inventories on packaging systems), we realised that this developing technique provided the basis for such a systematic approach.

People should undertake the journey because solid waste management is changing and evolving rapidly and decisions need to be taken about the future directions to take. Society needs to strive towards the total quality goal of sustainable development. Industry has a role to play in this by developing products that meet the needs of the world's population whilst using less of the earth's resources, including energy, and creating less waste emissions. One aspect of this 'more with less' approach is to manage solid waste more efficiently so that there is less of it, and so that the waste that is inevitably created has usefulness (value) restored to it where possible. The inevitable emissions to air, water and land must be reduced as far as possible. This book provides a logical approach to this challenge that can ensure all possible efforts are also taken to manage the economic costs.

Procter & Gamble is concerned with solid waste because some of our products and most of our packages enter the solid waste stream. Our products are found in almost every household in countries where we have an established business. Our consumers want us to do everything we can to make sure that our products and packages are environmentally responsible. This responsibility, which we readily accept, involves us in constantly seeking improvements in the design and manufacture of our products. All of our packages and some of our products enter the solid waste stream. We have been working with others in various countries to help develop environmentally improved solid waste management systems that are affordable. The creation of such systems will help to satisfy the needs of our consumers and, because we understand the emerging systems, we will be able to develop products and packages that provide better value (including reduced environmental impact) for our future consumers.

To be sure that the approaches we are espousing for the development of these improved solid waste management systems are soundly based, we sought for an analysis of the environmental impact of waste management systems and for methods that could be used to evaluate possible improvements. We found lots of data in all sorts of places. We found no adequate compilation; no detailed critical review; no logical framework within which to study and perform operations on the data. We determined that we needed to undertake this work to ensure that we were basing our views on the best available information. This work will be of most value to us if others review it and build upon it. To achieve this we are making it publicly available.

Where to?

What is the aim of this book?

By the final page, the reader will be ready to consider how solid waste management systems can be designed to be environmentally and economically sustainable. These terms are defined as follows:

- *Environmentally sustainable* solid waste management systems must reduce as much as possible the environmental impacts of waste disposal, including energy consumption, pollution of land, air and water and loss of amenity.
- *Economically sustainable* solid waste management systems must operate at a cost acceptable to the community, which includes private citizens, businesses and government. The costs of operating an effective solid waste system will depend on existing local infrastructure, but ideally should be little or no more than current waste management costs.

The need for safety in waste management operations is paramount. Beyond that, however, there will always be the need to balance the environmental quality of a waste management system and its economic cost. The precise balance struck by communities will vary with geography, and over time, so there will never be one unique and universal solution as to what is 'the best waste management system'. This book argues that a waste management system that reduces the overall environmental impact compared to the current situation, at little or no extra cost, represents a reasonable balance.

We believe that the reader will be convinced that the best way of achieving this goal is to use an integrated approach to solid waste management. An integrated approach deals with all types and sources of non-hazardous solid waste, using a range of treatment options, in line with local needs. This approach contrasts with approaches requiring special systems for specific elements of waste.

Those readers who so wish will be in a position to use and interpret results from the lifecycle inventory spreadsheet that is provided for use with Excel™ and Lotus 1-2-3™ software.

The route?

How is the book organised?

Chapters 1–4 set out the principles of the book. They cover the key concepts of environmental and economic sustainability, integrated waste management, total quality management, system modelling and lifecycle inventory techniques in sufficient detail for a full understanding of the rest of the book.

Chapters 5–11 provide a compilation and critical review of existing data on waste management processes. This is drawn from many European sources and also from North America. Some of these data are previously unpublished.

Chapter 12 shows how the lifecycle inventory (LCI) system can be used to study solid waste management systems. This chapter takes the form of case studies. They are not published elsewhere. It also shows how the systematic approach advocated by the book is the best way of seeking answers to the challenge of dealing with solid waste in environmentally improved ways that are affordable.

The Appendices provide detailed information that will be of use to waste management specialists.

Signposts?

What techniques are used in the book to make it easy to read?

The text of the book is mainly descriptive. There is little detailed use of data within the text. The sections of each chapter are headlined and numbered. The headlines and numbers are included in the contents list. Data are found in Tables and Figures. Summaries of key points in the form of diagrams or bullet points are contained in each chapter as Boxes. These can provide an easy way for refreshing key points and concepts.

Chapters 5–11 contain an LCI Box and an LCI Data Box at the end of the chapter. The LCI Box shows the questions that must be answered in the IWM-1 computer spreadsheet dealing with that part of the system. The LCI Data Box summarises how the data in the chapter are used in the computer model. These Data Boxes can be ignored by those not wishing to use the computer model; they are an easy reference for those using the model.

The book and the computer model are built upon the system represented schematically in the frontispiece of the book. This Scheme is explained in detail in Chapter 4. Chapters 5–11 deal with specific parts of the Scheme. The appropriate part is depicted at the start of each chapter.

Warnings?

What should the reader be careful about?

We do not claim that this book is the final word on how to develop improved solid waste management systems. It seems to us, however, to be the right moment to set out a logical, holistic approach to waste management, together with a critical review of the available environmental and economic data. We expect and hope that knowledge will rapidly accumulate that allows us all to do an even better job of cooperating to develop improved systems.

The learnings developed in this book challenge perceived wisdom in some quarters. With data on the table, it is difficult to maintain dogmatic positions. We all have to be prepared to challenge our own wisdom. This can be uncomfortable. It can also be uncomfortable trying to use data in discussions when others want to rely on the telling phrase 'No, it is obvious that . . .'. We make no apology for placing the data on the table and trying to use them to do things in a better way.

The fascination of the data and the model must not become a cause of inaction. We do not want 'paralysis by analysis'. It can be argued that more sophisticated models and more comprehensive data are needed and will lead to even better waste management systems. We agree. We hope that this book will stimulate such improvements. In the meantime there is a need to take some hard decisions about improvements in waste management systems that can be put into place today. These improvements must be made. Total quality teaches us that there is a need for continuous improvement. As we make today's improvements, we must measure the effects that they have and use those data to stimulate us into further improvements.

The book contains the usual disclaimers of responsibility for actions taken as a result of information in this book and, especially, in the computer model.

Acknowledgements

Many people have helped us develop our understanding which has allowed us to write this book.

One of us (MF) studied environmental technology and engineering at university, specialising in waste management, and has worked in the area for over a decade. The others (PRW, PH) have come to this subject in the past three years.

We have all had the benefit of discussing issues with many people inside and outside our company. It is impossible to define specifically how each of those discussions fashioned our thinking. Those specialists who sense that they find their view reflected in the book are probably right. We hope that they will accept our thanks for the stimulus and insights they have given us.

The European Recovery & Recycling Association (ERRA) has been a particular source of inspiration. Like any association, it is the people who make the difference. Those working in the Brussels headquarters and on the demonstration projects have been particularly helpful. We have found the many discussions that we have had, in working groups and informally, to have been of immense value.

The Organic Reclamation & Composting Association (ORCA) and its publications were especially valuable in the preparation of Chapters 6 and 9.

Lifecycle inventory forms the basis for the treatment of the data in the book. We are indebted to the Society Of Environmental Toxicology and Chemistry (SETAC) for the leadership it has given in this emerging scientific discipline. Two of us (PRW and MF) have especially profited from involvement in some of the SETAC workshops. One of us (MF) has profited from involvement in lifecycle inventory and assessment work groups in Germany. We have all benefited from many discussions with Procter & Gamble colleagues on this important topic. We also acknowledge comments we have received on some of our earlier publications and presentations in this area.

We owe special thanks to colleagues within Procter & Gamble who have critiqued early drafts of the book and who have given unstintingly of their knowledge, data and insights. We mention their names and the countries where they are based to demonstrate the global perspective into which we have been able to tap.

Derek Gaskell, UK; Mariluz Castilla, Spain; Dr Rolf Seeboth, Germany; Klaus Dräger, Germany; Dr Roland Lentz, Germany; Phillipe Schauner, France; Chris Holmes, Belgium; Willy van Belle, Belgium; Nick de Oude, Belgium; Dr Celeste Kuta, USA; Keith Zook, USA; Karen Eller, USA; Dr Bruce Jones, USA; Tom Rattray, USA; Dr Eun Namkung, Japan.

P.W.
M.F.
P.H.

Contents

1 Introduction

Summary

The concept of waste as a by-product of human activity and the current environmental concerns over waste disposal are discussed. From these, environmental objectives for waste management are formulated. Current approaches to reaching these objectives rely on legislation, both end-of-pipe and strategic. The principles of, and difficulties with, present legislation are discussed. An alternative approach, integrated management, is introduced as the basic theme of this book.

1.1 What is waste?

Definitions of 'waste' invariably refer to lack of use or value, or 'useless remains' (Concise Oxford Dictionary). Waste is a by-product of human activity. Physically, it contains the same materials as are found in useful products; it only differs from useful production by its lack of value. A basic way to deal with waste, therefore, is to restore value to it, at which point it will cease to be 'waste'. The lack of value in many cases can be related to the mixed and, in many cases, unknown composition of the waste. Separating the materials in waste will generally increase their value if uses are available for these secondary materials. This inverse relationship between degree of mixing and value is an important property of waste, and is a recurring theme in this book (Box 1.1). Defining exactly what is a 'waste' and what is not is not merely of academic interest. It will determine which materials are subject to the increasing amount of legislation on waste handling and which are not.

Waste can be classified by a multitude of schemes (Box 1.1): by physical state (solid, liquid, gaseous) and then within solid waste by original use (packaging waste, food waste, etc.); by material (glass, paper, etc.); by physical properties (combustible, compostable, recyclable); by origin (domestic, commercial, agricultural, industrial, etc.) or by safety level (hazardous, non-hazardous). While there is benefit in addressing the whole issue of waste, it is not possible to deal with all of these different areas of solid waste management within a single volume. This book therefore concentrates on management of household and commercial waste (referred to together as municipal solid waste or MSW). This only accounts for a

---

Box 1.1   WASTE: SOME KEY CONCEPTS.

1. The relationship between waste and value

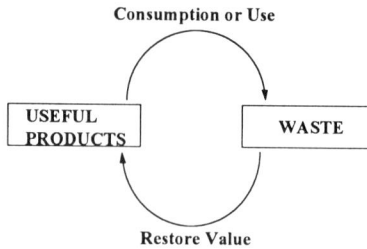

Consumption or Use

USEFUL PRODUCTS   WASTE

Restore Value

2. The relationship between value and mixing

$$\text{Value} = f \frac{1}{\text{degree of mixing}}$$

3. Possible classifications of waste

by   – physical state
      – original use
      – material type
      – physical properties
      – origin
      – safety level

---

relatively small part of the total solid waste stream; every year the countries of the developed world produce over 5 billion tonnes of municipal and industrial solid waste (OECD, 1993).

There are good reasons for addressing MSW. As the waste that the general public (and therefore voters) have contact with, management of MSW has a high political profile. Additionally, household waste is, by nature, one of the hardest sources of waste to manage effectively. It consists of a diverse range of materials (glass, metal, paper, plastic, organics) totally mixed together, with relatively small amounts of each. MSW composition is also variable, both seasonally and geographically from country to country, and from city to country areas. In contrast, commercial, industrial and other solid wastes tend to be more homogeneous, with larger quantities of each material. Thus, if a system can be devised to deal effectively with the materials in household waste, it should be possible to apply such lessons to the management of other sources of solid waste. In more concrete terms, as will be seen in later chapters, a system set up to collect, separate and deal with household waste can accept other large sources of similar wastes, and benefit from this addition.

Thus, although this book principally addresses municipal solid waste, the approach should be applicable to all areas of solid waste management.

1.2 Environmental concerns

Historically, health and safety have been the major concerns in waste management. These still apply; wastes must be managed in a way that minimises risk to human health. Today, society demands more than this; as well as being safe, waste management also needs to look at its wider effects on the environment. Environmental concerns about the management and disposal of waste can be divided into two major areas: conservation of resources and pollution of the environment.

1.2.1 Conservation of resources

In 1972, the best-selling book *Limits to Growth* (Meadows *et al.*, 1972) was published. It argued that the usage rates (in 1972) of the earth's finite material and energy resources could not continue indefinitely. Now, 20 years on, the sequel, *Beyond the Limits* (Meadows *et al.*, 1992), tells the same story, but with increased urgency; raw materials are being used at a faster rate than they are being replaced, or alternatives are being found. As the result of such reports, it is now becoming clear that the future of the planet lies in the concept of sustainable development. This is defined in the Brundtland Report 'Our Common Future' (WCED, 1987) as 'development that meets the needs of the present without compromising the ability of future generations to meet their own needs.' (Figure 1.1a). Sustainability requires that natural resources be efficiently managed, and where possible conserved.

Against this backdrop, the production and disposal of large amounts of waste is seen by many to represent squandering of the earth's resources. Putting the waste into holes in the ground certainly smacks of inefficient management of materials. It needs to be remembered, however, that although the earth is an open system regarding energy, it is essentially a closed system for materials. Whilst they may be moved around, used, dispersed or concentrated, the total amount of the earth's elements stays constant (with the exception of unstable radioactive elements). Thus, although resources of 'raw materials' may be depleted, the total amount of each element present on Earth remains constant. In fact, the concentration of some useful materials is higher in landfills than in their original raw material ores. Such materials could be dug up at a later date, and 'mining of landfills' has already begun in some countries.

Putting waste in holes in the ground, i.e. landfilling, could there-fore be considered as long-term storage of materials rather than actual

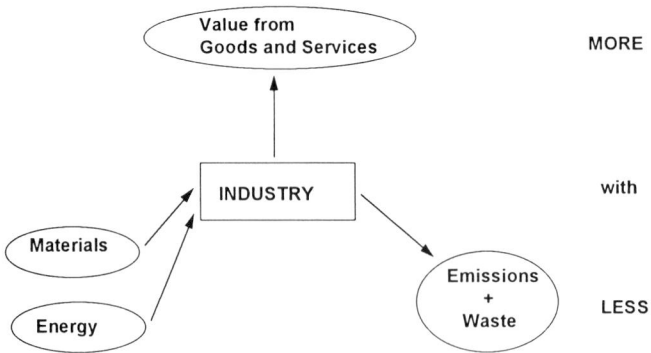

(a)

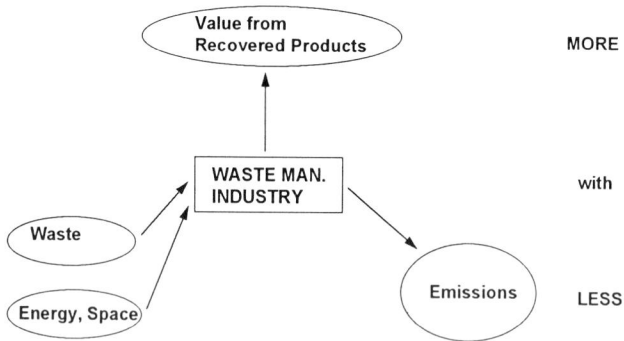

(b)

Figure 1.1 'More with less' and sustainability.
(a) Sustainable development. The Brundtland Report on Sustainable Development (WCED, 1987) introduced the concept of 'more with less', i.e. the need to produce more value from goods and services with less raw material and energy consumption, and less waste and emission production.
(b) Sustainable waste management. This also calls for 'more with less', i.e. more valuable products recovered from the waste with less energy and space consumption and less emissions.

disposal. Is this the most efficient way to manage such materials, however?

Concerns over conservation of resources have led to calls for, firstly, general reductions in the amount of waste generated, i.e. waste minimisation or waste reduction, and secondly, for ways to recover the materials and/or energy in the waste, so that they can be used again. Raw materials and energy carriers used by society can be divided into renewable and non-renewable resources. Recovery of resources from waste should slow down the exhaustion of non-renewable resources, and help to lower the use of renewable resources to the rate of replenishment.

1.2.2 Pollution

Potential or actual pollution is the basis for most current environmental concern over waste management. Historically, the environment has been considered as a sink for all wastes produced by human activities. Materials have been released into the atmosphere or watercourses, or dumped into landfills and allowed to 'dilute and disperse'. At low levels of emissions, natural biological and geochemical processes are able to deal with such flows without resulting changes in environmental conditions. However, as the levels of emissions have increased with exponential rises in human population and activity, natural processes do not have sufficient turnover to prevent changes in environmental conditions (such as the level of atmospheric CO_2). In extreme cases of overloading (such as gross sewage pollution in rivers), natural processes may break down completely, leading to drastic changes in environmental quality.

Just as raw materials are not in infinite supply, the environment is not an infinite sink for emissions. Environmental pollution produced by human activity will come back to haunt society, by causing deterioration in environmental quality. Consequently there is growing awareness that the environment should not be considered as an external sink for wastes from society, but as part of the global system that needs careful and efficient management.

As well as such broad concerns over environmental pollution, specific concerns emerge at the local level whenever new facilities for waste treatment are proposed. Planned incinerators raise concerns over likely emission levels in general, and recently, of dioxin levels in particular. Similarly, landfill sites are known to generate landfill gas. At the global level, this has a high global warming potential (GWP), but more immediately at a local level it can seep into properties, with explosive consequences. There is also increasing concern of the risk of groundwater pollution from the leachate generated in landfill sites. Such local environmental concerns (plus, no doubt, concern over the effects on property prices) have given rise to several acronyms describing the attitudes facing waste management planners (Box 1.2).

Box 1.2

THE CHALLENGE OF WASTE MANAGEMENT

```
┌──────────┐                    ╭──────────────╮
│ Human    │ ───────────────▶  │  Solid Waste │
│ Activity │                    ╰──────────────╯
└──────────┘
```

BUT

Common Attitudes:-

NIMBY	- Not In My Back Yard
NIMET	- Not In My Elected Term
BANANA	- Build Absolutely Nothing Anytime, Near Anybody

THE RESPONSE

AN OVERALL STRATEGY THAT REDUCES
ENVIRONMENTAL IMPACT

Reduce the amount of waste generated
Manage the inevitable waste produced

Whilst such attitudes are understandable, they ignore our common responsibility for waste management. All human activities generate waste. Each person in Western Europe generates an average of 318 kg of municipal solid waste each year (Elsevier, 1992). This waste has to be dealt with somehow, somewhere. Whilst all the methods for treating and disposing of waste are known to have environmental impacts, the waste must still be dealt with. Waste cannot be moved to 'other people's back yards' indefinitely. What is needed is an overall strategy to manage waste with reduced environmental impact, at an affordable cost.

1.3 Environmental objectives

Clearly, the first objective is to reduce the amount of waste generated. However, even after this has been done, waste will still be produced. The second objective, therefore, is to manage the waste in an environmentally sustainable way, by minimising the overall environmental impacts associated with the waste management system (Figure 1.1b). Environmental sustainability addresses both resource conservation and pollution concerns, but can it be assessed, and if so how?

Lifecycle assessment (LCA) is an emerging environmental management tool that attempts to predict the overall environmental impact of a product, service or function, and so can be applied to waste management systems (White *et al.*, 1993). In this book, the technique of lifecycle inventory (LCI), part of LCA (see Chapters 3 and 4), is used to look at whole waste management systems, to predict their likely environmental impacts. Then, by selecting appropriate options for dealing with the various fractions of solid waste, the environmental impacts of the whole waste management system can be reduced.

1.4 Current approaches: legislation

Waste management in developed countries is governed by legislation, the details of which could fill this volume several times over. Looking at the basic approach of legislative tools, they fall into two main categories: end-of-pipe regulations and strategic targets.

1.4.1 End-of-pipe regulations

These are technical regulations and relate to the individual processes in waste treatment and disposal. Emission controls for incinerators are a prime example. Such regulations may be set nationally (e.g. the T.A. Luft (1986) emission limit for Germany) or, increasingly, internationally by the European Commission. Municipal waste incinerator emissions are the subject of two EC Directives (89/369/EEC for new incinerator plants and 89/429/EEC for existing plants), and there is currently a draft EC Directive on Landfill. Such regulations are important to ensure safe operation of waste disposal processes. They operate as 'fine tuning' of the system, by promoting best available technology and practices, but as with other end-of-pipe solutions, they are not tools that will lead to major changes in the way the waste management system operates. These will be produced by strategic decisions between different waste management options rather than by refining the options themselves.

1.4.2 Strategic targets

Strategic legislation is becoming increasingly used to define the way in which solid waste will be dealt with in the future. At national levels, the German Töpfer Ordinance, Dutch Packaging 'Covenant' and UK

Environmental Protection Act (1990) all lay down either rules or guidelines on the waste management options (mainly recycling) that will be used for at least part of municipal solid waste. At the European level, the Draft Packaging and Packaging Waste Directive, if agreed, will require member states to set deadlines for meeting value recovery and recycling targets for packaging materials.

This legislation has several common threads: it builds (either explicitly or implicitly) on the 'hierarchy of solid waste management' and within this, it sets targets for recovery and recycling of materials. Much current legislation is also directed at specific parts, rather than the whole, of the municipal waste stream. Packaging, in particular, has been targeted.

The hierarchy of waste management options is headed by source reduction, or waste minimisation. This is the essential prerequisite for any waste management strategy – less waste to deal with. Next in the hierarchy come a series of options: re-use, recycling, composting, waste to energy, incineration without energy recovery and landfill, in some order of preference. It will be argued in this book that, although useful as a set of default guidelines, using this hierarchy to determine which options are preferable does not necessarily result in the lowest environmental impacts, nor an economically sustainable system. Different materials in the waste are best dealt with by different processes, so to deal effectively with the whole waste stream, a range of waste management options is desirable. Thus, there are no overall 'best' or 'worst' options; different options are appropriate to different fractions of the waste. Instead of relying on the waste management hierarchy, this book looks at the whole waste management system, and attempts to assess its overall environmental impacts and economic costs on a lifecycle basis.

Recycling targets, similarly, are useful ways to measure progress towards recycling goals, but they may not always measure progress towards environmental objectives. Typically recycling is used to reduce raw material consumption, and in many cases also energy consumption. It must be borne in mind, however, that it is a means to an end, not an end in itself. Rigidly set recycling targets may not produce the greatest environmental benefits. Environmental benefit (e.g. reduction in energy consumption) does not increase linearly with recycling rates (Boustead, 1992). At high levels of recovery, proportionately more energy is needed to collect used materials from diffuse sources, so there is little, if any, environmental gain. In such cases, lifecycle assessment could be used to determine the optimal recycling rate to meet defined environmental objectives. A sign that this more flexible approach may be gaining ground is the inclusion in the EU Draft Packaging Waste Directive (Article 4) of the provision that 'if scientific research (such as ecobalances) proves that other recovery processes show greater environmental advantages, the targets for recycling can be modified . . .'.

Packaging waste has been the focus of much recent strategic legislation which has concentrated attention on a specific part of the municipal waste stream. Consequently attention has been diverted away from the overall problem of dealing with all of the waste stream in an environmentally and economically sustainable way. Estimates of the contribution of packaging to household waste vary from around 27% (by weight) of the dustbin contents in the UK (WSL, 1992) to around 35% (by weight) for Germany. Setting targets for recycling of packaging alone, therefore, only deals, at maximum, with this 27–35% of the household waste. For example, a 50% reduction in packaging waste would only result in approximately, a 15% reduction in total household waste. Given that household waste represents less than 5% of total solid waste production (see Chapter 5), this means that special measures directed at packaging will have little impact on the overall solid waste management scenario.

At a more fundamental level, how waste is best recovered, treated, or disposed of depends on the nature of the materials in the waste, not on the original use of the discarded object. Whatever legislative instruments (e.g. recycling targets) are used, they should relate to the whole municipal waste stream. Otherwise, as has been seen in Germany, packaging-specific legislation can lead to the development of separate and parallel waste management systems for such packaging waste. Duplication of systems on the basis of original use of the material, or for any other reason, will give rise to increased environmental impacts and economic costs.

This book proposes that if real and sustainable environmental improvements are the objective, a single, integrated collection and sorting system, followed by material-specific recovery/treatment/disposal represents the most promising approach. It shows that a scheme designed to deal with all fractions of household waste can also deal with other sources such as commercial and some industrial waste. Fragmenting the household waste between different waste management systems rules out an integrated approach, and any economies and efficiencies of scale.

1.5 Economic costs of environmental improvements

Environmental improvements to waste disposal methods should be welcomed, where they are scientifically justifiable. Improvements, however, usually have economic costs associated with them. This is invariably the case with such end-of-pipe solutions as installing new technology to clean up emissions following tighter regulatory control. It can also be the case with introducing strategic solutions aimed at reducing the environmental impact of waste management, such as recycling. Take, for example, the 'Blue Box' schemes that have been introduced for kerbside collection of recyclable materials in many parts of North America, and recently in Sheffield and Adur, UK, or the Dual System operating in Germany for packaging materials. Whilst these systems can collect large quantities of high quality materials, the collection schemes operate in parallel with existing household waste collections. Two vehicles call at each property, where only one did before. These systems, as additions to residual waste collection, inevitably involve additional cost. Any other similar 'bolt-on' systems, whether for recovery of materials or collection of compostables will suffer from the same economic problems. This trade-off between economic cost and environmental impact has been seen as a major hurdle to environmental improvements in waste management. However, two concepts that are becoming established can help to overcome this obstacle.

Internalising external environmental costs. Environmental and social costs of waste disposal have historically been seen as external costs. For example, the effects of emissions from burning waste, or the leachate and gas released from landfill sites were not considered as part of the cost of these disposal methods. More recently, however, as emission regulations have become tighter, the costs of emission controls have been internalised in the cost of disposal. Similarly, as legislation (e.g. UK Environmental Protection Act, 1990) requires monitoring of waste disposal sites after closure, and the provision of insurance bonds to remediate environmental problems that may arise in the future, the real (i.e. full and inclusive) cost of each waste management option becomes apparent. Under such conditions, waste management options with lower environmental impacts that may have appeared more expensive can become economically viable.

Building environmental objectives into the waste management system. An additional 'bolt-on' system or an end-of-pipe solution will entail additional costs (Box 1.3). If, however, a waste management system is designed from scratch to achieve environmental objectives, it may include such features at little or no extra cost compared to existing practices. An integrated system that can deal with all the materials in the solid waste stream represents the total quality approach to waste management. Such a system would benefit from economies of scale; an integrated material-based system allows for efficient collection and management of waste from different sources. The total quality objective would be to minimise the environmental impacts of the whole waste management system, whilst keeping the economic costs to an acceptable level. The definition of acceptable will vary according to the group concerned, and with geography, but if the cost is

Box 1.3 ACHIEVING ENVIRONMENTAL IMPROVEMENTS IN SOLID WASTE MANAGEMENT.

1. The 'Bolt-on' Approach

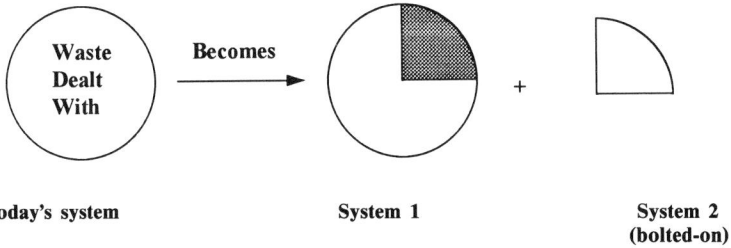

Today's system   System 1   System 2
                                                      (bolted-on)

Likely to result in: higher costs
                     higher lifecycle impact even if material is recycled or
                     composted

2. The Total Quality Approach

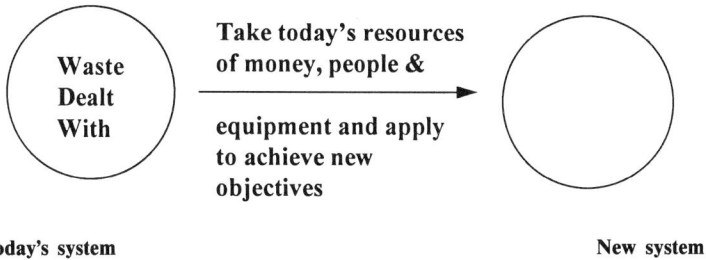

Take today's resources
of money, people &

equipment and apply
to achieve new
objectives

Today's system   New system

Will result in: a lifecycle impact improvement by definition
                lowest additional costs

little or no more than present costs, it should be acceptable to most (Box 1.3).

1.6 An integrated approach to solid waste management

The aim of this book, then, is to propose an integrated approach to solid waste management which can deliver both environmental and economic sustainability. It is clear that no one single method of waste disposal can deal with all materials in waste in an environmentally sustainable way. Ideally, a range of management options is required. The use of different options such as composting or materials recovery will also depend on the collection and subsequent sorting system employed. Hence, any waste management system is built up of many inter-related processes, integrated together. Instead of focusing on and comparing individual options (e.g. 'incineration versus landfill'), an attempt will be made to synthesize waste management systems that can deal with the whole waste stream, and then compare their overall performances in environmental and economic terms.

This approach looks at the overall waste management system, and develops ways of assessing overall environmental impacts and economic costs. As part of this, the various individual techniques for collection, treatment and disposal of waste are discussed. It is not intended, however, to be a technical manual for these processes. Techniques such as composting or thermal waste-to-energy can, and do, fill many volumes of their own. This book, in contrast, gives an overall vision of waste management, with a view to achieving environmental objectives using an economically sustainable system.

References

Boustead, I. (1992) The relevance of re-use and recycling activities for the LCA profile of products. *Proc. 3rd CESIO International Surfactants Congress and Exhibition*, London, pp. 218–226.

EC (1989a) 89/369/EEC. Council Directive on the Prevention of Air Pollution from New Municipal Waste Incineration Plants. *Off. J. Eur. Commun.* **L163**, 32–36.

EC (1989b) 89/429/EEC. Council Directive on the Prevention of Air Pollution from Existing Municipal Waste Incineration Plants. *Off. J. Eur. Commun.* **L203**, 50–54.

EC (1993a) Draft Landfill Directive.

EC (1993b) Draft Packaging and Packaging Waste Directive.

Elsevier (1992) *Municipal Solid Waste Recycling in Western Europe to 1996*. Elsevier, Science Publishers.

Meadows *et al.* (1972) *The Limits to Growth*. Universe Books, New York.

Meadows *et al.* (1992) *Beyond The Limits*. Earthscan Publications, London.

OECD (1993) *Environmental Data Compendium*. Organisation for Economic Cooperation and Development, Paris.

WCED (1987) *Our Common Future*. World Commission on Environment and Development, Oxford University Press, Oxford.

WSL (1992) Calculated from waste analysis data from Warren Spring Laboratory.

White, P.R, Hindle P. and Dräger, K. (1993) Lifecycle assessment of packaging, In *Packaging in the Environment*, ed. G. Levy. Blackie, Glasgow, pp. 118–146.

2 Integrated waste management

Summary

This chapter discusses the needs of society: less waste, and then an effective way to manage the inevitable waste still produced. Such a waste management system needs to be both environmentally and economically sustainable and is likely to be integrated, market-oriented, flexible and operated on a regional scale. The current hierarchy of waste management options is critically discussed, and in its place is suggested a holistic approach that assesses the overall environmental impacts and economic costs of the whole system. Lifecycle techniques are introduced for comparing the overall environmental impacts and economic costs.

2.1 Basic requirements

Waste is an inevitable product of society; managing this waste more effectively is a need that society has to address. In dealing with the waste, there are two fundamental requirements: less waste, and then an effective system for managing the waste still produced.

2.1.1 Less waste

The Brundtland report of the United Nations 'Our Common Future' (WCED, 1987) clearly spelled out that sustainable development would only be achieved if society in general, and industry in particular, learned to produce 'more with less'; more goods and services with less use of the world's resources (including energy) and less pollution and waste.

In this era of 'green consumerism' (Elkington and Hailes, 1988), this concept of 'more with less' has been taken up by industry which has, in turn, spawned a range of concentrated products, light-weight and refillable packaging and other innovations (Hindle *et al.*, 1993; IGD, 1994). Production as well as product changes have been introduced, with many companies using internal recycling of materials, or on-site energy recovery, as part of solid waste minimisation schemes.

All of these measures help to reduce the amount of solid waste produced, either as industrial, commercial or domestic waste. In essence, they are improvements in efficiency, whether in terms of materials or energy consumption. The costs of raw materials and energy, and rising disposal costs for commercial and industrial waste, will ensure that waste reduction continues to be pursued by industry for economic as well as environmental reasons.

There has been recent interest in promoting further waste reduction by the use of fiscal instruments. Pearce and Turner (1992) for example, suggest ways to reduce the amount of packaging used (and hence appearing as waste) by internalising the costs of waste disposal within packaging manufacture, by means of a packaging levy. It is not clear how effective such taxes would be, however, since they only affect a small section of the waste stream. (Packaging materials constitute approximately 1.5% by weight of the total European waste stream (Warmer, 1990).) Furthermore, economic incentives for waste reduction already exist.

Interestingly, there is one area where there are often no economic incentives for waste reduction, i.e. household waste generation. In some communities, notably in the USA (Schmidt and Krivit, 1992), and also in Germany, waste collection charges are on a scale according to the volume of waste generated, but in most communities, a flat rate collection fee applies. Charging according to waste generation could lead to a reduction in household waste, provided that unauthorised dumping or other alternative routes could be prevented. The issues to be resolved are the level of the charge, and an effective way of managing the system. Introducing such charging structures to societies not used to them can be politically unpopular.

'Waste minimisation', 'waste reduction' or 'source reduction' are usually placed at the top of the conventional waste management hierarchy. In reality, however, source reduction is a necessary precursor to effective waste management, rather than part of it. Source reduction will affect the volume, and to some extent, the nature of the waste, but there will still be waste for disposal (Figure 2.1). What is needed, beyond source reduction, is an effective system to manage this waste.

2.1.2 *Effective solid waste management*

Solid waste management systems need to ensure human health and safety. They must be safe for workers and safeguard public health by preventing the spread of disease. In addition to these prerequisites, an effective system for solid waste management must be both environmentally and economically sustainable.

- *Environmentally sustainable*. It must reduce as much as possible the environmental impacts of waste management, including energy consumption, pollution of land, air and water, and loss of amenity.
- *Economically sustainable*. It must operate at a cost acceptable to the community, which includes private citizens, businesses and government. The costs of operating an effective solid waste system will depend on existing local infrastructure, but ideally should be little or no more than existing waste management costs.

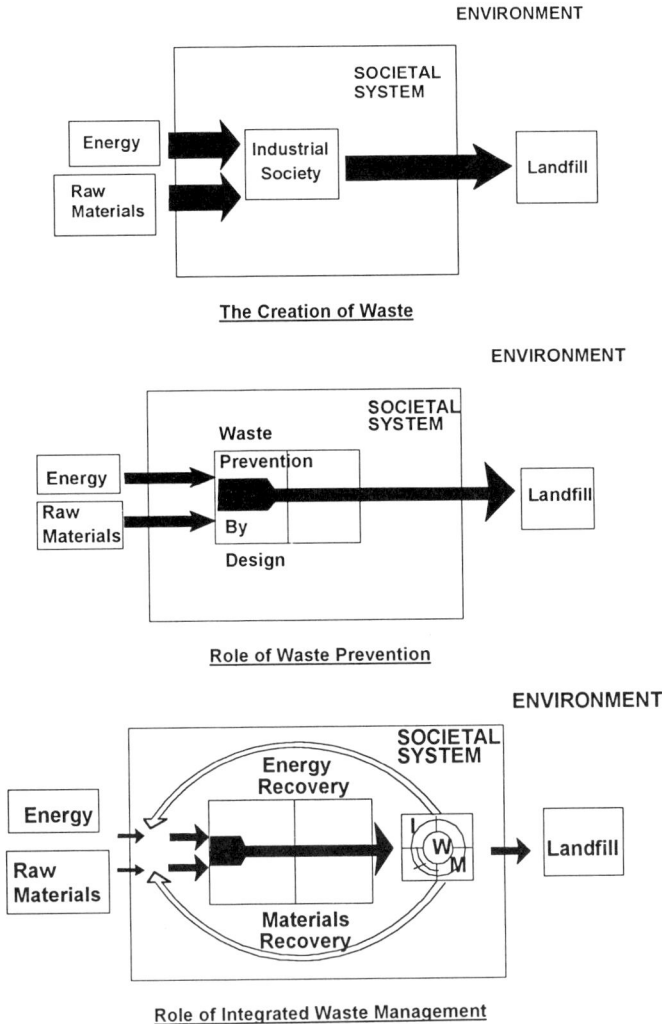

ENVIRONMENT

SOCIETAL SYSTEM

The Creation of Waste

ENVIRONMENT

SOCIETAL SYSTEM

Role of Waste Prevention

ENVIRONMENT

SOCIETAL SYSTEM

Role of Integrated Waste Management

Figure 2.1 The respective roles of waste prevention and integrated waste management. In life-cycle studies, a 'system' is defined (with boundaries indicated by lines). Energy and raw materials from the 'environment' are used in the system. Emissions, including solid waste, leave the system and enter the environment. (See Chapters 3 and 4 for more detailed discussion.)

Clearly it is difficult to minimise two variables, cost and environmental impact, simultaneously. There will always be a trade-off. The balance that needs to be struck is to reduce the overall environmental impacts of the waste management system as far as possible, within an acceptable level of cost. Deciding the point of balance between environmental impact and cost will always generate debate. Better decisions will be made if data on impacts and costs are available; such data will often prompt ideas for further improvements.

2.2 Waste management systems

2.2.1 Characteristics of an effective system

An economically and environmentally sustainable solid waste management system is likely to be integrated, market-oriented and flexible. The execution of these principles will vary on a regional basis. A key requirement, in total quality language, is to understand the 'customer-supplier relationships'.

An integrated system. 'Integrated waste management' is a term that has been frequently applied but rarely defined. Here it is defined as a system for waste management that deals with:

- *All types of solid waste materials.* The alternative of focusing on specific materials, either because of their ready recyclability (e.g. aluminium) or their public profile (e.g. plastics) is likely to be less effective, in both environmental and economic terms, than taking a multi-material approach.
- *All sources of solid waste.* Domestic, commercial, industrial, institutional, construction and agricultural. (Hazardous waste needs to be dealt with within the system, but in a separate stream.) Focusing on the source of a material (on packaging or domestic waste or industrial waste) is likely to be less productive than focusing on the nature of the material, regardless of its source.

An integrated system would include waste collection and sorting, followed by one or more of the following options:

- Recovery of secondary materials (recycling): this will require adequate sorting and access to reprocessing facilities.
- Biological treatment of organic materials: this will produce marketable compost or reduce volume for disposal. Anaerobic digestion produces methane that can be burned to release energy.
- Thermal treatment: this will reduce volume, render residues inert and may recover energy.
- Landfill: this can increase amenity via land reclamation but will at least minimise pollution and loss of amenity.

To handle all waste in an environmentally sustainable way requires a range of the above treatment options. Landfill is the only method that can handle all waste alone, since recycling, composting and incineration all leave some residual material that needs to be landfilled. Landfill, however, does not valorise any part of the waste. It can be used to reclaim land, but it can also cause methane emissions and groundwater pollution and consume space. Use of the other options prior to landfilling can both valorise significant parts of the waste stream, reduce the volume and improve the physical and chemical stability of the residue. This will reduce both the space requirement and the potential environmental impacts of the landfill.

Market-oriented. Any scheme that incorporates recycling, composting or waste-to-energy technologies must recognise that effective recycling of materials and production of compost and energy depends on markets for these outputs. These markets are likely to be sensitive to price and to consistency in quality and quantity of supply. Managers of such schemes will need to play their part in building markets for their outputs, working with secondary material processors, and helping to set material quality standards. They must also recognise that such markets and needs will change over time, so such standards should not be rigid and based on legislation, but be set as part of a total quality customer-supplier relationship.

Flexibility. An effective scheme will need the flexibility to design, adapt and operate its systems in ways which best meet current social, economic and environmental conditions. These will likely change over time and vary by geography. Using a range of waste management options in an integrated system gives the flexibility to channel waste via different treatments as economic or environmental conditions change. For example, paper can either be recycled, composted or incinerated with energy recovery. The option used can be varied according to the economics of paper recycling, compost production or energy supply pertaining at the time.

Scale. The need for consistency in quality and quantity of recycled materials, compost or energy, the need to support a range of disposal options, and the benefit of economies of scale, all suggest that integrated waste management should be organised on a large-scale, regional basis. The optimal size for such a scheme is considered in later chapters, but experience is beginning to suggest that an area containing upwards of 500 000 households is a viable unit (White, 1993). This may not correspond to the scale on which waste disposal is currently administered. In many cases, therefore, implementation of such schemes will require local authorities to work together.

2.2.2 The importance of an holistic approach

The operations within any waste management system are clearly interconnected. The collection and sorting method employed, for example, will affect the ability to recover materials or produce marketable compost. Similarly, recovery of materials from the waste stream may affect the viability of energy recovery schemes. It is necessary, therefore, to consider the whole waste management system in an holistic way. What is required is

---

BOX 2.1

DESIGNING AN EFFECTIVE SOLID WASTE MANAGEMENT SYSTEM

1. **Strive for both of the following:-**

    Environmental Sustainability: Reduce Environment Impact

    Economic Sustainability: Drive costs out

2. **To achieve these the system should be:-**

    Integrated: in waste materials

    in sources of waste

    in treatment methods

    - anaerobic digestion
    - composting
    - energy recovery
    - landfill
    - recycling

    Market Oriented: materials and energy have end uses

    Flexible: for constant improvement

3. **Take care to:-**

    Define clear objectives

    Design a total system against those objectives

    Operate on a large enough scale

4. **Never stop looking for improvements in environmental impacts and costs. There is no perfect system.**

---

an overall system that is both economically and environmentally sustainable (Box 2.1). A lot of recent effort has been put into schemes concentrating on individual technologies, e.g. recycling, or on materials from one source only (e.g. the German DSD system to collect packaging). From the perspective of the whole waste management system, such schemes may well involve duplication of efforts or other inefficiencies, making them both environmentally and economically ineffective.

The relevance of looking at the whole system could be challenged, since waste management is split up into so many different compartments. Collection is the duty of local authorities, although it may be contracted out to private waste management companies. Disposal often comes under the jurisdiction of another authority, and perhaps another private company. Different operators may contribute to recycling activities; in the case of material collection banks, these may be the material producers. Similarly, incineration, composting and landfill operations may all be under the control of different operating companies. Each company or authority only has control of the waste handling within its operation, so what is the feasibility of taking an overall systems approach when no-one has control over the whole system?

The holistic approach has several advantages:

1. It gives an overall picture of the waste management process. Such a view is essential for strategic planning. Handling of each waste stream separately is inefficient.
2. Environmentally, all waste management systems are part of the same system, the global ecosystem. Looking at the overall environmental burden of the system is the only rational approach, otherwise reductions in the environmental impacts of one part of the process may result in greater environmental impacts elsewhere.
3. Economically, each individual unit in the waste management chain should run at a profit, or at least break even. Therefore, within the boundaries controlled by each operator, the financial incomes must at least match the outgoings. By looking at the wider boundaries of the whole system, however, it is possible to determine whether the whole system operates efficiently and whether it could run at break-even, or even at a profit. Only then can all the constituent parts be viable, provided that income is divided up appropriately in relation to costs.

2.2.3 A total quality system

To achieve fully integrated waste management will require major system changes from the present situation. The objective of an integrated system is to be both environmentally and economically sustainable. This is a total quality objective (Oakland, 1989); it can never be reached, since it will always be possible to reduce environmental impacts further, but it will lead to continual improvements.

Application of total quality thinking can be of further use in waste management. To reach a total quality objective one builds a system to achieve this objective. To deliver environmentally and economically sustainable waste management requires putting together a system designed for

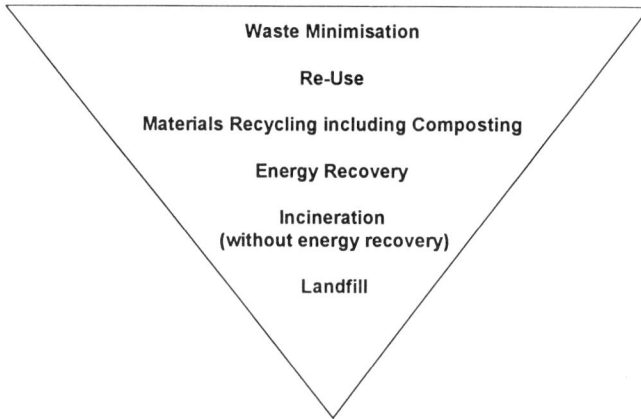

Figure 2.2 A hierarchy of solid waste management.

this purpose. This is a key point. Trying to improve present systems by using recycling or composting as 'bolt-on extras' is unlikely to work. Different components of the system are inter-connected so it is necessary to design a whole new system rather than tinker with the old one. Radical system changes may also allow previous economic inefficiencies to come to light, which can be used to offset any increased costs.

2.3 A hierarchy of waste management options?

Current thinking on the best methods to deal with waste is centred on a broadly accepted 'hierarchy of waste management' (Figure 2.2) which gives a priority listing of the waste management options available. Several variations of the hierarchy are currently in circulation, but they are essentially similar. The hierarchy gives important general guidelines on the relative desirability of the different management options, but it has limitations:

(a) Using the hierarchy rigidly will not always lead to the greatest reduction in the overall environmental impacts of a given system. Equally, its use will not necessarily lead to economically sustainable systems. The hierarchy itself makes no attempt to measure the impacts of individual options, nor of the overall system. Planning an integrated waste system for a given region should involve comparing the environmental impacts and economic costs of different schemes to determine which are environmentally and economically sustainable in that region. Only if data are not available for a case-by-case systems analysis should the guidelines of the hierarchy be used.

(b) A danger exists that the hierarchy will become accepted as dogma;

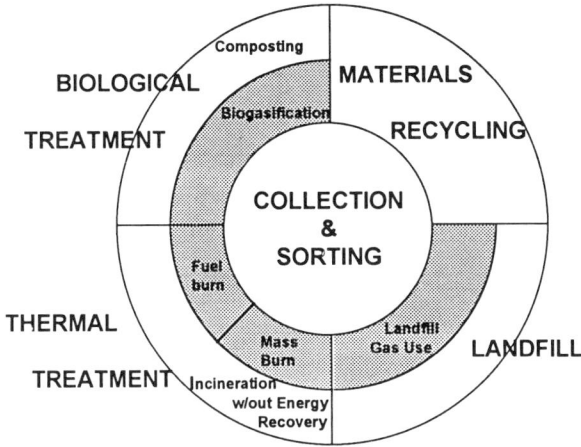

Figure 2.3 The elements of integrated waste management. Shaded area represents waste-to-energy.

that re-use will always be seen to be better than recycling, for example. Yet, if heavy bottles have to be transported long distances to be refilled, re-use may not be preferable over recycling on either environmental or economic grounds. The only way to ensure that environmental impacts and economic costs are kept to a minimum is to measure them throughout the lifecycle of the waste.

Rather than a hierarchy of preferred waste management options (Figure 2.2), an holistic approach is proposed which recognises that all disposal options can have a role to play in integrated waste management (Figure 2.3). This would include materials recycling, biological treatment (composting, anaerobic digestion/biogasification), thermal treatment (incineration with/without energy recovery, RDF burning) and landfill. The model stresses the inter-relationships of the parts of the system. Each option should be assessed using the most recent data available, but the overriding objective is to optimise the whole system, rather than its parts, to make it environmentally and economically sustainable.

Unlike the hierarchy, this approach does not predict what would be the 'best' system. There is no universal best system. There will be geographic differences in both the composition and the quantities of waste generated. Similarly there will be geographic differences in the availability of some disposal options (such as landfill), and in the size of markets for products derived from waste management (such as reclaimed materials, compost and energy). The economic costs of using different treatment methods will reflect the existing infrastructure (i.e. whether the plant already exists or needs to be built from scratch). This approach allows comparisons to be made between different waste management systems for dealing with the solid waste of the local regions. The best system for any given region will be determined locally to reduce overall environmental impact for an acceptable cost.

2.4 Modelling waste management

2.4.1 Why model?

Optimising the waste system to reduce environmental impacts or economic costs requires that these impacts and costs be predicted. Hence the need to model waste management systems. Modelling may at first glance appear to be a purely academic exercise, but delving further reveals that it has very practical uses:

1. The process of building a model focuses attention on missing data. Often the real costs, in either environmental or economic terms, for parts of the waste chain are not widely known. Once identified, missing data can either be sought out, or if not in existence, experiments can be run to obtain it.
2. Once completed, the model will define the *status quo* of waste management, both by describing the system, and by calculating the overall economic and environmental costs.
3. Modelling allows 'what if ...?' calculations to be made, which can then be used to define the points of greatest sensitivity in the system. This will show which changes will have the greatest effects in reducing costs or environmental impacts.
4. Last but not least, the model can be used to predict likely environmental impacts and economic costs in the future. Such forecasts will not be 100% accurate, but will give ball-park estimates valuable for planning future strategy. These are useful especially in such long-term processes as the development of markets for secondary materials. Market development is vital to ensure that higher levels of recycling can be sustained. Modelling the waste system will allow prediction of the likely amounts of reclaimed material available, which will in turn allow investment in the necessary equipment to proceed with confidence.

2.4.2 Previous modelling of waste management

Modelling of waste management is not a new idea. Clark (1978) reviewed the use of modelling techniques then available to optimise collection methods, predict the most efficient collection routes and define the optimal locations for waste disposal facilities. Such models concentrated on the detailed mechanics of individual processes within the waste system. Other models have attempted to take a broader view and have compared alternative waste disposal strategies from an economic perspective (e.g. Greenberg *et al.*, 1976).

More recently, detailed models have been developed to model the economics of materials recovery for recycling, and some of its environmental impacts (Boustead, 1992), as well as broader models including cost, public acceptance, environmental impact and ease of operation and maintenance of waste disposal alternatives (Sushil, 1990). The model constructed here attempts to predict both the overall environmental impacts and economic cost, since both are crucial, for an integrated waste management system.

2.4.3   Using lifecycle assessment for waste

Modelling may be divided into two areas: model structure (which will determine how the model will work) and data acquisition (for insertion into the model). Recent developments in each of these two key areas have made this work both possible and timely.

Models.   Several models for predicting environmental impacts have been produced within the discipline of lifecycle assessment (LCA). This is a relatively new branch of applied science that has considerable potential as an environmental management tool. The modelling technique used here is essentially a lifecycle inventory of waste. The methodology of a lifecycle inventory and its part in lifecycle assessment is discussed in the following two chapters.

Data.   Even the best models are useless without accurate, relevant and accessible data to plug into them. Whilst data have been available for many of the technical processes in waste disposal, such as incinerator emissions, information on the costs and impacts of collection and sorting systems has not been readily available. Until very recently, many source separated collection and sorting systems existed only on the planning table; operational data could not be obtained because in many cases the schemes did not exist on the ground. Fortunately, through the efforts of a range of bodies, many different pilot schemes, both on small and large scales, have been set up, and data are becoming available. The European Recovery and Recycling Association (ERRA) has assisted in the setting up of eight kerbside collection schemes in seven European countries (ERRA, 1991). Data from these schemes, and others, are being collated in the ERRA database, and are cited in this book. The DSD (Dual System Deutschland) has been set up to collect packaging materials across Germany, and there too, data are beginning to emerge on collection rates. Many cities and towns have independently set up recycling, composting or, in some cases, such as Leeds (UK) and Bonn, Coburg, Rhein-Sieg county and Lahn-Dill county (Germany), fully integrated schemes. For the first time, it is now possible to model integrated waste management schemes based on actual data.

The model developed in this book therefore attempts to combine recent developments in lifecycle inventory methodology with the stream of hard data beginning to emerge. This fusion should allow prediction of the overall environmental impacts and economic costs of different executions of integrated waste management.

References

Boustead, I. (1992) The relevance of re-use and recycling activities for the LCA profile of products. *Proc. 3rd CESIO International Surfactants Congress and Exhibition*, London, pp. 218–226.

Clark R.M. (1978) *Analysis of Urban Solid Waste Services – A Systems Approach*. Ann Arbor Science Publishers, Ann Arbor, MI.

Elkington, J. and Hailes, J. (1988) *The Green Consumer Guide*. Victor Gollancz.

ERRA (1991) Resource. Report of European Recovery and Recycling Association, 1991. ERRA, Brussels.

Greenberg, M., Caruana, J. and Krugman, B. (1976) Solid-waste management: a test of alternative strategies using optimization techniques. *Environ. Planning* **8**, 587–597.

Hindle, P., White, P.R. and Minion, K. (1993) Achieving real environmental improvement using value: impact accessment. *Long Range Planning* **26**(3), 36–48.

IGD (1994), Environmental impact management. Report by A. Hindle for Policy Issues Programme of the Insitute for Grocery Distribution, Watford, UK, 65 pp.

Oakland. J.S. (1989) *Total Quality Management*. Heinemann, Oxford, UK, 316 pp.

Pearce, D.W. and Turner, R.K. (1992) Packaging waste and the polluter pays principle – a taxation solution Working paper. Centre for Social and Economic Research on the Global Environment, University of East Anglia and University College London, 20 pp.

Schmidt, S. and Krivit, D. (1992) Variable fee systems in Minnesota. *Biocycle* September, 50–53.

Sushil (1990) Waste management: a systems perspective. *Industrial Management and Data Systems* **90** (5), 7–66.

Warmer, (1990) The packaging puzzle. *Warmer Bull.* **25**, 14–15.

WCED (1987) *Our Common Future*. World Commission on Environment and Development, Oxford University Press, Oxford.

White, P.R. (1993) Waste-to-energy technology within integrated waste management. In *Proc. Cost Effective Power and Steam Generation from the Incineration of Waste*. Institute of Mechanical Engineers Seminar, London.

3   Lifecycle inventory: a part of lifecycle assessment

Summary

Lifecycle assessment (LCA) is introduced. It is a tool to predict the overall environmental impact of a product or service. It consists of four stages: goal definition, inventory, impact analysis and valuation, of which the first two are well developed and the latter two in need of further work. In view of this, the analysis of solid waste management will be limited to a lifecycle inventory (LCI) (comprising goal definition and inventory stages) and not involve a detailed impact analysis or valuation. Economic lifecycle assessment is an equally important, although separate, part of an overall assessment.

3.1   What is lifecycle assessment?

Lifecycle assessment (LCA) is an environmental management tool increasingly used to predict and compare the environmental impacts of a product or service, 'from cradle to grave' (for recent reviews see White *et al.*, 1993 and Elkington *et al.*, 1993). The technique examines every stage of the lifecycle, from the winning of raw materials, through manufacture, distribution, use, possible re-use/recycling and then final disposal. For each stage, the inputs (in terms of raw materials and energy) and outputs (in terms of emissions to air, water and as solid waste) are calculated, and these are aggregated over the lifecycle. These inputs and outputs are then converted into their effects on the environment, i.e. their environmental impacts. The sum of these environmental impacts then represents the overall environmental effect of the lifecycle of the product or service. Conducting LCAs for alternative products or services thus allows comparison of their overall environmental impacts. This will not necessarily determine that one option is 'environmentally superior' to another, but it will allow the trade-offs associated with each option to be assessed.

Similar assessments have been conducted for some time under a range of different titles, including Ecobalances, Ecoprofiles, Resource and Environmental Profile Analysis (REPA), and Produktlinienanalyse (PLA). An explanation of these and other terms and their important differences is given in Table 3.1.

Table 3.1 Lifecycle terminology

Term	Description
Lifecycle assessment	A process to predict the overall environmental impact of a product or service, over the whole lifecycle 'from cradle to grave', i.e. from raw material mining to final disposal; currently considered to consist of 4 stages: goal definition; inventory; impact analysis; valuation
Goal definition	Stage at which the functional unit for comparison is defined (normally per equivalent use), as well as the purpose, boundaries and scope of the assessment
Inventory	Process of accounting for all the inputs and outputs in the lifecycle; will result in a list of raw material and energy inputs, and of individual emissions to air, water and as solid waste
Impact analysis	Sometimes called 'classification'; conversion of the inventory of materials and energy into their actual environmental effects, e.g. ozone depletion; includes aggregation within such categories of environmental effect
Valuation	Evaluation of the overall environmental impact of the lifecycle by attempting to compare or aggregate the different categories of environmental concern resulting from the impact analysis; this stage is the least well accepted or defined, and involves subjective value judgements
Lifecycle analysis	Sometimes used synonymously with lifecycle assessment; however, early lifecycle analyses involved only the first two of the presently recognised stages, goal definition and inventory, so current usage usually relates to these stages only
Cradle to grave analysis	Essentially any analysis that encompasses the whole lifecycle, but may or may not include impact analysis; little used nowadays
Lifecycle inventory	Terms used by different practitioners for the first two stages of a lifecycle assessment, i.e. goal definition and inventory; they produce an inventory of materials and energy inputs and outputs
Resource and environmental profile analysis (REPA)	Terms used in two different ways; either equivalent to a lifecycle inventory (goal definition plus inventory stages) or to a lifecycle inventory with some subsequent attempt at impact analysis/valuation
Ecobalance/Environmental balance	Equivalent to a lifecycle assessment, although usually combine impact analysis and valuation in one single phase with an expert making the value judgement
Ecoprofile/Environmental profile	A broader assessment procedure, incorporating social and economic assessments as well as an environmental lifecycle assessment
Produktlinienanalyse (PLA)	

To date, lifecycle assessments have been mainly used by companies to predict and compare the environmental performance of their products and packages. Whilst mainly used as an internal tool to guide development and identify areas for environmental improvement, results of lifecycle assessments have been communicated to the general public or published in environmental journals (e.g. ENDS, 1991). Because of its perceived high profile, packaging has received considerable attention. Major studies have been published on the comparison of different packaging alternatives for beverages (Franke, 1983; Thomé-Kozmiensky and Franke, 1988; Boustead and Hancock, 1989; Boustead, 1993), and of the environmental impacts of different packaging materials (Tillman et al., 1991).

Several LCA computer models and spreadsheets for packaging are available commercially. Procter and Gamble (1992) has made its own life-cycle inventory spreadsheet (LCI1) freely available in order to stimulate widespread usage and improvements of lifecycle techniques.

Recently, lifecycle assessment requirements have also been included in legislation. The EC Ecolabelling Regulation (1992) requires that the whole lifecycle be considered when setting labelling criteria, and provision for life-cycle assessment (or Ecobalance) is also included in the latest draft of the EC Packaging and Packaging Waste Directive (1993). As it is developed further, it is likely that lifecycle assessment will find many additional applications.

3.2 Benefits of the lifecycle approach

Lifecycle assessment is an inclusive tool.   The inventory portion is essentially an accounting process, by which all necessary inputs and emissions in the lifecycle are considered. This includes not only direct inputs and emissions for production, distribution, use and disposal, but also indirect inputs and emissions, such as from the initial production of the energy used. It is essential that all of the processes are included to assess the overall environmental impact. Similarly, the analysis integrates over time, i.e. all inputs and emissions over the whole lifecycle, and includes all of the sites involved. If real environmental improvements are to be made, it is important that any changes do not cause greater environmental deteriorations at another time or another location in the lifecycle.

LCA offers the prospect of assessing the environmental impact of the total system, i.e. it is an holistic approach to environmental assessment. Comparing such impacts for different options, whether for different products or waste management systems, allows real environmental improvements to be identified.

LCA goes beyond single environmental issues.   Concern over the environment is sometimes expressed in terms of single issues. With regard to waste management, criteria such as compostability, recyclability or level of recycled content of products come to mind. Concentrating on one issue alone, however, ignores and may even worsen other environmental impacts. The power of LCA is that it allows us to expand the debate on environmental concerns beyond the single issue, and address the overall environmental effects. By using a quantitative methodology, it also gives an objective basis for decision-making, and so can take some of the emotional element out of environmental debates.

3.3 Structure of a lifecycle assessment

As a result of intensive recent efforts to define LCA structure and harmonise the various methodologies used (notably by the Society of Environmental Toxicology and Chemistry, SETAC), LCA is now considered to have four distinct stages: goal definition, inventory, impact analysis (or classification) and valuation (SETAC, 1992) (see Figure 3.1 and Box 3.1).

3.3.1 Goal definition

This first stage sets the goals for the LCA study, defines the options that will be compared and the intended use of the results. As will be made apparent, the intended use for the LCA will influence the type of study carried out, and the type of data required.

A fundamental part of this stage is the definition of the 'functional unit'. This is the unit of the product or service whose environmental impacts will be compared. The importance of defining the most appropriate functional unit cannot be over-emphasised. It is often expressed in terms of amount of product (e.g. per kilogram or litre), but should really be related to the amount of product needed to perform a given function, i.e. per equivalent use. For a car, for example, the functional unit could be per kilometre travelled. As all of the impacts are calculated per functional unit, any alteration in the size of the functional unit, e.g. to allow for higher performance of one of the products compared, will have a major effect on the outcome of the assessment.

Another essential requirement is the definition of system boundaries, i.e. what is included within the assessment, and what is omitted. Both the 'length' and 'breadth' of the study need to be defined.

The 'length' involves defining exactly where the cradle and the grave of the lifecycle lie. Should the mining of the raw materials be included? Similarly should the grave include the emissions from used materials after they have been buried in the ground, i.e. landfilled? The 'breadth' of the study will determine how much detail will be included at each stage of the

---

BOX 3.1
THE STAGES OF A LIFECYCLE ASSESSMENT

1. **Goal Definition** (now a routine procedure)
   Define:-
   - Options to be compared
   - Intended use of results
   - The functional unit
   - The system boundaries

2. **Inventory** (now a routine procedure)
   Account for:-
   - All materials and energy
   - Inputs
   - Outputs
        across the whole lifecycle.

3. **Impact Analysis** (undergoing active development)
   - converts the LCI into environmental effects
   - methodology to aggregate according to effects being developed

4. **Valuation** (currently little agreement on how to perform)
   - The process of balancing the importance of different effects
   - No agreed scientific method
   - Requires public debate

   Note: A lifecycle inventory (LCI) - the basis of this book - includes the goal definition and inventory stages

---

lifecycle. For example, the impacts of the manufacturing process for the product itself will be included in the assessment. Should the impacts of making the factory and equipment that make the product also be included? In this instance, previous studies have shown that such 'second level' impacts, when divided between the number of units that the factory produces, are insignificant, so can be omitted from most analyses.

It should be stressed that a lifecycle assessment can be done on any system, whatever the boundaries defined. There are no right or wrong boundaries to choose, but some are more appropriate to the defined goal

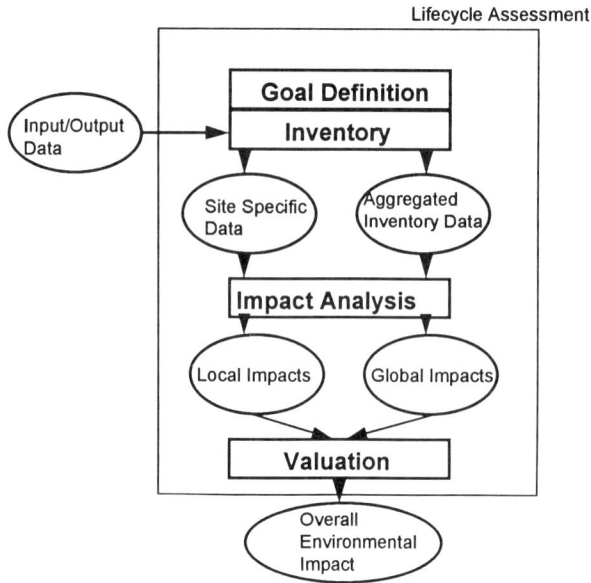

Figure 3.1 Structure of a full lifecycle assessment (from Hindle *et al.*, 1993).

than others. Where valid comparisons are required, then the same boundaries must be used in each case.

In any lifecycle assessment, defining the functional unit and the boundaries of the system assessed are both important steps. Reporting the functional unit and the boundaries used are as important as giving the actual results.

3.3.2 Inventory

This stage, formerly referred to as lifecycle analysis (White *et al.*, 1993), consists of accounting for all of the material and energy inputs and outputs over the whole lifecycle of the product or service. This entails describing the lifecycle as a series of steps, and then calculating the inputs and outputs for each of these steps (Figure 3.2). This amounts to constructing a materials and energy balance for each step in the lifecycle, a process that will be familiar to chemical engineers. Inventories for each stage in the lifecycle can then be combined to give the overall lifecycle inventory.

Such analyses have been conducted for many years, initially as energy analyses and more recently including materials balances as well (Hunt *et al.*, 1974, 1992). Consequently, the methodology of this stage is well established. A few areas of disagreement still remain, however, notably

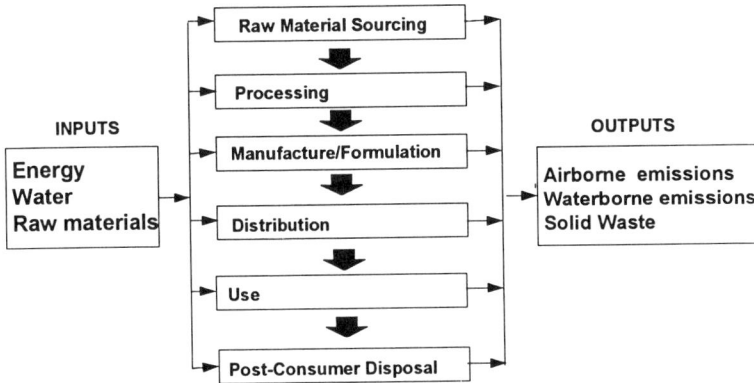

Figure 3.2 The stages of a product or package lifecycle.

around the problem of co-allocation of emissions and energy consumption. Where two or more products arise from the same process, it is necessary to divide the emissions and energy consumption associated with the process between the products, but there are several ways in which this allocation can be made (by relative mass, energy content or even economic value) (Huppes, 1992). This is an area where clear conventions need to be agreed to standardise future studies.

These fine details aside, the methodology for the inventory stage is generally accepted and well used. A lifecycle inventory describes all of the relevant material and energy flows. The large amount of data generated makes decision-making challenging. There is a need to find methods of aggregating data but this is full of dangers. There is also a need to translate the material and energy flows (especially the outputs) into environmental impacts.

Notwithstanding these limitations, an LCI provides enormous knowledge about and insights into the environmental consequences of a given system. A lifecycle inventory (LCI), as used in this volume, involves both the goal definition and inventory stages.

3.3.3   Impact analysis

Alternatively known as 'classification', this involves converting the lifecycle inventory of materials and energy into their environmental effects. Past attempts at impact analysis have often aggregated all emissions to each environmental compartment, i.e. to air, to water, as solid waste. In some cases, the emissions have simply been added together to give a total mass of emissions (e.g. Hunt *et al.*, 1974), regardless of the nature of the materials emitted. An improvement on this has been to weight the emissions, according to their toxicity, before aggregating them. This system of calculating 'critical volumes' of polluted air and water has been extensively used (BUS, 1984; Habersatter, 1991), but this approach also has its limitations. Emissions to air can contribute to a range of different environmental problems, such as ozone depletion, global warming and acid-rain formation in addition to toxic effects, as can emissions to water. Furthermore, one emitted chemical can contribute to more than one problem (CFC-11 causes ozone depletion and has a potent greenhouse effect). Indeed, the basic concept of aggregating according to different physical media is in itself questionable, since some outputs can have effects in more than one medium. Organic solid waste in landfills, for example, can produce methane and leachate, so can affect soil, air and water.

It is now generally accepted that the most meaningful aggregation of impacts will be according to their effects on the environment (SETAC, 1992). Thus, all releases that contribute to ozone depletion, for example, will be aggregated together, each weighted by their relative contribution to the problem. Impact analysis can then be considered to consist of four steps (Guinée, 1992):

- a classification of all effects into a surveyable number of environmental problems, e.g. contribution to global warming;
- the definition of units to measure these classes of problems, e.g. units of global warming potential;
- a conversion of the inventory inputs and outputs into these units, e.g. amounts of CO_2 and methane into their global warming potential;
- an aggregation of the units within each effect, e.g. adding up the total global warming potential of all the emissions.

The result of the impact analysis will be the contribution of the lifecycle to a surveyable number of environmental problems. The details of different impact analysis schemes differ slightly (Finnveden *et al.*, 1992; Guinée, 1992) but are in general agreement on the main effects to be considered (Table 3.2).

3.3.4 Valuation

Whilst there are significantly fewer impact categories than inventory categories, there are still many environmental issues to be considered. Comparing one lifecycle option with another will not normally show which is 'environmentally superior' (except in the case where one has a lower impact in all categories), but will demonstrate the trade-offs between the two options.

To facilitate decision-making between options, attempts have been made to aggregate the outputs of a lifecycle assessment further, ultimately to a

Table 3.2 Classification of lifecycle impacts

Heading	Scale of effect
Depletion	
– Non-renewable resources	Global
– Energy carriers	
– Others (minerals)	
– Scarce renewable resources	Global/local
– Space consumption	Regional/local
Pollution	
– Global warming	Global
– Ozone depletion	Global
– Human toxicity	Continental/regional
– Ecotoxicity	Continental/regional
– Acidification	Continental/regional
– Photo oxidant formation	Regional
– Nutrification/eutrophication	Regional
– Final solid waste	Regional/local
– Radiation	Regional/local
– Dispersion of heat	Local
– Occupational health	Local
– Noise	Local
– Smell	Local
Disturbances	
– Desiccation	Regional/local
– Landscape degradation	Regional/local
– Ecosystem degradation	Regional/local

From: Udo de Haes (1992) with amendments.

single environmental score. This requires some form of weighting of the importance of the different environmental problems in Table 3.2. This is very difficult at present, since there are no generally accepted environmental objectives against which to make such weightings. As a result, attempts so far have relied on scoring systems from 'expert panels' (which are subjective) (e.g. Landbank, 1991), comparing emissions against Government targets (which are political and not global) (e.g. the Ecopoints approach, Ahbe *et al.*, 1990), or by converting all impacts into monetary units. None of these schemes has been widely accepted. For the foreseeable future, decision-making on the basis of the impact analysis results should be done by open public debate as part of the democratic process. Any scheme that aggregates the impact categories to a single score may appear to make decision-making easier, but in the process obscures the assumptions and priorities upon which the decision will be based. Thus, broad acceptance of the outcome of the decision is uncertain. That verdict will only be passed when the results become evident to those affected.

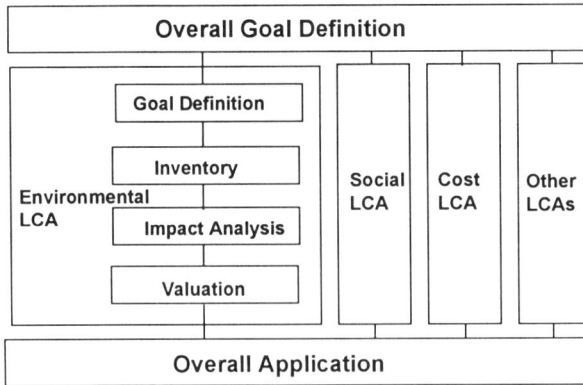

Figure 3.3 The relationship of different types of lifecycle assessment.
The whole process, consisting of environmental, social and economic analyses, is known as
product line analysis (PLA) (from Assies, 1992).

3.4 Current state of development

As discussed above, the first two stages of a lifecycle assessment are now established and can be performed routinely. Impact analysis is still undergoing active development, whilst there is little agreement about the form, or even the need, for a formal valuation stage. Consequently, at present, many studies concentrate on producing a lifecycle inventory (LCI), and this is the approach taken in this book. Where methods for converting the inventory into actual effects are available, they will be applied, but this will not be possible in all cases. Methods for aggregating global effects, such as ozone depletion, are already available; methods for aggregating local effects such as toxic releases to aquatic systems, noise and smell (many of which are especially relevant to waste management) have yet to be agreed.

3.5 Environmental and economic lifecycle assessments

This discussion has been limited to environmental lifecycle assessment, sometimes abbreviated to LCA_{env}. Other lifecycle considerations need to be addressed in any overall assessment, in particular, economic and social factors. Some schemes have tried to include these within an environmental lifecycle assessment, but they make the whole procedure complex and unwieldy, and may limit the application of LCA.

In waste management, as in other activities, an economic lifecycle assessment is an essential part of any overall assessment. It should be done in parallel with an environmental LCA, and the outcomes of the two assessments evaluated together (see Figure 3.3).

3.6 Lifecycle inventory in reverse

A lifecycle inventory (and assessment) may be used to improve the environmental performance of a product or service. Given that the product will be used and disposed of in a fixed disposal system (i.e. relative levels of recycling/incineration/landfill), it is possible to determine how changes in the product affect its overall environmental impact. However, it is possible to conduct 'lifecycle (inventory) in reverse' (White *et al.*, 1993), by keeping the product constant and changing the disposal conditions to see how this affects the overall environmental impact.

This is essentially what is involved in a lifecycle inventory of waste. Assuming that the composition of waste produced is fixed, and details of its composition are known, at least for household waste, it is possible to determine how the use of different options for waste management affect the overall environmental impact. This technique is used throughout this book, and the general format of an LCI for waste is explored in Chapter 4.

References

Ahbe, S., Braunschweig, A. and Müller-Wenk, R. (1990) Methodik für Oekobilanzen auf der Basis ökologischer Optimierung. Bundesamt für Umwelt, Wald und Landschaft (BUWAL), Report No. 133. Bern, Switzerland.

Assies, J.A. (1992) State of art. In *Life-Cycle Assessment*. Society of Environmental Toxicology and Chemistry – Europe (SETAC), Brussels, pp. 1–20.

Boustead, I. (1993) Resource Use and Liquid Food Packaging. EC Directive 85/339: UK Data 1986–1990. A report for INCPEN.

Boustead, I. and Hancock , G.F. (1989) EEC Directive 85/339. UK Data 1986. A report for INCPEN (Industry Council for Packaging and the Environment). The Open University, UK.

BUS (1984) EMPA (1984). Oekobilanzen von Packstoffen. Bundesamt für Umweltschutz, Report No. 24. Bern, Switzerland.

EC (1992) Ecolabelling regulation. *Off. J. Eur. Commun.* **L99** (1992).

EC (1993) Draft Packaging and Packaging Waste Directive.

Elkington *et al.* (1993) *The LCA Sourcebook*. SustainAbility, SPOLD and Business in the Environment, 112 pp.

ENDS (1991) P & G throws down gauntlet on aerosols with propellant-free refillable sprays. *Environ. Data Services Rep.* **199**, 23–24.

Finnveden, G., Andersson-Sköld, Y., Samuelsson, M.-O., Zetterberg, L. and Lindfors, L.-G. (1992) Classification (impact analysis) in connection with lifecycle assessments – a preliminary study. In *Product Lifecycle Assessment – Principles and Methodology*. Nordic Council of Ministers, 172–231.

Franke, M. (1983) *Umweltauswirkungen durch Getraenkeverpackungen – Systematik zur Ermittlung von Umweltauswirkungen am Beispiel der Mehrweg-Glasflasche und des Getraenkekartons*. E. Freitag Verlag, Berlin.

Guinée, J.B. (1992) Headings for classification. In *Life-Cycle Assessment. Report of the SETAC Workshop*, Leiden. Society of Environmental Toxicology and Chemistry, Brussels, pp. 81–85.

Habersatter, K. (1991) Oekobilanz von Packstoffen Stand 1990, Bundesamt für Umwelt, Wald und Landschaft (BUWAL) Report No. 132, Bern, Switzerland.

Hindle, P. White, P.R. and Minion, K. (1993) Achieving real environmental improvement using value: impact assessment. *Long Range Planning* **26**(3), 36–48.

Hunt, R.G., Franklin, W.E., Welch, R.O., Cross, J.A. and Woodal, A.E. (1974) *Resource and Environmental Profile Analysis of Nine Beverage Container Alternatives.* Midwest Research Institute for US Environmental Protection Agency, Washington DC.

Hunt, R.G., Sellers, J.D. and Franklin, W.E. (1992) Resource and environmental profile analysis: a life cycle environmental assessment for products and procedures. *Environ. Impact Assessment Rev.* **12**, 245–269.

Huppes, G. (1992) Allocating impacts of multiple economic processes in LCA. In *Life-Cycle Assessment. Report of the SETAC Workshop*, Leiden. Society of Environmental Toxicology and Chemistry, Brussels, pp. 57–70.

Landbank (1991) Packaging – An Environmental Perspective. A Gateway Foodmarkets report prepared by Landbank Consultancy, UK.

Procter & Gamble (1992) Lifecycle inventory for consumer goods packages. A copy of this spreadsheet can be obtained from Procter and Gamble European Technical Center, Temselaan 100, B-1853 Strombeek-Bever, Belgium.

SETAC (1992) *Life-Cycle Assessment. Report of the SETAC Workshop*, Leiden, Society of Environmental Toxicology and Chemistry, Brussels, pp. 57–70.

Thomé-Kozmiensky, K.J. and Franke, M. (1988) *Umweltauswirkungen von Verpackungen aus Kunststoff und Glas.* EF-Verlag fuer Energie- und Umwelttechnik, Berlin.

Tillman, A.-M., Baumann, H., Eriksson, E. and Rydberg, T. (1991) Lifecycle analyses of selected packaging materials, report commissioned by Swedish National Commission on Packaging (draft).

Udo de Haes, H.A. (1992) Workshop conclusions on classification session. In *Life-Cycle Assessment. Report of the SETAC Workshop*, Leiden. Society of Environmental Toxicology and Chemistry, Brussels, pp. 95–98.

White, P.R., Hindle, P. and Dräger, K. (1993) Lifecycle assessment of packaging. In *Packaging in the Environment*, ed. G. Levy. Blackie, Glasgow, pp. 118–146.

4   A lifecycle inventory of solid waste

Summary

The lifecycle inventory technique described in Chapter 3 is applied to waste management. The possible uses for an inventory of different waste management options are discussed. The functional unit for the comparison is defined, as are the system boundaries. This includes defining the 'cradle' and 'grave' for waste. The general structure of waste management systems, which forms the basis of the LCI model, is mapped out, and the computer spreadsheet developed to conduct the LCI is introduced.

4.1   Integrated waste management and lifecycle inventory

The objective of integrated waste management is to deal with society's waste in an environmentally and economically sustainable way (Chapter 2). To assess such sustainability, we need tools that can predict the likely overall cost and environmental impacts of any system. Lifecycle assessment is an emerging environmental management tool that allows prediction of the likely environmental impacts associated with a product or service over the whole lifecycle, from 'cradle to grave'. This technique can be usefully applied to waste management to assess environmental sustainability. At the same time, a parallel economic lifecycle assessment can determine the economic sustainability of waste management systems, a criterion equally crucial to their successful implementation.

As described in Chapter 3, a lifecycle assessment is comprised of four stages: goal definition, inventory, impact analysis and valuation (Box. 3.1). Currently the goal definition and inventory stages (which together comprise a lifecycle inventory study) are routinely carried out in various applications; impact analysis and valuation present significant challenges. This book and the associated computer spreadsheet are based on a lifecycle inventory of solid waste. This chapter addresses the concept and practicalities of using lifecycle assessment to compare waste management systems.

Whilst the lifecycle technique has been used to compare specific options for waste disposal (e.g. Kirkpatrick, 1992), it has not previously been used to assess complete integrated waste systems. Consequently, it is necessary first to address such basic questions as where is the cradle of waste, and where is its grave?

4.2   A lifecycle inventory of waste

4.2.1   Goal definition

The first stage of a lifecycle inventory, the 'goal definition stage', addresses three major questions, namely, What is the purpose of the study?, What will be compared, i.e. what is the functional unit for comparison? and What are the boundaries of the system? (Box 4.1) This last question defines what will be included in the study and what will be omitted, and specifies the 'length' and 'breadth' of the study.

What are the purposes of the lifecycle inventory?

- To predict both environmental performance (in terms of emissions and energy consumption) and economic costs of an integrated waste management system. Because specific data for all parts of the lifecycle are not available, and thus generic (typically averaged) data will be frequently used, the result of the inventory will not be 100% accurate. However, it will provide a 'first cut' and will provide rough comparisons between different system options. The objective of predicting environmental performance of waste management systems can be met in two ways. Detailed lifecycle inventory studies can be run for several individual waste management systems, and general conclusions extrapolated from the results. The alternative is to construct a generic, flexible tool that can be applied to any waste management system to assess the overall environmental performance. Constructing a model that is flexible enough to describe all possible waste management scenarios is a very challenging task, but is the option attempted in this book. A general model such as this will rely on generic data, so will not give such accurate results as specific studies that describe particular waste management systems. However, the flexibility to apply the same model to most waste management systems, both existing or planned, is considered to outweigh this.
- To demonstrate the interactions that occur within waste management. As it attempts to model the whole waste system, a lifecycle model will show how different parts of the system are inter-connected and will aid understanding of the system's behaviour.
- To clarify the objectives of the waste management system. Above, it has been argued that the objective of waste management is environmental and economic sustainability, i.e. minimising the environmental impacts for an acceptable cost. Because it specifically calculates both the cost and individual environmental impacts (i.e. energy consumption, air emissions, water emissions, landfill requirements, etc.) it focuses attention on which parameters need to be maximised or

---

Box 4.1

A LIFECYCLE INVENTORY OF WASTE: GOAL DEFINITION.

1. Options to be compared:
- Different systems for managing solid waste

2. Purposes:
- To predict environmental performance (emissions and energy consumption)
- To predict economics costs
- To allow "What if..?" calculations
- To support achieving environmental and economic sustainability
- To demonstrate interactions within integrated waste management
- To supply waste management data for use in individual product LCIs

3. Functional unit:
- To manage the household and similar commercial waste from a given geographical area.

4. System boundaries (environmental LCI):
- Cradle (for waste): when material ceases to have value and becomes waste (e.g. the household dustbin)
- Grave: when waste becomes inert landfill material or is converted to air and/or water emissions or assumes a value (intrinsic if not economic)
- Breadth: "second level" effects such as building of capital equipment ignored. Indirect effects of energy consumption included.

5. System boundaries (economic LCI):
- Cradle: as for environmental LCI
- Grave: as for environmental LCI
- Breadth: capital equipment costs included

---

minimised. N.B. A lifecycle assessment will not in itself decide this, but will provide the data on which these societal/political decisions can be based.
- To allow for 'What if . . .?' calculations. The use of a simple computer spreadsheet for the lifecycle inventory allows any user to compare a number of hypothetical waste management systems and find their relative economic costs and environmental impacts.
- To provide data on waste management methods that can be used in LCI studies of individual products and packages.

Defining the functional unit: what will be compared?   The functional unit is the unit of comparison in a lifecycle inventory. Most LCI studies conducted to date have been related to products, such as washing machines (DoE/DTI, 1991) or detergents (Stalmans, 1992), or packages (White *et al.*, 1993). The functional unit in such cases relates to the product/package made, and the comparisons are made on the basis of per amount or per equivalent use of the product. The functional unit is therefore expressed in terms of the system's output (Box 4.2). Such studies are usually run to see how changes to the product will affect their overall environmental impacts.

The function of a waste management system, in contrast, is not to produce anything, but to deal with the waste of a given area. Therefore, the functional unit in an LCI of waste is the waste of the geographical area under study. In this study, this functional unit is refined further into the household and similar commercial waste of the specified geographical area. This is a key difference in approach: in a product LCI, the functional unit is defined by the output (product) of the system; in an LCI of waste, the functional unit is defined in terms of the system's input, i.e. the waste.

Using this LCI method, different systems for dealing with the solid waste of a given area can be compared. The geographical area under study, and the waste produced by this area, are defined by the user, and the LCI model will calculate the overall cost and environmental impacts of different options chosen for dealing with this waste. The waste input from any given area will be constant, and the LCI can be used to assess the overall performance, both environmental and economic, of different waste handling systems.

The impacts and costs for the whole system can be broken down further, however. For economic costs, it is useful to have the cost attributed per household, since revenue is usually collected in this way for domestic solid waste. Alternatively, costs can be attributed per tonne of waste collected, which is especially relevant to commercial/industrial waste where charges are levied on such a basis. As well as overall system costs and impacts, costs and impacts per tonne and per household will also be calculated, but neither alone is satisfactory.

A figure that is often quoted in reports of recycling schemes is the cost per tonne of recovered or recycled material (usually high!). This will not be used as the functional unit in this assessment, since it is not the function of an integrated waste management scheme to produce recycled material. The objective is to deal with waste in an economically and environmentally sustainable way; recovering and recycling materials is a means to this end but not the end in itself.

System boundaries: where is the cradle of waste and where is the grave?
All lifecycles run from cradle to grave, hence the often interchanged terms

---

BOX 4.2

KEY DIFFERENCES between LCI STUDIES FOR PRODUCTS and an LCI STUDY OF WASTE.

1. Functional Unit

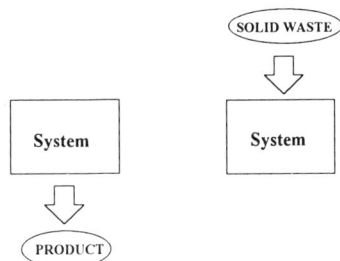

In a product LCI the functional unit is defined in terms of the system's output i.e. the product. For example, per kg of product made, or per number of laundry loads washed.

In the LCI of waste the functional unit is defined in terms of the system's input, i.e. the waste. The functional unit could be the management of the waste of one household or the total waste of a defined geographical region.

2. Boundaries - Definition of Cradle and Grave.
Product LCI studies consider the whole lifecycle of one particular product, from raw material extraction, through manufacture, distribution and use, to final disposal. The last part of the lifecycle will be spent as waste in a waste management system (Fig. 4.1). This book considers the lifecycle of waste, from the moment it becomes waste by losing value, to the moment it regains value or leaves the waste management system as an emission.
A Product LCI considers the whole lifecycle of a single product; an LCI of waste includes part of the lifecycles of all products.

3. Using What if...? Scenarios.
Product LCIs are normally used to determine the environmental effect of changes to the product. This LCI of waste can be used to determine the environmental effect of changes to the waste management system.

'lifecycle inventory' and 'cradle to grave inventory'. When considering the lifecycle of products, inventories usually go back to the source of the raw materials, by mining for example, to define the product's 'cradle'. The 'grave' is the final disposal of the product, often back into the earth as land-fill. Whilst it shares the same grave as individual products, the lifecycle of 'waste' does not share the same cradle (Figure 4.1). Waste only becomes waste at the point at which it is thrown away, i.e. ceases to have any value

(a)

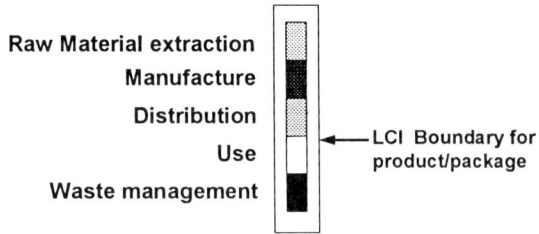

Raw Material extraction
Manufacture
Distribution
Use
Waste management

LCI Boundary for product/package

(b)

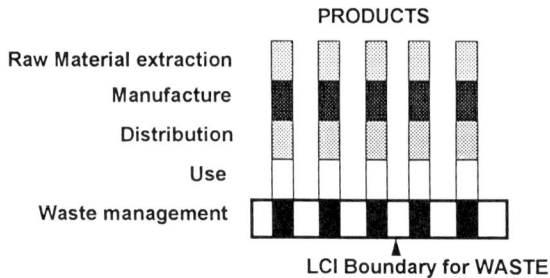

PRODUCTS

Raw Material extraction
Manufacture
Distribution
Use
Waste management

LCI Boundary for WASTE

Figure 4.1 The lifecycle of a product (a) versus the lifecycle of waste (b). The LCI for solid waste considers part of the lifecycle of all products and packages.

to the owner. Thus, the 'cradle' of waste, in households at least, is usually the dustbin. This is another key difference between an LCI for a product and an LCI for waste. Every product spends part of its lifecycle as waste. Conversely a lifecycle study of waste includes part (but only part) of the lifecycle of every product or package.

Household and similar commercial waste can be either collected or delivered (e.g. to bottle banks) in a variety of ways, so for a practical boundary, the cradle of such waste in this study is taken as the point at which it leaves the household or commercial property.

Prior to this stage, individual items in the waste stream will be affected by source reduction, waste minimisation and other processes for environmental improvement. Whilst these are valuable contributions, they occur during product manufacture, distribution or use and are thus upstream of waste management. As a result, they are not included here. Other waste treatment methods can also be used to reduce the amount of solid waste within households, prior to collection. Home composting of organic material or the burning of waste on domestic fires would be two such examples. Such treatment methods will similarly not be considered within the boundaries of this LCI, since they occur on-site, and prior to the materials leaving the household or commercial property.

The 'grave' of the waste lifecycle is its final disposal back into the environment. Incineration and landfilling are often described as 'final disposal' options, but neither represents the true end of the lifecycle for the materials involved. Incineration produces ash, for example, which then needs disposal, often by landfilling. Similarly, landfills are not the final resting place for some of the materials contained, since they can in turn release gas emissions and leachate. It has been suggested that landfills should not be considered as environmental impacts that should be measured, but as waste management processes that in turn produce gas and water emissions with measurable impacts (Finnveden, 1992).

The 'grave', or end of the waste lifecycle used here, is considered to be when it becomes inert landfill material, or is converted into air or water emissions. Alternatively, the waste can regain some value (as compost, secondary material or fuel) and thus ceases to be waste. 'Value' normally implies positive economic value.

Defining this exact point when waste acquires value and ceases to be waste has important practical consequences, as well as being essential in this analysis. Different regulations generally apply to handling and transporting raw materials rather than waste, and different emission levels are allowed when burning fuels rather than burning waste. Thus, a clear definition of exactly when waste becomes a secondary raw material or a fuel is essential.

One area where difficulties arise is the recovery of materials for recycling. Once materials are collected and sorted, they regain value as a secondary raw material. This value equates to positive economic value in many cases, where sorted materials are sold from material collection banks or from material recovery facilities (MRFs). Consequently, this has been chosen as the boundary of the waste management system for such materials. In some other schemes, however, where collected material supply greatly exceeds recycling capacity, the collected material may still have a negative economic value at this point. This describes the present situation in Germany, for example, where plastic material is offered for recycling along with a subsidy for the recycling process. Such material would not then regain value until after it had been converted into recycled resin pellets. Choice of this point as the boundary would mean including the whole recycling industry within

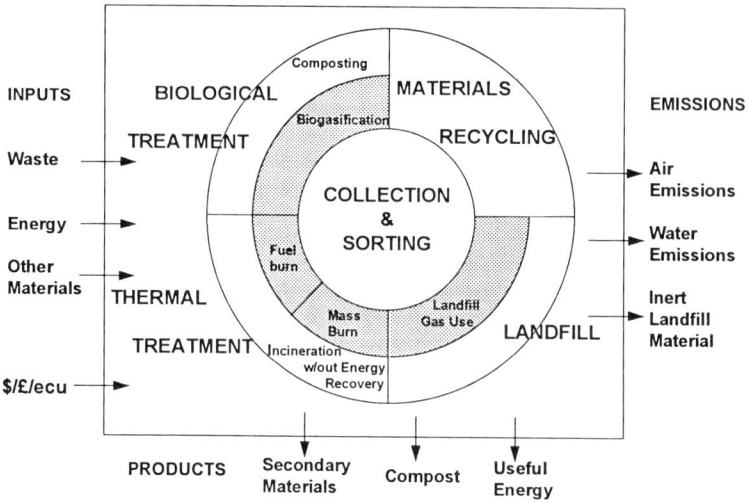

Figure 4.2 System boundaries for the environmental lifecycle inventory of solid waste. Shaded area represents waste-to-energy.

the waste management system. Since many recycling processes, e.g. for steel, aluminium, paper, etc. are integrated within the virgin material industries, this would necessitate taking most of industry into the waste management system, which would make system modelling unwieldy, if not impossible.

Since all industries are inter-related, defining hard and fast boundaries is not easy, but some boundaries must be chosen. For this study, recovered materials are considered to leave the waste management system when they leave the MRF or are collected from material banks. They then enter the plastics/paper/metal/glass industry system. These reclaimed materials will eventually replace other, virgin, materials in the manufacture of new products. In many cases this will lead to environmental improvements via energy savings, and reduced raw material consumption. These potential savings (or costs) are not included within the boundaries of this LCI study, but are calculated in Chapter 8 so that their significance can be assessed.

The general system boundaries for an integrated waste management system are shown diagrammatically in Figure 4.2. Along with the waste itself, there are energy and other raw material inputs (e.g. petrol, diesel) to the system. The outputs from the system are useful products in the form of reclaimed materials and compost, emissions to air and to water and inert landfill material. Energy will also be produced in waste to energy options (which also includes use of recovered landfill gas); combining this with the energy inputs to the system gives a net value for energy consumption/ generation. The exact boundaries at which materials and energy enter or leave the waste management system studied here are defined in Box 4.3.

Box 4.3 BOUNDARY DEFINITIONS FOR THE LCI OF WASTE.

	System boundary	Units
Inputs		
Energy	Extraction of fuel resources	GJ thermal energy
Outputs		
Energy	Electric power cable leaving waste-to-energy facility (The electrical energy generated is subtracted from the energy consumed, so is effectively used within the system, and not exported)	
Recovered materials	Material collection bank *or* Exit of Material Recovery Facility *or* Exit of RDF plant *or* Exit of biological treatment plant	tonnes
Compost	Exit of biological treatment plant	tonnes
Air emissions	Exhaust of transport vehicles *or* Stack of thermal treatment plant i.e. after emission controls *or* Stack of power station (for electricity generation) *or* Landfill lining/cap	kilograms
Water emissions	Outlet of biological treatment plant *or* Outlet of thermal treatment plant *or* Outlet of power station (electricity production) or landfill lining	kilograms
Final Solid Waste	Content of landfill at end of biologically active period	tonnes or cu. metres of inert material

---

For the energy inputs into the systems (as fuels, petrol, diesel, gas or as electricity), in addition to the energy content delivered, energy will have been expended in drilling, mining, and/or production. These processes would also have generated emissions to air and water, and solid waste. Consequently these impacts associated with energy production will also be included whenever energy is consumed within this lifecycle inventory. Effectively, the energy production industries are included within the system boundaries of this LCI study, although they are not depicted in Figures 4.4–4.6.

For any energy recovered by the waste management system, it is assumed that this is converted into electrical power only. Subtracting the amount of electrical energy produced from that consumed gives the net amount of electrical energy consumed by the waste management system.

The system boundaries for the parallel economic lifecycle assessment are shown in Figure 4.3. Economic outputs of a waste management system include costs for collection, sorting, various forms of treatment, transport

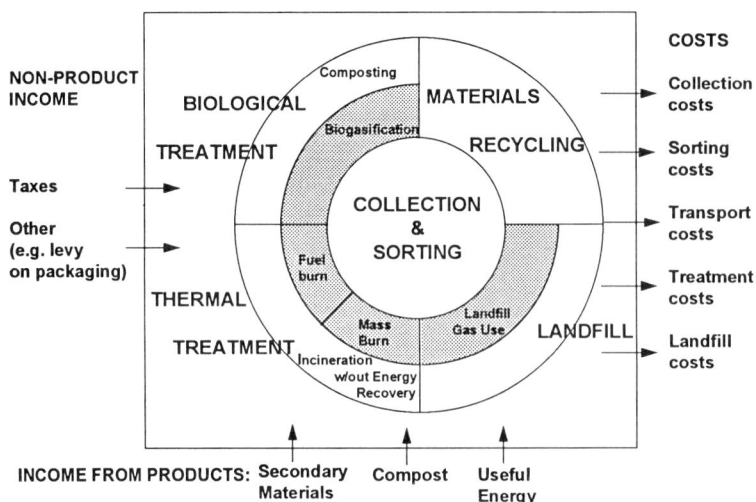

Figure 4.3 System boundaries for an economic lifecycle inventory of solid waste.

and for final disposal to landfill. Revenues produced by the system come from the sale of reclaimed materials, compost and energy. Subtracting the revenues from the costs will give the net cost of operating the system.

As discussed above, reclaimed materials will displace virgin materials, so if the reclaimed materials are produced for less than virgin materials, further economic savings occur. As with the corresponding environmental savings, these cost savings are not included within the boundaries of the basic LCI system defined here (Figure 4.3), but they are calculated in Chapter 8.

System boundaries: what is the 'breadth' of the present study? The breadth of an LCI study determines the level of detail included. Should the impacts of building the trucks that carry the waste be included in the overall assessment? What about the impacts of constructing incinerators?

In most lifecycle inventories to date, such 'second level' impacts are insignificant, when spread over the lifecycle of the truck/plant, and so are normally omitted. This practice will be followed in this study for the environmental impacts of waste management. It cannot be justified, however, when considering the economic costs of waste management. The capital cost of installing plant and vehicles is clearly significant in comparison to running costs, since it must be financed and interest must be paid. When considering the economic costs, therefore, the full inclusive cost of waste management will be addressed, including purchase of capital equipment, depreciated over a suitable time span.

4.2.2 The inventory stage

The inventory stage looks at all of the inputs and outputs in the lifecycle of waste. The first step, however, is to define the lifecycle. Since the objective of the LCI is to be able to describe the majority of waste management systems, existing or planned, all possible processes and combinations of processes need to be possible. The main stages, and their interconnections in the lifecycle of solid waste are shown diagrammatically in Figure 4.4, comprising pre-sorting and collection, central sorting, biological treatment, thermal treatment and landfilling. It is then necessary to consider the processes within each stage (Figure 4.5), and list all materials and energy entering and leaving each process. By linking up all processes within each stage and then all stages in the lifecycle, it is possible to define the overall waste management system. This is shown in Figure 4.6 and Box 4.4, which also define the boundaries of the system, and is the basis on which the lifecycle inventory is performed. Materials enter the system primarily as the waste, and leave the system when they are converted into recovered materials or compost, emitted to air or water, or are deposited as solid waste in landfills.

A lifecycle inventory for solid waste management consists of two main steps. Firstly, the waste management system to be considered must be described. This involves choosing between the different possible waste treatment options. Such variable data are needed to show how the waste is treated in any given system, and what route materials take through the system, e.g. how much of the household waste is separated at the kerbside, how many fractions it is sorted into, whether organic material is composted separately or left with the residue, what proportion is burned as opposed to landfilled directly and so on. Answers to these questions must be chosen by the user, and can be altered freely to carry out 'What if...?' calculations.

Secondly, the inputs and outputs of the chosen processes must be calculated, using fixed data for each process. Such fixed data are dependent on the performance of the equipment and technologies involved and are expressed relative to the amount of material treated. For example, the energy generation and emissions per tonne of waste treated in a waste-to-energy plant.

Fixed data are generally lacking for most routine manufacturing processes, let alone for waste management. The lack of quality data is a serious problem in any LCI of waste, and has probably been one of the reasons why this has not been attempted before. Suitable data sources are now beginning to emerge, however, and will be used in this study.

For data on collection and sorting of waste, ERRA, the European Recovery and Recycling Association, has helped set up eight multi-material collection and sorting schemes across Europe, and another two are at an

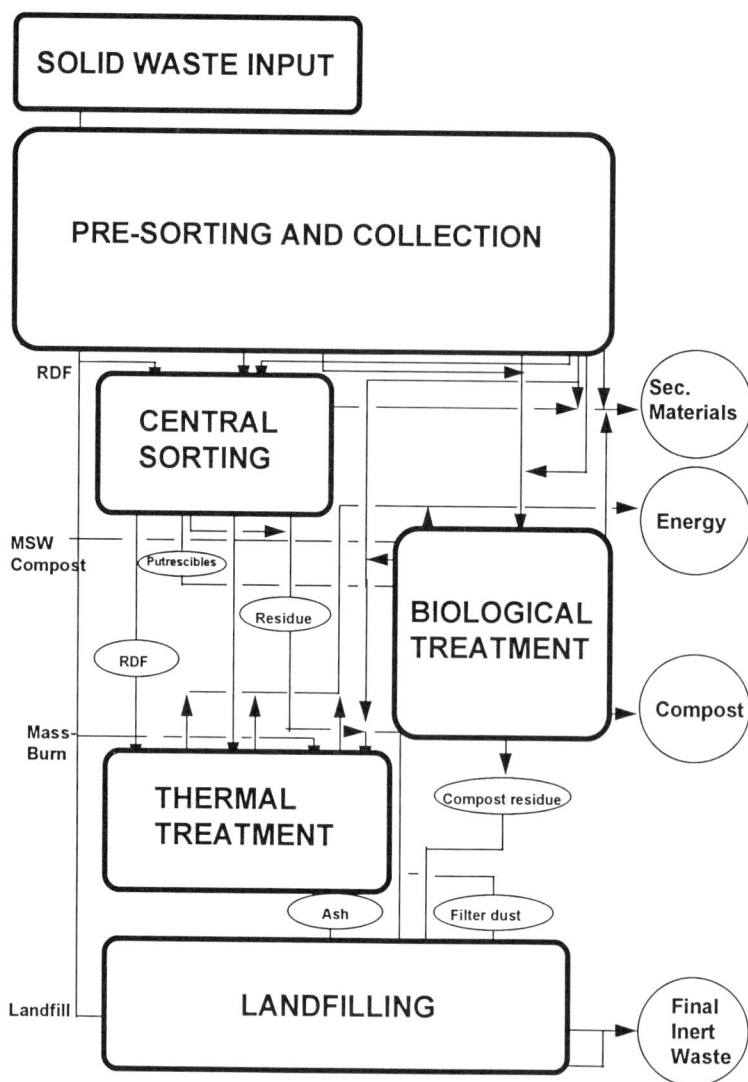

Figure 4.4 Components of an integrated waste management system.

advanced planning stage (ERRA, 1991). Details of the amounts of materials collected and the costs of the schemes are systematically collected into the ERRA database, along with details of other non-ERRA schemes. This extensive database will be made available to interested parties for a signing-on fee. Similarly, in Germany, the Dual System operated by DSD (Duales System Deutschland), collects a wide range of packaging materials, and

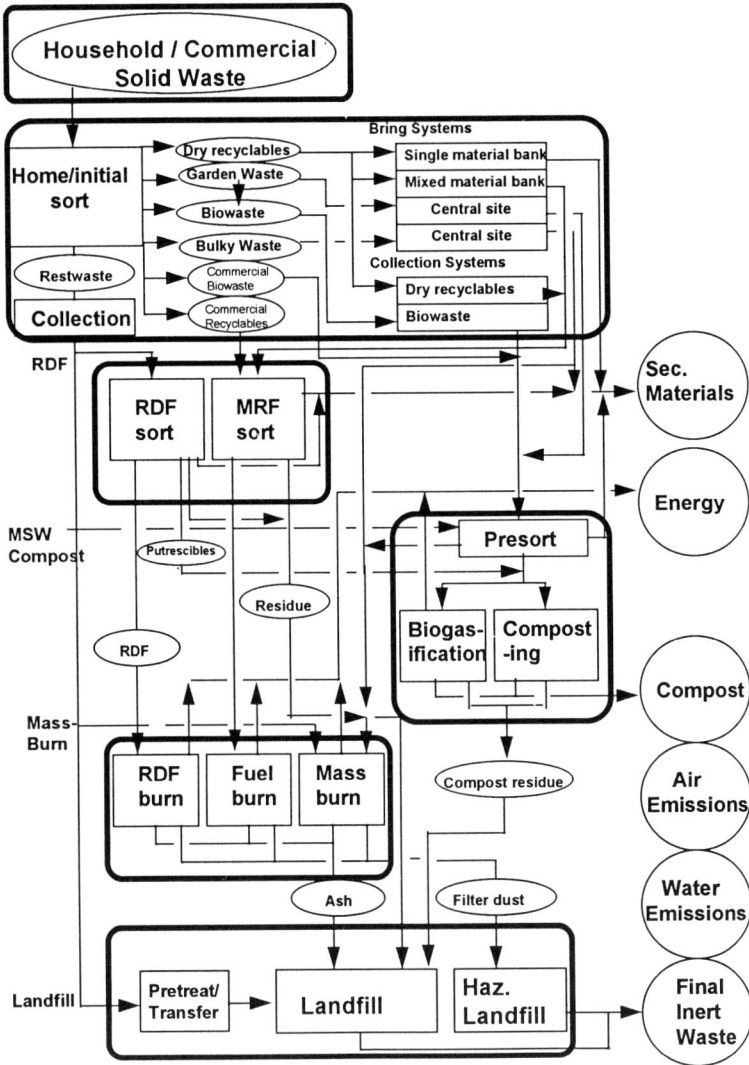

Figure 4.5 Detailed structure of an integrated waste management system.

useful data on the quality and quantities of different materials are beginning to emerge.

Data on general processes such as transportation are available from other LCI reports (e.g. Habersatter, 1991). Data on the performance of waste treatment processes are available from the technical literature (e.g. Vogg, 1992; Clayton *et al.*, 1991, for incineration/waste to energy; Coombs, 1990, for anaerobic digestion; Ogilvie, 1992 for recycling processes). There is one

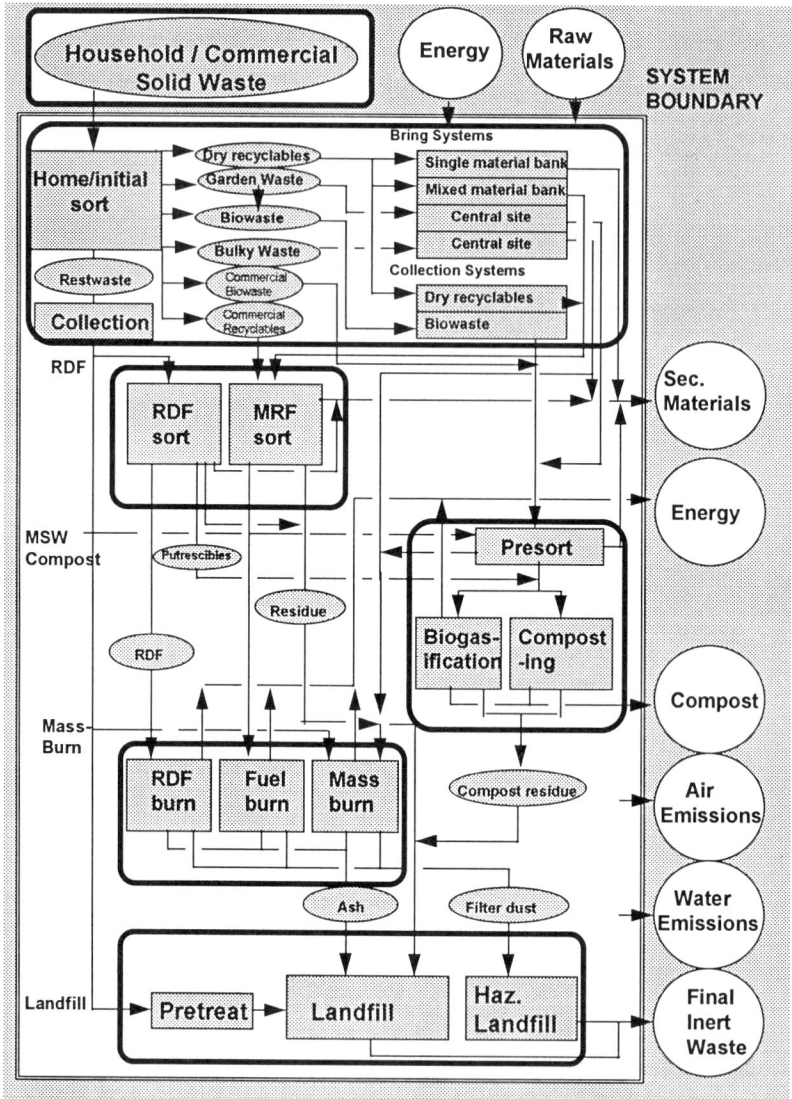

Figure 4.6 Detailed system boundaries and inputs/outputs for the lifecycle inventory of solid waste.

Box 4.4 LIFECYCLE INVENTORY OF WASTE: INVENTORY STAGE.

1. System studied
See Frontispiece

2. Data inputs

a) Types of data – Variable data (inserted by user) which define the waste management system considered.
– Fixed data (embedded in spreadsheet) which define inputs and outputs for unit waste management processes.

b) Sources of data: Waste composition and amounts: International agencies, technical literature
Collection and sorting: ERRA and other local schemes
Treatment processes: technical literature, company reports.

c) Data quality: Overall poor quality, especially for waste quantity and composition.
Data available for treatment of mixed waste stream
Little/no data on treatment of individual fractions of waste.

3. Data outputs

System Inputs: Net Energy Consumption
System Outputs:
    Products: Amounts of recovered materials
Amount of compost produced
    Emissions: Emissions to air
Emissions to water
Final solid waste
Performance Indicators: Materials recovery rate
Overall recovery rate
Diversion rate from landfill

---

area in which current fixed data sources are particularly lacking, however: the allocation of impacts to the different parts of the waste stream. Data are available for most waste management treatment processes, such as composting or incineration, in terms of inputs and outputs per tonne of waste treated, but solid waste varies not only in quantity, but also in composition. In an integrated waste management system, different materials within the waste stream are separated and treated in different ways. Thus, the residual waste that enters a mixed waste treatment method, e.g. incinerator or landfill, will vary in composition depending on what has been removed for other treatment methods. Without data allocating the inputs and outputs to the particular fractions of waste, it is not possible to reliably predict the inputs and outputs of any mixed waste treatment method. Where process inputs and outputs can be allocated to individual materials in this LCI, this will be applied, otherwise average data for mixed waste streams will be used.

The important distinction between generic and specific data should also be borne in mind. Generic data are generally averages, and although use-ful to give a rough 'ball-park' measure of the environmental impacts associated with a system, can never be more specific than that. To assess the actual impacts of a given system requires measurement of all the processes in that actual system. Clearly this is costly both in terms of money and time; lifecycle inventories collecting and using specific data have taken several years to complete. Most of the data used in this LCI model for waste will be generic data, but this is often the best available. As more specific and appropriate data sources become available, they can be incorporated.

4.2.3   Results of the model: system inputs and outputs

Net energy consumption. All energy inputs to the system and the energy produced during certain treatment processes are considered. The inherent energy of the waste is not included since it is common to all possible options for dealing with the same amount of waste. The energy delivered by fuels and electricity is included, plus the indirect energy consumption during fuel and electricity production.

Air emissions, water emissions. These will be considered separately rather than aggregating them (e.g. into 'critical volumes' (Habersatter, 1991)). This will allow subsequent impact analysis methods to be applied as they are developed. Since there are a large number of different emissions from many single processes, let alone complete waste management systems, it is necessary to define which emissions will be considered. Not surprisingly, previous studies have varied in the categories that they have considered. The list of emissions to air and water considered most important, and thus included in this LCI study, are given in Box 4.5. This list of emissions is used for each of the processes included within the waste system boundaries, but since a range of data sources is used, not all data sets contain all the required information, nor use the same categories. Therefore, it has some-times been necessary to convert data from other sources into the categories required.

Landfill volume (inert). Since landfills fill up rather than get too heavy, volume rather than weight is the key measure of concern. This volume needs to reflect the level of compaction that occurs naturally in the landfill.

Recovered materials. Recovered materials are products, and therefore outputs of the system. It is important to predict the types and amounts of materials that are likely to be produced by any system, since industry will

---

Box 4.5 AIR AND WATER EMISSION CATEGORIES USED IN THE LCI OF SOLID WASTE.

Emissions to air	Emissions to water
Particulates	BOD/COD
CO	Suspended solids
CO_2	Total organic compounds
CH_4	AOX (adsorbable organic halides)
NO_x	Chlorinated HCs
N_2O	Dioxins/furans (TEQ)
SO_x	Phenol
HCl	Ammonium
HF	Total metals
H_2S	Arsenic
Total hydrocarbons (HC)	Cadmium
Chlorinated hydrocarbons	Chromium
Dioxins/furans (TEQ)	Copper
Ammonia	Iron
Arsenic	Lead
Cadmium	Mercury
Copper	Zinc
Lead	Chloride
Mercury	Fluoride
Nickel	Nitrate
Zinc	Sulphide

---

need to help build both the re-processing capacity and markets to deal with such amounts.

Compost. Again, it is essential to be able to predict the amount and quality of compost that would be produced since large scale markets might need to be developed.

Other statistics. Although descriptors of the system rather than inputs or outputs, it is useful to generate certain statistics since Government targets are usually set in these terms.

- *Material recovery rate*: the percentage of the waste stream that is recovered as usable secondary materials.

$$\text{Material recovery rate} = 100\% \times \frac{\text{Amount of recovered recyclable materials leaving the system}}{\text{Total amount of waste entering waste management system}}$$

- *Overall recovery rate*: this would include both 'dry recyclables' and compost.

$$\text{Overall recovery rate} = 100\% \times \frac{\text{Total amount of recovered recyclable materials and compost produced}}{\text{Total amount of waste entering waste management system}}$$

- *Landfill diversion rate*: the percentage of the waste stream that is diverted away from final disposal in a landfill. This is not the same as the material recycling rate, as diverted material may be released as emissions, e.g. during compost production.

$$\text{Landfill diversion rate} = 100\% \times \left(1 - \frac{\text{Amount of waste entering landfill}}{\text{Total amount of waste entering system}} \right)$$

4.2.4 Fuel and electricity consumption in the lifecycle of solid waste

Wherever fuels or electricity are used within the waste management systems, there will be resultant impacts not only due to their actual use, but also due to the mining, drilling, transport and production of the fuels and electricity. By using generic data throughout the lifecycle inventory, every time that fuel or electricity are consumed, the relevant consumption of thermal energy (including 'pre-combustion' energy consumption), emissions to air and water and production of solid waste can be added to the overall inventory totals. Different sources of data for energy production and use are given in Table 4.1, and the data selected for use in this LCI spreadsheet are given in LCI Data Box 0.

Electricity consumption. The overall thermal energy consumption, emissions and solid waste generated by electricity production will vary according to the method used, and its efficiency. Clearly, the environmental impact of generating 1 kW-h of electricity from fossil fuels will differ markedly from hydro-electric generation of the same amount. Since the mix of electricity generating methods varies from country to country, the impacts of consuming 1 kW-h of electricity will also vary. To overcome this variability, Habersatter (1991) uses an average European scenario with a mix of 42.9% thermal power generation, 36.9% nuclear power and 20.2% hydro-electricity (based on the UCPTE (Union for the Connection, Production and Transport of Electricity) 1988 model). Ideally, the energy mix for the country under study should be used to give the most accurate results. Table 4.1 gives the inputs and outputs for electricity generation for the overall European model, and country specific data (Boustead, 1993) for comparison. Using country-specific mixtures of electricity generation methods would increase the complexity of the model, and so for this first version, the European UCPTE 88 data will be used.

The amount of electrical power generated by waste-to-energy methods within the waste management system is subtracted from the electrical energy used, to give the overall net electrical energy consumption. This amount is then used to calculate the thermal energy consumption, emissions and solid waste that are due to electricity generation, using the UCPTE 88 data. If the amount of electrical energy recovered from waste should exceed the amount consumed by the system, there will be a net export of electrical energy from the waste management system, and a net saving of the emissions and solid waste that would have been associated with the production of this amount of electricity using the conventional generation methods.

Petrol and diesel consumption. The delivered energy content of petrol and diesel are around 36.0 and 37.7 MJ/l respectively (Boustead, 1993). Additional energy has been expended, however, to drill, transport and process the crude oil to produce these fuels. Similarly, air and water emissions and solid waste will have been generated. For example, the energy efficiency of the diesel production industry in Europe is estimated as 91.6% (Habersatter, 1991). Thus, to supply 1 MJ of energy as diesel requires a gross energy input of 1.09 MJ. Such 'pre-combustion' impacts of petrol and diesel production are given, per litre, in Table 4.1. (Litres are used as the unit since this is how fuel consumption is totalled over the lifecycle). The use of petrol and diesel will generate further emissions to air. The actual amounts generated, per litre of fuel used, will vary according to the vehicle or machine involved. Since most of the petrol and diesel consumption in the waste management system will be by private cars and heavy goods vehicles, respectively, data for these vehicles are used as the basis for calculating the emissions per litre used. (Note: by calculating emissions in this way, the fuel consumption of different vehicle types will be accounted for, since this will determine the amount of fuel consumed). Air emission data for fuel usage are given in Table 4.1 where they are added to the pre-combustion impacts to give the total impacts due to fuel production and use. Boustead (1993) has similarly calculated the total impacts (gross energy consumption, air emissions, water emissions and solid waste) associated with consumption of both petrol and diesel for the UK, and these data are also included in Table 4.1 for comparison.

Natural gas consumption. Data for production and use of natural gas are also presented in Table 4.1. Although the energy content of natural gas will vary along with its composition, the data are given per cubic metre.

Table 4.1 Fuel and electricity production and use: lifecycle impact data for pre-combustion and use impacts

	Electricity production (Europe) (per kW-h) (Habersatter, 1991)	Electricity production (UK) (per kw-h) (Boustead, 1993)	Petrol production and use (per litre) (Boustead, 1993)	Diesel production (per litre) (Habersatter, 1991)	Diesel production and use (per litre) (Boustead, 1993)	Natural gas production (per m³) (Habersatter, 1991)	Natural gas production and use (per m³) (Boustead, 1993)
Energy consumption (MJ)	9.5	13.6	42.0	39.0	44.1	42.5	42.1
Air emissions (mg)							
Particulates	197	4245	2446	184	2564	6.7	3220
CO	349	1202	25 323	210	26 548	129	39
CO_2	441 657	1 084 594	2 491 318		3 036 258		2 061 211
CH_4							
NO_x	1236	4504	32 301	1598	33 901	376	29 604
N_2O	70			41		43	
SO_x	2502	13 140	9640	3245	10 106	1653	660
HCl		234	<36		<38		
HF	0.01	7.2	<36		<38		
H_2S							
Total HC	2112	572	10 395	5700	10 898	23 976	53 932
Chlorinated HC							
Dioxins/furans (TEQ)							
Ammonia	0.49	<3.6		17		0	
Arsenic							
Cadmium							
Chromium							
Copper							
Lead			144				
Mercury							
Nickel							
Zinc							

Water emissions (mg)						
BOD	0.15	<3.6	<36	5	<38	0
COD	0.44	<3.6	36	15	38	0
Suspended solids	0.15	338	<36	5	38	0
Total org. compounds	4.7	21.6	396	137	415	31
AOX						
Chlorinated HCs						
Dioxins/furans (TEQ)						
Phenol	0	<3.6	<36	0	<38	0
Ammonium	0.62	140	<36	0	<38	0
Total metals						
Arsenic						
Cadmium						
Chromium						
Copper	0.003					
Iron						
Lead						
Mercury						
Nickel						
Zinc	0.02					
Chloride	1.335	<3.6	<36		<38	
Fluoride	1.32	<3.6	<36		<38	
Nitrate						
Sulphide						
Solid waste (g)	49.1	80.5	5.3	1.3	5.7	3.0

4.3   The economic LCI

The parallel economic LCI which assesses the overall costs of the waste management system also requires both variable and fixed data. The variable data will be the same as for the environmental LCI, but details for the cost of each process in the lifecycle are required. These also need to be entered by the user, since generic data on costs are of little use. Costs for waste management processes are extremely variable, both between and even within countries, and reflect availability of local facilities, salaries and land prices. In each of the following chapters, examples of average costs for a range of countries in Europe are given, but local cost estimates are needed for any reliability in the result. For consistency, all costs in this book will be quoted in European Currency Units (ecu), using the exchange rates current at the time of writing (March 1994). These conversion rates are given at the end of the book in Appendix 5.

The methods used for waste management costings vary. Collection costs for domestic waste depend on the number of households visited, and so are often calculated per household per year (IGD, 1992). Collection costs for commercial and industrial waste, and processing costs for waste treatment and disposal relate to the amount of waste handled, so are calculated per tonne. As the aim of this model is to compare the costs of the whole waste management system, the cost will be calculated as the overall cost per year. For comparison with other literature, this can be converted into cost per household per year and cost per tonne of throughput.

4.4   The computer spreadsheet

Both the environmental and economic LCIs are included in one computer spreadsheet, which operates in Excel 4.0 and Lotus 1-2-3 v 2.01. Based on Figure 4.5, the spreadsheet follows the waste stream through its lifecycle. Each of the stages in the lifecycle of waste (large boxes in Figure 4.4) is represented in the spreadsheet by a box containing input questions. The answers to these questions define the waste management system considered. The first box will define the amount and composition of waste as it enters the waste management system from both household and commercial sources. Since the effectiveness of any treatment process, e.g. composting, thermal waste-to-energy, will depend on what is in the waste stream entering the process, it is necessary to keep the different materials separate in the spreadsheet, even though they may be physically commingled. By doing this, it is possible to characterise the material composition of the waste, and hence also its calorific value, at any point in the lifecycle.

The following boxes represent pre-sorting and collection, central sorting, materials recycling, biological treatment, thermal treatment and landfilling, respectively. This structure mirrors the following chapters in this book: the operation of the module for each stage and a copy of the input box are presented early in the relevant chapter. Within each stage, as materials are recovered, they are subtracted from the waste stream and entered into a reclaimed materials stream. Other outputs from processes are entered into the relevant columns for emissions or energy, where they accumulate. Total costs for the system accumulate through the lifecycle to produce the economic LCI.

By the end of the lifecycle, all of the materials will have left the waste stream columns and have been entered into either the products or emissions columns. This emulates the definition of the cradle and grave of solid waste discussed above (Section 4.2.1); the cradle is the point at which the material is thrown away (i.e. ceases to have value), the grave is the point at which the material regains value (i.e. as secondary products) or is released as an emission to land, air or water. The spreadsheet then totals the energy consumption, energy production, recovered materials, compost, emissions to air, emissions to water and final solid waste to produce the lifecycle inventory for the waste of the chosen region.

4.5 The relationship between a lifecycle inventory for waste and product or packaging lifecycle inventories

This chapter has emphasised the clear difference between the lifecycle of a product (or package) and the lifecycle of solid waste (Figure 4.1). Similarly, LCI studies of solid waste systems and LCIs of products or packages fulfil different functions. An LCI of solid waste aims to optimise the waste treatment system for a given input of waste. This will be of use to waste managers, whether in local or national governments or private waste management companies. It will not, however, predict whether one form of a product or package is better or worse for the environment than another. A product or package only spends a part of its lifecycle in the waste management system, and its compatibility with waste management processes may be offset by environmental burdens in raw material sourcing, manufacture, distribution or use. Any comparison should be on a cradle to grave basis for the product or package lifecycle, i.e a product specific LCI. Similarly, this spreadsheet is not designed to answer questions such as whether one-way packaging or returnable packaging is preferable from an environmental point of view. An example often used is the comparison between returnable bottles and single-use cartons for milk packaging. This comparison needs to look at the packaging LCIs for the two options, including the initial manufacture of the bottles and cartons, use, and subsequent refilling, recycling or disposal processes as appropriate.

Product LCIs and waste stream LCIs are complementary; the waste

LCI Data Box 0 Input/output data attributed to fuel and electricity consumption (including production and use) in the LCI of solid waste.

	Electricity consumption (per kW-hr)	Petrol consumption (per litre)	Diesel consumption (per litre)	Natural gas consumption (per m³)
Energy consumption (MJ)	9.5	42.0	44.1	42.1
Air emissions (mg)				
Particulates	197	2446	2564	3220
CO	349	25 323	26 548	39
CO_2	441 657	2 491 318	3 036 258	2 061 211
CH_4				
NO_x	1236	32 301	33 901	29 604
N_2O	70		41	43
SO_x	2502	9640	10 106	660
HCl		36	38	
HF	0.01	36	38	
H_2S				
Total HC	2112	10 395	10 898	53 932
Chlorinated HC				
Dioxins/furans				
Ammonia	0.49			0
Arsenic				
Cadmium				
Chromium				
Copper				
Lead		144		
Mercury				
Nickel				
Zinc				
Water emissions (mg)				
BOD	0.15	36	38	0
COD	0.44	36	38	0
Suspended solids	0.15	36	38	0
Total organic compounds	4.7	396	415	31
AOX				
Chlorinated HCs				
Dioxins/furans (TEQ)				
Phenol	0	36	38	0
Ammonium	0.62			0
Total metals		36	38	
Arsenic				
Cadmium				
Chromium				
Copper				
Iron	0.003			
Lead				
Mercury				
Nickel				
Zinc				
Chloride	0.02	36	38	
Fluoride	1.335	36	38	
Nitrate	1.32			
Sulphide				
Solid Waste (g)	49.1	5.3	5.7	3.0

stream LCI can be used to optimise the overall waste management system for all waste materials; product-specific LCIs can optimise the individual items that end up within the waste stream. Although there is an area of overlap in the processes, the two different objectives are distinct.

References

Boustead, I. (1993). Resource Use and Liquid Food Packaging. *EC Directive 85/339: UK Data 1986–1990*. A report for INCPEN, 1993.

Clayton *et al.* (1991) Review of municipal solid waste incineration in the UK. Warren Spring Laboratory Report LR 776 (PA), Department of the Environment Research Programme.

Coombs J. (1990) The present and future of anaerobic digestion. In: *Anaerobic Digestion: a Waste Treatment Technology*, Elsevier Applied Science, London, pp. 1–42.

DoE/DTI (1991) Ecolabelling of washing machines. *Environmental Labelling*; Newsletter from the National Advisory Group on EcoLabelling, Issue 2.

ERRA (1991) Resource. Report of European Recovery and Recycling Association, 1991. ERRA, Brussels.

ERRA (1993) Project summary sheets. European Recovery and Recycling Association, 83 Ave E. Mounier, Box 14, Brussels 1200, Belgium.

Finnveden, G. (1992) Landfilling – a forgotten part of lifecycle assessment. In: *Product Life Cycle Assessment-Principles and Methodology*. Nordic Council of Ministers, pp. 263–288.

Habersatter, K. (1991) Ökobilanz von Packstoffen Stand 1990. Bundesamt für Umwelt, Wald und Landschaft (BUWAL) Report No. 132, Bern, Switzerland (1991).

IGD (1992) Sustainable waste management: the Adur project. Report by the Institute for Grocery Distribution, Letchmore Heath, Watford, UK. 85 pp.

Kirkpatrick, N. (1992) Choosing a waste disposal option on the basis of a lifecycle assessment. *Proc. PIRA Conf.: Lifecycle Analysis-Protecting Your Market Share*, Gatwick, UK.

Ogilvie, S.M. (1992) A review of the environmental impact of recycling. Warren Spring Laboratory report LR 911 (MR).

Stalmans, M. (1992) LCA studies for chemical substances: major detergent surfactants and their raw materials. *Proc. 3rd CESIO Congress*, London, pp. 237–250.

Vogg, H. (1992) Arguments in favor of waste incineration. *Annual European Toxicology Forum*, Copenhagen.

White, P.R., Hindle P. and Dräger, K. (1993). Lifecycle assessment of packaging. In: *Packaging in the Environment*, ed. G. Levy. Blackie, Glasgow, pp. 118–146.

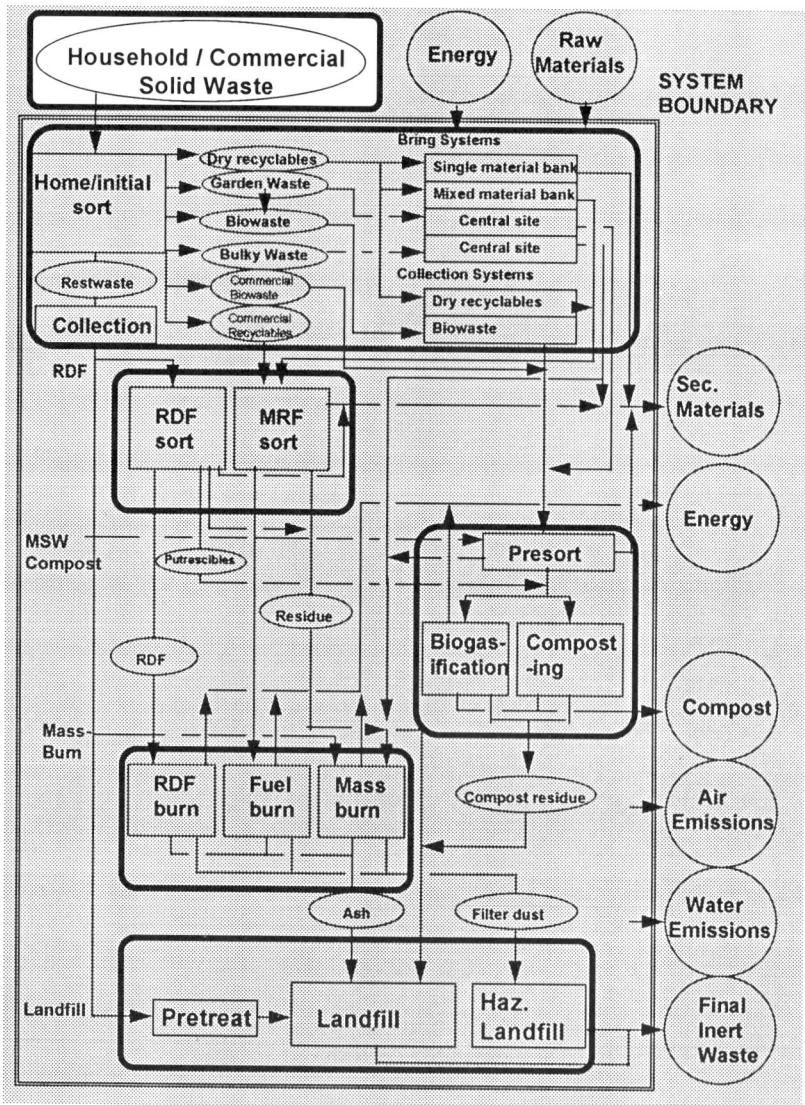

Figure 5.1 The solid waste inputs to an integrated waste management system.

5 Solid waste generation

Summary

This chapter starts the construction of the LCI model for solid waste. It attempts to assess the amount and composition of solid waste likely to be generated in a given area. The lack of comprehensive and standardised data collection is one of the limiting factors in this process, and in the development of effective solid waste management in general. This chapter presents the data currently available on the generation and composition of solid waste in general, and of municipal solid waste (MSW) in particular, for Europe. These data are limited; they are incomplete and are based on different definitions of waste categories. Definitions are given for the type of waste that will be dealt with in this book, namely MSW comprising of household (collected and delivered), commercial and institutional waste. The limitations of present classification schemes are discussed and new developments in waste analysis outlined. The first module of the computer LCI spreadsheet (IWM-1), which defines the waste entering a solid waste management system, is described.

5.1 Introduction

The functional unit of the LCI study of solid waste has been defined as the amount of household and similar solid waste generated in the specified geographical area. The first requirement, therefore, is to determine both the amount and composition of the waste generated in the region considered. Ideally this should be data specific to this region, collected from local waste analyses. There is increasing interest in collecting such information, but at present, for most areas, it is not available. Assessment of waste generation rates and composition must consequently be based on generic data, presented in the form of country-specific averages. This information suffers from being at best an overall average, at worst an estimate, but is often all that is available.

Accurate definition of the waste stream is essential in an LCI study of waste, since the waste itself represents the majority of material inputs into the system. Whatever enters the system in the waste, whether energy or contaminants, has to leave the system somewhere. Thus, most of the total emissions from the system reflect what was in the incoming waste. This underlines the need to reduce the amount of solid waste produced in the first place, and the need to eliminate any potentially harmful materials from the waste. Alterations to the waste management system can change where the waste materials leave the system (e.g. in compost or as air emissions), or whether energy is harnessed for use or dissipated, but will not change the total amount of waste materials or energy that arise from the waste.

5.2   Solid waste generation in Europe

Total solid waste generation in Europe is in the region of five billion (10^9) tonnes per year. Table 5.1 gives the amounts generated by each country, broken down according to waste source. Statistics for waste generation are notoriously unreliable, however. Although Table 5.1 represents one of the most recent pan-European sets of data, it is clearly incomplete. Furthermore, the figures it does contain in many cases represent only estimates of the amounts of waste generated. National governments have openly stated that many figures quoted for waste generation are based on estimates (e.g. DoE, UK 1990).

Two key factors contribute to this paucity of reliable data: absence of systematic data collection and the lack of a standard classification for waste.

Historically, wastes have been measured in tonnages, on disposal, rather than when and where they are generated. Thus, where wastes are produced and dealt with *in situ*, as with many agricultural wastes, these are not measured or included in statistics. Also, since waste disposal has not been high on the political agenda in the past, efforts to maintain up-to-date national statistics on waste disposal, let alone waste generation, have been limited.

Waste classification has traditionally been by source rather than by composition. Because of the different administrative methods used in the countries of Europe, however, no universal classification has been adopted. The category 'Municipal Solid Waste' typifies the confusion (Carra and Cossu, 1990). In some countries, figures are collected for household waste only (the MSW figure for the UK in Table 5.1 refers only to household waste), whereas other countries include waste derived from commercial and sometimes light industrial sources. Similarly, figures for solid waste generated during energy production may be quoted separately, or, alternatively, included under the umbrella of industrial wastes. Clearly, 'Solid Waste' is a very diverse category; it has been said that every consignment of waste is unique. Although classification of waste types is difficult, the lack of consistent categories renders comparisons across countries

Table 5.1 Solid waste generation (in thousands of tonnes/year) in Europe

Country	Year	MSW	Agriculture	Mining/ quarrying	Manufacturing	Energy production	Demolition	Dredge spoils	Sewage sludge	Others
Austria	1990	4783	880	21	31 801	1150	18 309	111	365	
Belgium	1988	3410			27 000	1069	680	4805	687	2830
Bulgaria	1990	2562	9028	1 506 755	370 757	195 560		333 885	776	
Czech Republic	1987	2600	451	533 373	39 604	25 774	2677	23 071	2750	
Denmark	1985	2430			2304	1532	1747		1263	
Finland	1990	3100	23 000	21 650	10 160	950	7000	3000	1000	150
France	1990	20 320	400 000	100 000	50 000				600	9800
Germany	1990	27 958		19 296	81 906	29 598	120 394		1750	
Greece	1990	3000	90	3900	4304	7680				
Hungary	1989	4900	62 000		45 000				30 000	
Iceland	1990	80			135					
Ireland	1984	1100	22 000	1930	1580	130	240		570	860
Italy	1991	20 033			34 710		34 374		3428	
Luxembourg	1990	170			1300		5240		15	
Netherlands	1990	7430	19 210	391	7665	1553	12 390	17 500	320	
Norway	1990	2000	18 000	9 000	2000		2000		1000	
Poland	1991	13 300	81 000	85 200	27 000	18 800			1000	
Portugal	1990	2538		202	662	165				15
Slovak Republic	1987	1901	942	4276	22 602	3128	5977		1041	497
Spain	1990	12 546	112 102	70 000	13 800		22 000		10 000	
Sweden	1990	3200	21 000	28 000	13 000	625	1200		220	3850
Switzerland	1990	3000			1000		2000		260	
Turkey	1989	19 500								
UK	1990	20 000	80 000	107 000	56 000	13 000	32 000	21 000	1000	0

Source: EUROSTAT (1994b)

Table 5.2 Categories of solid waste

Category	Description
Agricultural	Waste arising from agricultural practices, especially livestock production; often either used (applied to land) or treated *in situ*
Mining and quarrying	Mainly inert mineral wastes, from coal mining and other mineral extraction industries
Dredging spoils	Organic and mineral wastes from dredging operations
Construction and demolition	Building waste, mainly inert mineral or wood wastes
Industrial	Solid waste from industrial processes; sometimes will include energy production industries
Energy production	Solid waste from the energy production industries, including fly ash from coal burning
Sewage sludge	Organic solid waste, disposed of by burning, dumping at sea (soon to cease), application to land or composting; may result from industrial or domestic waste water treatment
Hazardous/special waste	Solid waste which can contain substances that are dangerous to life is termed 'special waste' in UK, or 'hazardous waste' in EC directives
Commercial	Solid waste from offices, shops, restaurants, etc. often included in MSW
Municipal solid waste (MSW)	Solid waste collected and controlled by municipality; typically comprised of: household waste, commercial waste, institutional waste

fraught with problems. The most commonly used categories are listed in Table 5.2.

The lack of reliable statistics for waste generation has resulted in wide variability in reported figures. Such variability is graphically shown in Figure 5.2, which gives different data for UK waste generation over the same time period. These figures vary by 300 million tonnes per year (some 75% of the lowest estimate). Most of this variability can be accounted for by differences in the estimates for agricultural and mining wastes generated. Both may be treated *in situ*, so their generation is hard to assess. There has also been a re-evaluation of what constitutes a 'waste' in agriculture (DoE, 1992).

Discussion of waste management is also hampered by uncertainty over data sources. Lack of reliable data has resulted in the proliferation of reports which, while quoting data, do not reveal their sources. In this work,

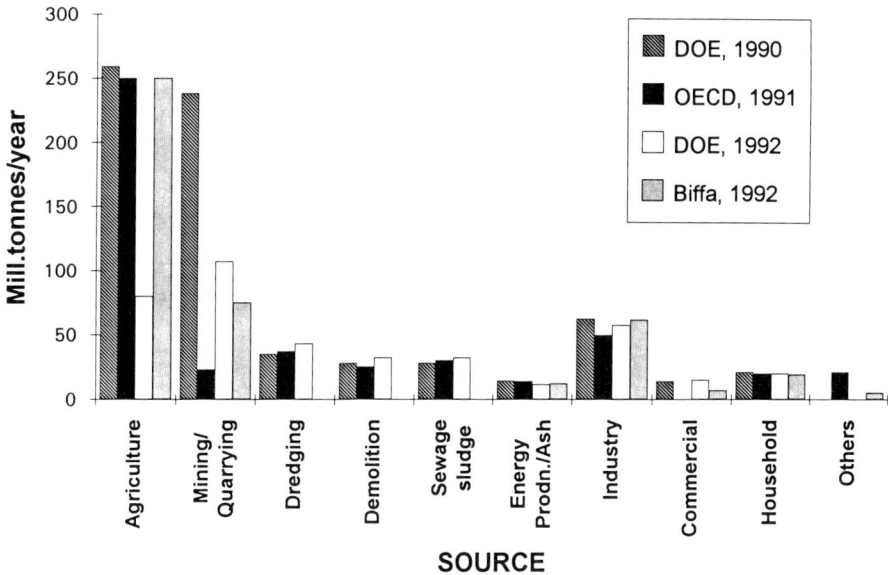

Figure 5.2 Variability in waste generation data (UK waste, late 1980s).

data that can be traced back to their original source will be used wherever possible.

Lack of consistent definitions and reliable statistics on a national scale does not necessarily preclude the planning of local or regional integrated waste management schemes, however, since these require accurate local rather than national data. Most regional authorities will have weighbridge data on the amount of selected wastes that are produced. They will often have to rely on national averages for the composition of the waste, however, as waste analyses are more expensive and labour-intensive to conduct.

5.3 Solid wastes dealt with in this study

Although it is widely used in waste management, Municipal Solid Waste (MSW) is not a naturally defined category. It is defined simply as the waste collected and controlled by the local authority or municipality. Consequently, there is no uniformity in material composition, merely in how it has been collected, or more precisely, by whom. MSW is potentially the most diverse category of waste, as it comprises waste from different sources, each of which is heterogeneous. For the purposes of this study these sources are defined as:

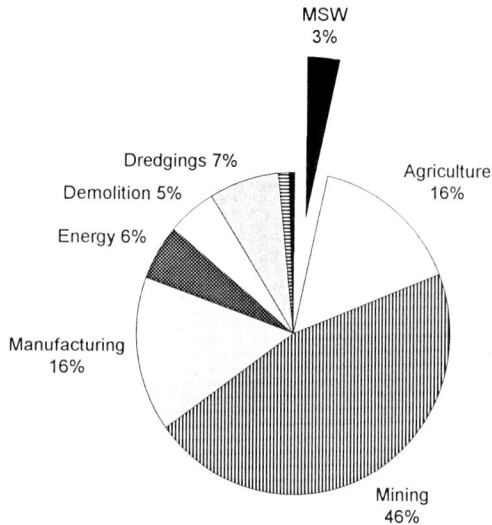

Figure 5.3 Solid waste generation in Western Europe, by source.

- *Household waste*: generated by individual households. It will include all solid wastes originating on the property, including garden waste. It may also be termed domestic waste. This is further subdivided into:
  - *Collected household waste*: household waste collected from the property by the waste collection service. Essentially this is the combined contents of the dustbin, bin bag, blue box, etc.
  - *Delivered household waste*: household waste delivered to a collection point by the householder. This may include bulky waste items (e.g. cookers, refrigerators) and garden waste (in some areas), plus recyclable materials deposited in bring systems (e.g. bottle banks, waste paper collections). This category of waste would not be included in analyses of dustbin contents, and may be overlooked in some statistics.
- *Commercial waste*: waste generated by commercial properties, in particular shops, restaurants and offices.
- *Institutional waste*: waste generated in schools, leisure facilities, hospitals (excluding clinical waste), etc. This is often included within the commercial waste category.

5.4 Quantities of municipal solid waste (MSW) generated

MSW represents a small but significant proportion of solid waste, accounting for about 3-4% of total solid waste production in Europe

Table 5.3 Generation of municipal solid waste by countries

Country	MSW generation (kg/person/per year)
Austria	325
Belgium	343
Czech Republic	251
Denmark	475
Finland	624
France	328
Germany	350
Greece	296
Hungary	463
Iceland	314
Ireland	312
Italy	348
Luxembourg	445
Netherlands	497
Norway	472
Poland	338
Portugal	257
Slovak Republic	359
Spain	322
Sweden	374
Switzerland	441
Turkey	353
United Kingdom	348
EC average	350
USA	720
Japan	410

Data for year 1990
Sources: EUROSTAT (1994a,b); OECD (1993)

(Table 5.1 and Figure 5.3). When broken down by country, it can be seen that there are regional differences in the amounts generated per person (Table 5.3 and Figure 5.4). Given possible discrepancies between countries in what is included within MSW, Finland, Netherlands, Denmark and Norway have the highest per capita generation of MSW in Europe, although all Western European countries have lower generation rates than Canada and the USA.

5.5 Composition of MSW

5.5.1 By source

Data on the generation of MSW are difficult to interpret because of the different definitions used for MSW. National data breaking down MSW according to the source of the material, i.e. into collected household waste, delivered household waste, commercial waste and institutional waste are

kg/ person/ yr

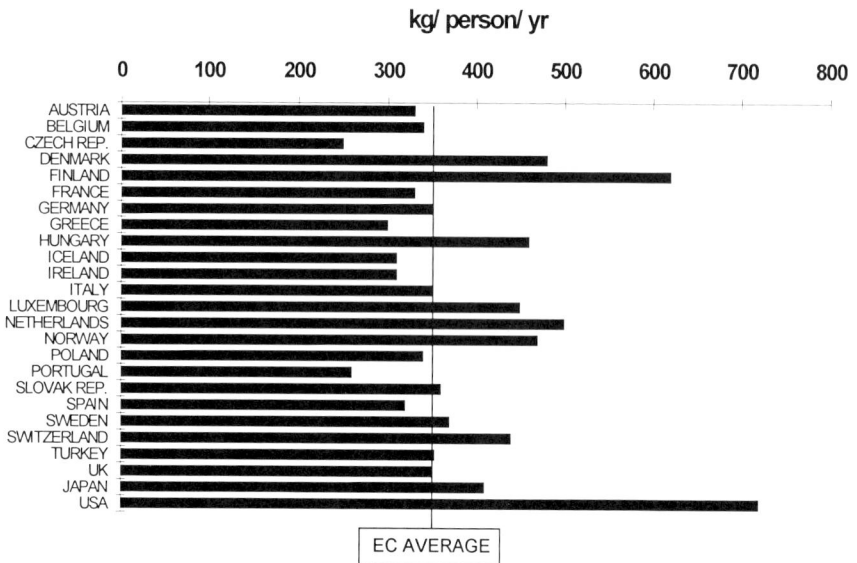

Figure 5.4 Generation of municipal solid waste. *Source*: EUROSTAT (1994a).

equally hard to obtain. Collected household wastes are normally the best documented, with poorer records available for delivered household and commercial wastes (Table 5.4).

Complete and up-to-date figures on this split into MSW sources are essential when planning an integrated MSW management system, since the source of the MSW will determine the collection strategy necessary. Such data may be available in a geographical area, although not consolidated into a national figure.

5.5.2 By materials

Just as data on the sources of MSW within a defined geographical area are necessary to design effective collection systems, knowledge of the material composition is essential for effective management and disposal. The composition of MSW by weight, compiled from various sources for European countries is given in Table 5.5. Again, the data are not fully comparable, since while most of the figures relate to MSW, some relate specifically to household waste, which is only one part of MSW in most countries. There are clear differences in the composition of collected household waste, delivered household waste and commercial waste (see Figures 5.5 and 5.6) so it is important to specify precisely the type of waste analysed for any data quoted. As with the above data for amounts generated, household waste (and in particular collected household waste) is the best documented

Table 5.4 Breakdown of MSW by source for selected countries

Country	Collected household (%)	Delivered household (%)	Commercial (%)	Reference
Denmark	32	22[a]	46[b]	Christensen (1990)
France	30	6[c]	64[d]	Barres *et al.* (1990)
Germany (West)	62.5	8.5[e]	29[d]	Stegmann (1990)
Netherlands	66	8.5[e]	25.5[f]	Beker (1990)
UK	65	20	15	DoE/DTI (1992)

[a] Bulky waste plus garden/park waste.
[b] Commercial plus industrial waste.
[c] Bulky household waste, plus car wrecks and tyres.
[d] Industrial waste similar to household refuse.
[e] Bulky waste.
[f] Shop/office/service waste.

part of MSW in terms of materials composition. Data for the commercial part of MSW are harder to obtain, and are likely to vary more from area to area. Commercial waste often forms a significant part of MSW, however (see Table 5.4), and can have an important role to play in improving the economics of recovery and recycling operations (IGD, 1992). There is thus an equally urgent need to understand the amount and material composition of this portion of MSW.

Despite these differences in the way data have been compiled, some trends in MSW composition can be seen from Table 5.5. The two major fractions in all countries are paper/board and food/garden waste. Plastics, glass and metals occur at much lower levels. There is, however, evidence of geographical variability. Southern European countries (e.g. Spain, Portugal, Italy) generally have a higher level of food/garden waste than northern countries (e.g. Finland, Denmark, France, UK), whereas paper and board show the opposite trend (Figure 5.7). Patterns are less easy to detect in the proportion of plastics, glass and metals in MSW, although there are interesting individual points, such as the high proportion of plastic waste in Ireland and Switzerland, high levels of glass in France and Germany, and high levels of metals in Denmark.

5.5.3   By chemical composition

A third valid way to classify MSW is by its chemical composition. Since the solid waste represents the largest input into the overall solid waste system, the composition of the waste will determine the majority of the emissions of the overall system. If the chemical composition of the incoming waste, and of individual fractions, is known, some of the emissions from waste treatment processes can be predicted. This is particularly

Table 5.5 Composition of MSW (by weight) for European countries

Country	Waste type	Year[a]	Paper/board (%)	Plastics (%)	Glass (%)	Metals (%)	Food/garden (%)	Textiles (%)	Other (%)	References[b]
Austria	MSW	1990	21.9	9.8	7.8	5.2	29.8	2.2	23.3	1,2
Belgium	MSW	1990	30.0	4.0	8.0	4.0	45.0		9.0	2
Bulgaria	MSW	1990	8.6	6.9	3.8	4.8	36.7		39.2	2
Czech Republic	MSW	1990	9.5	5.9	7.6	6.4	7.2		63.4	2
Denmark	MSW	1985	29.0	5.0	4.0	13.0	28.0		21.0	1
Finland	Household	1985	51.0	5.0	6.0	2.0	29.0	–	5.0	3
France	MSW	1990	31.0	10.0	12.0	6.0	25.0	2.0	12.0	2,4
Germany	MSW	1990	17.9	5.4	9.2	3.2	44.0	4.0	20.3	2
Greece	MSW'	1990	22.0	10.5	3.5	4.2	48.5		11.3	2
Hungary	MSW	1990	21.5	6.0	5.5	4.5			62.5[c]	2
Iceland	MSW	1990	37.0	9.0	5.0	6.0	15.0		28.0	2
Ireland	MSW	1992	34.0	15.0	5.0	4.0	24.0	3.0'	15.0	5
Italy	MSW	1990	23.0	7.0	6.0	3.0	47.0		14.0	6
Luxembourg	MSW	1990	17.0	6.0	7.0	3.0			67[c]	2
Netherlands	MSW	1990	24.7	8.1	5.0	3.7	51.9	2.1	4.5	2,7
Norway	MSW	1990	31.0	6.0	5.5	4.5	30.0		23.0	2
Poland	MSW	1990	10.0	10.0	12.0	8.0	38.0		22.0	2
Portugal	MSW	1990	23.0	4.0	3.0	4.0	60.0		6.0	6
Spain	MSW	1992	20.0	7.0	8.0	4.0	49.0	1.6	10.4	8
Sweden	MSW	1990	44.0	7.0	8.0	2.0	30.0		9.0	2
Switzerland	MSW	1990	31.0	15.0	8.0	6.0	30.0	3.1	6.9	2,9
Turkey	MSW	1990	37.0	10.0	9.0	7.0	19.0		18.0	2
UK	Household	1992	34.8	11.3	9.1	7.3	19.8	2.2	15.5	10

a Where more than one source is used, year gives latest date.
b References: 1, Carra and Cossu (1990); 2, OECD (1993); 3, Ettala (1990); 4, Barres et al. (1990); 5, ERL/UCD (1993); 6, Elsevier (1992); 7, Beker (1990); 8, MOPT (1992); 9, Gandolla (1990); 10, Warren Spring Laboratory (see Table 5.9).
c Includes food/garden waste

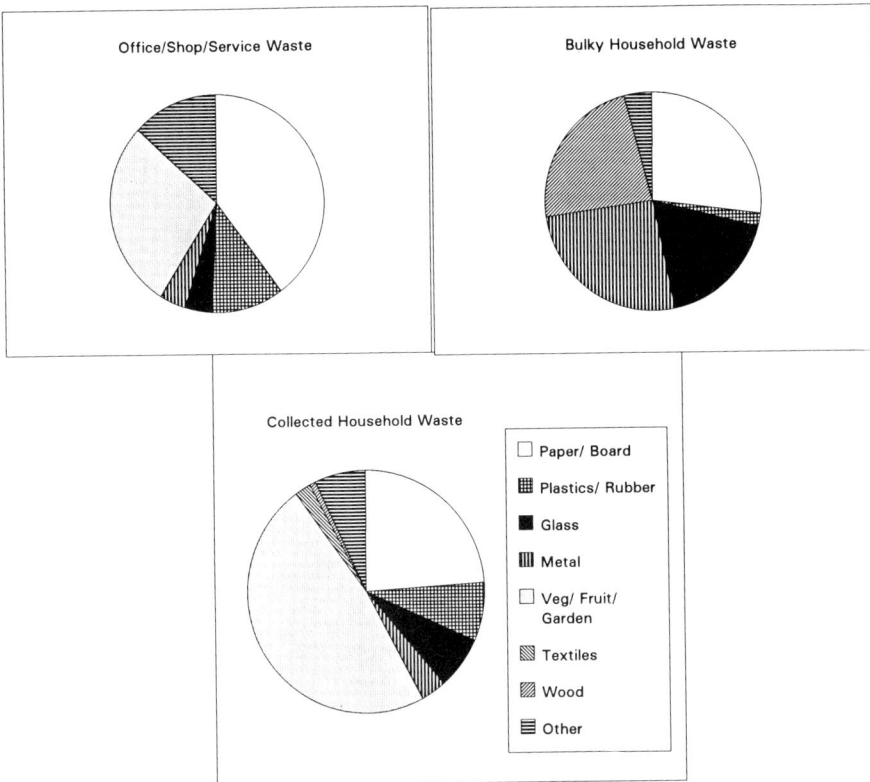

Figure 5.5 Composition of different parts of MSW in the Netherlands. *Source*: Beker (1990).

true for inorganic trace contaminants such as heavy metals (Table 5.6) which pass generally unchanged through the waste management process. Consequently, the total amount released will reflect the total input in the waste. Knowing how the heavy metals enter the waste stream allows efforts to be made either to reduce the levels of these contaminants, or to ensure that they are effectively handled. Two other useful characteristics of waste fractions are their carbon content (which allows calculation of emissions of carbon dioxide, methane, etc.), and their water content (which varies markedly between different waste fractions and will affect their calorific value) (see LCI Data Box 1).

5.6 Variability in MSW generation

When planning an MSW management system for any given region, it is important to have relevant and recent data for that region, since waste generation will vary geographically, both between and within countries. The composition of commercial waste will clearly vary according to the nature

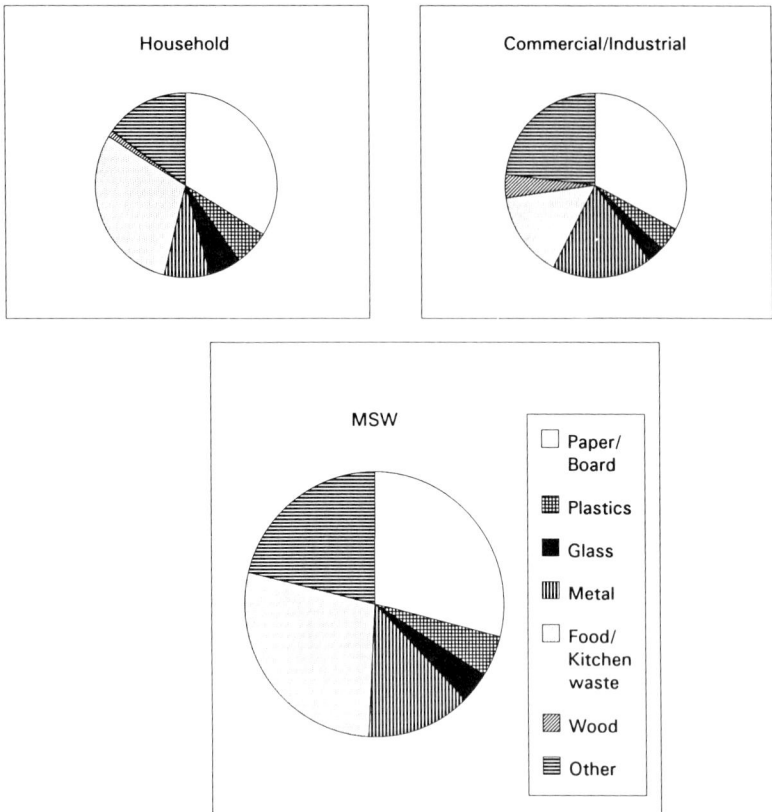

Figure 5.6 Composition of different parts of MSW in Denmark. *Source*: Christensen (1990), Carra and Cossu (1990).

of local commerce. Additionally, both the amount and composition of household waste will vary with population density and housing standards. Rural areas, for example, are likely to have a greater amount of vegetable, fruit and garden waste than inner city areas (Table 5.7). Differences have also been measured between areas depending on the type of domestic heating installed. In former East Germany, for example, areas with domestic open fires burning brown coal produced up to 190 kg of waste per person per year, whereas areas with central heating produced up to 260 kg/person per year (Bund, 1992). Differences in waste composition also occurred, as might be expected, with less paper waste in houses with open fires, but more fine material, i.e. ash (Table 5.8).

Even within a given area, there will also be seasonal effects. The composition and amount of household waste generated will vary, especially over holiday periods, and the amount of garden waste included will clearly vary with the seasons. Thus, while the figures quoted here will give a guide

(a)

(b)

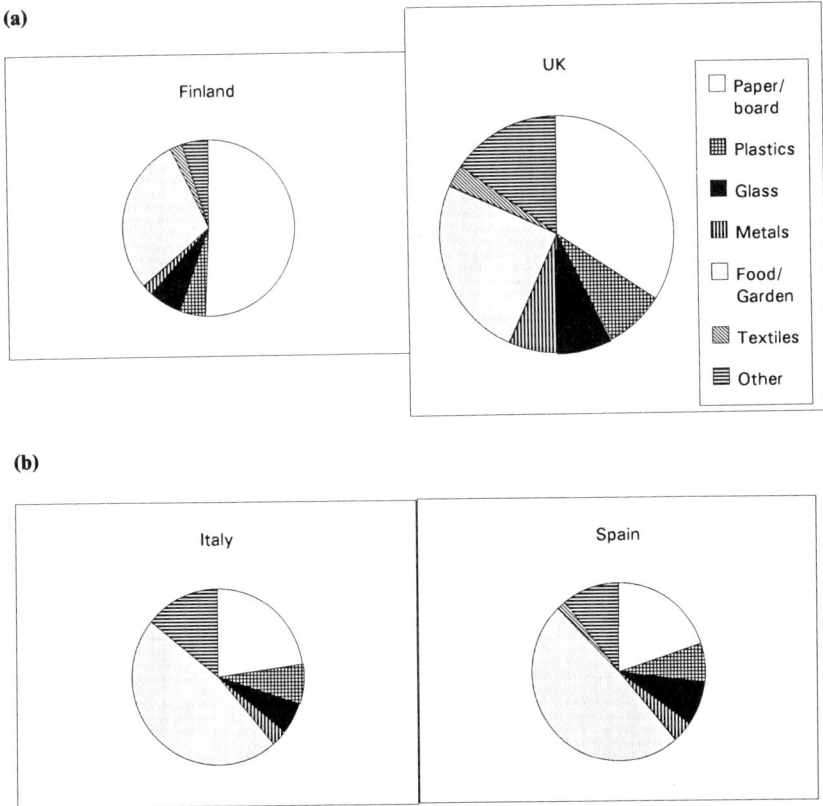

Figure 5.7 Geographical trends in MSW composition. (a) Northern Europe. (b) Southern Europe (see Table 5.5).

to both amounts and composition, they will not always reflect local conditions, nor a particular time of the year.

5.7 Effects of source reduction

The amount and composition of solid waste generated by a region will also be affected by any attempts to promote source reduction. The above example of domestic burning of some parts of household waste is a form of source reduction, since less waste remains to be collected and managed. The practice is not allowed in some countries, however, because of concern over the resulting gaseous emissions. Another form of source reduction that is on the increase, however, is the promotion of home composting. A variety of home composting units are commercially available which are capable of dealing with both garden and kitchen wastes, and some

Table 5.6 Presence of heavy metals (in % of total load) in household waste fractions

Total load (mg/kg dry)	Cadmium (3–15)	Nickel (80)	Zinc (1000–2000)	Copper (200–600)	Lead (400–1200)	Mercury (4–5)	Chromium (250)
Fines <10 mm	1–2	12–13[a]	5	7	5–7		2–3
Fines 10–20 mm	1–2		5	33–39	13–16	1	
Organics	2–3	16–19	5	4–6	5–13	2	4–6
Paper/carton	1–2	9–11	8–9	7–8	18–19	2–4	8–12
Textiles	2	3–4	1	1–2	1	1	3–4
Leather	4	3	1–2	3–8	1		39–50
Rubber			11–13		2		
PVC	36–40						
Other plastic	13–14	24–25	3–4	4–7	8–9	1	8–9
Glass		6–10	1	1	2		9–12
Ferrous metal			1	27–31	35–41	1	3–26
Non-ferrous	6–7		12–13	2–31	1		1–2
Batteries	39–48	20–22	44–47	12	1	93	

Source: Rousseaux (1988).
[a] Figures in bold type indicate significant load.

Box 5.1 CLASSIFICATION OF SOLID WASTE USED IN THE IWM-1 LIFE-CYCLE INVENTORY SPREADSHEET.

Category	Description
MSW FRACTIONS	
Paper (PA)	Paper, board and corrugated board, paper products.
Glass (GL)	Glass bottles (all colours), sheet glass.
Metal (ME)	All metals including cans. Further subdivided into: Ferrous (ME-Fe) and Non-Ferrous (ME-nFe)
Plastic (PL)	All plastic resin types, including bottles, films, laminates. Further subdivided into rigid plastics (PL-R) and plastic film (PL-F)
Textiles (TE)	All cloth, rag etc., whether of synthetic or natural fibres
Organic (OR)*	Putrescible kitchen and garden waste, food processing waste
Other (OT)	All other materials, including fines material, leather, rubber, wood.
WASTE TREATMENT RESIDUES	
Compost (CO)	Residues from biological treatment (composting or anaerobic digestion, that cannot be marketed as products due to contaminant levels or lack of suitable markets.
Ash (AS)	Bottom ash, clinker or slag from incinerators, RDF or alternate fuel boilers.
Filter dust (FD)	Fly ash and residues from gas cleaning systems from incinerators, RDF or alternate fuel boilers.

* Paper and plastic fractions are also strictly of organic origin, but to maintain alignment with the ERRA classification system, the term 'organic' is used here to describe putrescible kitchen and garden waste only.

---

municipalities have offered these to residents in an effort to reduce the amount of such wastes that need collection and treatment. The Adur Home Composting Scheme (in Adur District, West Sussex, UK) for example, has a district wide participation rate of 22% (Adur District Council, personal communication). An initial survey suggests that around 13% of household waste (by volume) can be diverted from the normal collection system in this way.

Since the LCI boundary used in this study is the waste leaving the householder's property, home composting is not considered within the waste management system modelled in the spreadsheet. However, it is an effective means of source reduction, and can reduce the amount of waste requiring treatment.

5.8 MSW classification: need for standardisation

It is clear from the above sections that to approach waste management on a European scale requires considerable standardisation of terminology.

Table 5.7 Effect of population density on waste composition

Population density (persons/km^2)		Waste fractions (%)		
		Paper	Glass	Fruit/veg/garden
Inner city	2000	24	20	28
Suburban	1000	20	14	34
Urban	500	16	11	39
Rural	150	12	8	52

Source: Rheinland-Pfalz, Ministry of Environment (1989).

Table 5.8 Variation of household waste composition in former East Germany with domestic heating type

Waste fraction	Regions with central heating (%)	Regions with open coal fires (%)
Metal	3.2	3.2
Glass	12.3	10.2
Plastics	5.5	3.6
Textiles	3.5	2.8
Paper/board	24.5	10.9
Wood	0.5	0.2
Bread	6.4	3.4
Fine material (< 16 mm)	11.7	41.0
Other	32.4	24.7
Total	100	100

Source: von Schoenberg (1990). Data for Dresden, 1988.

Whilst waste management practices will continue to vary from country to country, it is important that lessons from one area can be disseminated widely and implemented elsewhere. For this a common understanding of what is included under the terms household waste (collected and delivered), commercial waste and MSW is required.

Standardisation of waste material categories is similarly essential, but more detailed classifications are also required. The most effective method of dealing with any particular waste item will depend on what it is made of, so detailed knowledge of the composition of waste is a prerequisite for effective waste management. Whilst the classification scheme used in Table 5.5 gives useful general trends, there is insufficient detail for waste management purposes. Plastics, for example, may exist as thin films, rigid bottles or as a multitude of other objects. Knowing that they are made of plastic may give a guide to their suitability for waste-to-energy schemes, but further knowledge of their form is necessary to determine their suitability for material recycling.

To fill this need, more detailed waste classification schemes have been devised. In the UK, Warren Spring Laboratory have analysed household

Table 5.9 Household waste classification scheme developed by Warren Spring Laboratory (UK)

Category	% by weight[a]
Newspapers	12.29
Magazines	4.99
Other paper	10.08
Liquid containers	.64
Card Packaging	3.81
Other Card	2.98
Refuse sacks	1.15
Other plastic film	4.27
Clear plastic beverage bottles	.65
Coloured plastic beverage bottles	.12
Other plastic bottles	1.15
Food packaging	1.96
Other dense plastic	2.04
Textiles	2.17
Disposable nappies	3.87
Other miscellaneous combustibles	3.56
Miscellaneous non-combustibles	1.66
Brown glass bottles	.36
Green glass bottles	1.08
Clear glass bottles	1.32
Clear glass jars	1.55
Other glass	4.83
Garden waste	3.15
Other putrescible material	16.62
Steel beverage cans	.50
Steel food cans	3.86
Batteries	.05
Other steel cans	.39
Other ferrous metal	.99
Aluminium beverage cans	.40
Foil	.46
Other non-ferrous metal	.64
< 10 mm fines	6.39
Total	100.00

Source: Warren Spring Laboratory, personal communication.
[a] Figures give average of UK waste analyses conducted using these categories to date (January 1993).

waste using a 33 category classification (Table 5.9). This gives a much more detailed picture of what items are in the waste, but it still does not define the composition of all items (e.g. plastic resin type is not specified), To meet this requirement, the European Recovery and Recycling Association (ERRA) have proposed a hierarchical classification system which specifies not only the form of waste items (films, bottles, cans, etc.) but also the material (Figure 5.8; Appendix 1; ERRA, 1992a). If this proposal is adopted as a standard, information flow on waste composition will be significantly improved. A simplified version of this ERRA classification is used as the basis for the LCI model in this book (see Box 5.1).

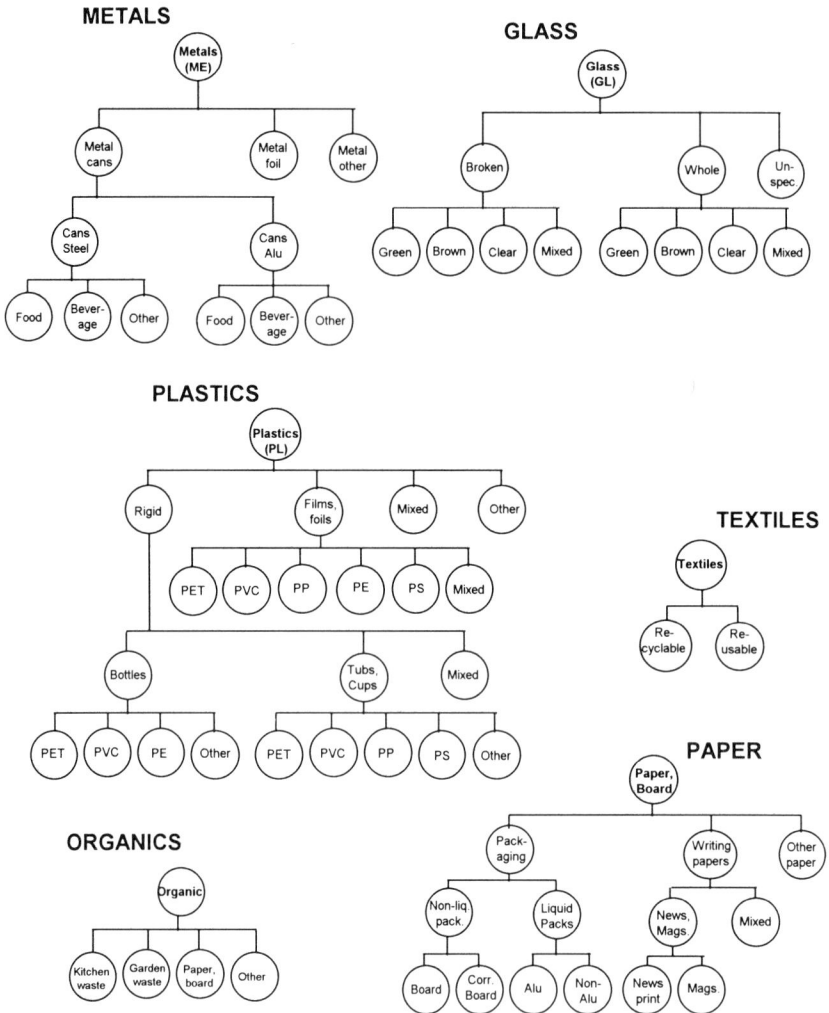

Figure 5.8 Proposed classification of waste materials (ERRA, 1993).

5.9 MSW analysis methods

An integrated approach to waste management clearly requires better waste statistics in the future. Use of a more detailed and standardised classification scheme is only part of the solution. Standardised and appropriate sampling and analysis techniques are also required.

Methods for household waste analysis have been developed from techniques originally used to sample mineral cores (e.g. Poll, 1991). Such methods suffer from two major limitations. Firstly, MSW is much more heterogeneous in particle size than mineral samples, so different techniques need to be applied in the taking of samples. Secondly, many sampling procedures require the use of specialised equipment. The need for widespread sampling at many waste collection and treatment facilities precludes the routine use of such equipment; simple but robust sampling techniques are required. In line with this, ERRA have proposed a simplified waste analysis procedure (ERRA, 1993; Appendix 2), that gives guidelines on both the techniques to be used, and the number of samples required for statistical analysis.

Sampling methods have been applied mainly to collected household waste rather than the more bulky and heterogeneous delivered household waste. There is little evidence of systematic sampling and analysis of commercial waste, although this tends to be more homogeneous than household waste.

Information on waste composition also comes from various collection and recovery schemes (bottle banks, kerbside collection schemes, material recovery facilities). These are valuable sources of data, but here again there are problems in the interpretation of data due to lack of standardisation in the terminology. The terms capture rate, recovery rate and recycling rate can vary widely in their intended meaning. To overcome this, ERRA has also published proposed definitions of such 'programme ratios' (Appendix 3; ERRA, 1992b) to clarify the confusion existing in many reports in the literature.

5.10 Defining the waste input for the LCI computer spreadsheet

5.10.1 Data sources

The first step in carrying out an LCI for household and associated commercial solid waste is to define the amount and composition of such waste generated by the area in question. As discussed above, local weighbridge data on amounts and waste analysis results are needed for an accurate estimate. The quantity and quality of household waste will depend on factors such as population density, levels of affluence, housing types and efforts at source reduction. Commercial waste will reflect both the types and level of local commercial activity.

If locally sourced data are not available, generic average data can be used, but with caution. It is important to know what the average data used reflect. Does household waste include garden waste and bulky waste, or does it only reflect 'collected household waste', i.e. the dustbin contents? Most national waste composition figures are based on collected household waste; data on delivered household waste often do not exist. Similarly, few good data on commercial waste exist. Thus, while generic information can be used in the absence of local data, it is not a satisfactory substitute.

5.10.2 Classification of solid waste used in the lifecycle inventory

The categories of waste used in this analysis, and their definitions, are given in Box 5.1. It will be seen that only the most basic level of the proposed ERRA classification (Figure 5.8; Appendix 1) has been used. For this first attempt to model the waste management system, this level of detail is considered sufficient. Using more detailed categories will lead to a more accurate prediction of the overall environmental impacts and economic costs of a waste management system, but will considerably increase the complexity of the model. It is hoped, however, that more sophisticated models will be developed in the future.

Some basic parameters of these categories of waste, which will be used throughout this analysis, are given in LCI Data Box 1.

5.10.3 Operation of the waste input module of the LCI spreadsheet

LCI Box 1 shows the data required on the amount and composition of solid waste entering the system. The instructions given in the box are listed below, with additional comments on the type of data needed, and how the spreadsheet handles the input.

WASTE INPUTS
Country:
The user first enters the country in the first shaded box. This calls up generic national data from the body of the spreadsheet on the average amounts and composition of waste generated (from Tables 5.3 and 5.5 above).

System Area
  Population:
  Average number of persons/household: *No. of households served:*
Either the population or the number of households in the specified area need to be entered, along with the average number of persons per household. This allows both per capita and per household data to be used.

Waste Composition
(a) Household
  Average generated in system area (kg/person/yr):
  Composition (% by weight):
    Detailed composition %:   *Metals*
                                 Plastics
If region-specific data are available on the amounts of household waste generated per capita, and its composition, these should be inserted. The

LCI BOX 1

1. WASTE INPUTS

Country: [   ]

System Area
Population [   ] thousand
Average number of persons/household [   ]   # of households served [   ] thousand

Waste Composition
        FILL IN SHADED BOXES USING LOCAL DATA. IF NO DATA AVAILABLE, LEAVE BLANK AND DEFAULT VALUES WILL BE USED

(a) Household
        Amount Generated in System area [   ] kg/person/yr
                                    Paper (PA)   Glass (GL)   Metal (ME)   Plastic (PL)   Textiles (TE)   Organics (OR)   Other (OT)
        Composition (% by wt) [   ]

        National Average (default value) [   ] kg/person/yr
                                    Paper (PA)   Glass (GL)   Metal (ME)   Plastic (PL)   Textiles (TE)   Organics (OR)   Other (OT)
        National Average Composition(% by wt) [   ]
                                                Ferrous   Non-Fe
        Detailed composition %   Metals [   ]
                                                Film   Rigid
                                  Plastics [   ]

(b) Commercial
Average generated in area   [   ] 000 tonnes/yr
                                    Paper (PA)   Glass (GL)   Metal (ME)   Plastic (PL)   Textiles (TE)   Organics (OR)   Other (OT)
Consisting of (% by weight) -   [   ]
                                                Ferrous   Non Fe
        Detailed composition %   Metals [   ]
                                                Film   Rigid
                                  Plastics [   ]

Go to Box 2a.

LCI Data Box 1 Physical parameters of the waste categories used in the Lifecycle Inventory (LCI) analysis.

	PA	GL	ME-Fe	ME-nF	PL-R	PL-F	TE	OR	OT	Source
Moisture content (%) as received	30	10	15	10	15	25	25	65	30	1
Calorific value (MJ/kg) (dry)	12	0	0	0	30	27	15	5.6	5.2	1
Calorific value (MJ/kg) wet-as received	10.5	−0.5	−0.5	−0.5	28	25	13.5	3.7	4	1
Carbon content (%) of dry weight	44	0	0	0	80	85	44	44	40	2
Ash content (%) as received	8.4	90	85	90	6.0	9.0	7.5	7.7	42	1

For definitions of categories, see Box 5.1.
Sources: 1. Barton (1986); 2. Calculated.

spreadsheet will enter generic values taken from national averages, which can be used as default values in the absence of specific local data.

(b) Commercial

Average generated in area (000 tonnes/yr):
Consisting of (% by weight):
    Detailed composition %:   *Metals*
                                       Plastics

Estimates of amount generated and composition of commercial waste in the region need to be entered in the shaded boxes. No reliable generic data on commercial waste generation are available for use as default values.

The spreadsheet then calculates the amounts of each waste material entering the system, using the categories in Box 5.1, and then adds these totals to the waste stream columns of the LCI spreadsheet.

References

Barres, M., Grenet, Y., Millot, N. and Meisel, A. (1990) Sanitary Landfilling in France. In: *International Perspectives on Municipal Solid Wastes and Sanitary Landfilling*, eds. J.S. Carra and R. Cossu. Academic Press, London, pp. 78–93.

Barton, J. (1986) The application of mechanical sorting technology in waste reclamation: options and constraints. Warren Spring Laboratory. Paper presented at the Institute of Waste Management and INCPEN Symposium on Packaging and Waste Disposal Options, London.

Beker, D. (1990) Sanitary landfilling in the Netherlands. In: *International Perspectives on Municipal Solid Wastes and Sanitary Landfilling*, eds. J.S. Carra and R. Cossu. Academic Press, London, pp. 139–155.

Biffa (1992) Biffa Waste Services. Personal communication.

BUND (1992) Bioabfallkompostierung vorrangige Abfallverwertung. Report by von Lossau, E, Krauss, M. and Neidhardt, R. Bund Hessen.

Carra, J.S. and Cossu, R. (1990) Introduction. In: *International Perspectives on Municipal Solid Wastes and Sanitary Landfilling*, eds. J.S. Carra and R. Cossu. Academic Press, London, pp. 1–14.

Christensen, T.H. (1990) Sanitary landfilling in Denmark. In: *International Perspectives on Municipal Solid Wastes and Sanitary Landfilling*, eds. J.S. Carra and R. Cossu. Academic Press, London, pp. 37–50.

DoE (1990) Our common inheritance. Environmental Protection Act White Paper UK. Department of Environment.

DoE (1992) *The UK environment*. Department of Environment *and* Government Statistical Office Report, ed. A. Brown. HMSO, London, Ch. 13.

DoE/DTI (1992) *Economic Instruments and Recovery of Resources from Waste*. Department of Trade and Industry and Department of the Environment, 75 pp.

ERL/UCD (1993) *Waste Management Statistics for Ireland*. Environmental Resources Ltd. and Environmental Institute of University College Dublin.

ERRA (1992a) Nomenclature: secondary materials. Reference report of the ERRA Codification Programme. Available from European Recovery and Recycling Association, 83 Ave E. Mounier, Box 14, Brussels 1200, Belgium. Reproduced, with permission, as Appendix 1.

ERRA (1992b) Programme ratios. Reference report of the ERRA Codification Programme. Available from European Recovery and Recycling Association, 83 Ave E. Mounier, Box 14, Brussels 1200, Belgium. Reproduced, with permission, as Appendix 3.

ERRA (1993) Waste analysis procedure. Reference report of the ERRA Codification Programme. Available from European Recovery and Recycling Association, 83 Ave E. Mounier, Box 14, Brussels 1200, Belgium. Reproduced, with permission, as Appendix 2.

Elsevier (1992) *Municipal Solid Waste Recycling in Western Europe to 1996.* Elsevier Science Publishers, 1992.

Ettala, M.O. (1990) Sanitary landfilling in Finland. In: *International Perspectives on Municipal Solid Wastes and Sanitary Landfilling*, eds. J.S. Carra and R. Cossu. Academic Press, London, pp. 67–77.

EUROSTAT (1994a) Report of the Statistical Office of the European Community. News Release, 15th February, 1994.

EUROSTAT (1994b) *Europe's Environment 1993: Statistical Compendium.* Statistical Office of the European Community, June 1994.

Gandolla, M. (1990) Sanitary landfilling in Switzerland. In: *International Perspectives on Municipal Solid Wastes and Sanitary Landfilling*, eds. J.S. Carra and R. Cossu, Academic Press, London, pp. 190–198.

IGD (1992) Sustainable waste management: the Adur project. Report by the Institute for Grocery Distribution, Letchmore Heath, Watford, UK, 85 pp.

MOPT (1992) Residuos sólidos urbanos. Report by Luis Ramon Otero Del Peral for Ministerio de Obras Publicas y Transportes, Madrid (ISBN 84-7433-820-4).

OECD (1993) *Environmental Data Compendium.* Organisation for Economic Cooperation and Development, Paris.

ORCA (1992) *Information on Composting and Anaerobic Digestion.* ORCA Technical Publication No. 1. Organic Reclamation and Composting Association, Brussels. 74 pp.

Poll, A.J. (1991) Sampling and analysis of domestic refuse – a review of procedures at Warren Spring Laboratory. Warren Spring Laboratory Report LR 667 (MR).

Rheinland-Pfalz Ministry of Environment (1989) Leitfaden zur Kompostierung organischer Abfälle. Cited in ORCA, (1992).

Rousseaux, P. (1988) Les métaux lourds dans les ordures ménagères origines, formes chimiques, teneurs. R and D programme on recycling and utilization of waste, EEC, General Directorate XII.

Stegmann, R. (1990) Sanitary landfilling in the Federal Republic of Germany. In: *International Perspectives on Municipal Solid Wastes and Sanitary Landfilling*, eds. J.S. Carra and R. Cossu. Academic Press, London, pp. 51–66.

von Schoenberg, A. (1990) Waste disposal in East Germany – an overview. *Warmer Bull.* **27**, 4–5.

Household / Commercial Solid Waste

Energy

Raw Materials

SYSTEM BOUNDARY

Home/initial sort

Dry recyclables
Garden Waste
Biowaste
Bulky Waste

Restwaste

Collection

Commercial Biowaste
Commercial Recyclables

Bring Systems

Single material bank
Mixed material bank
Central site
Central site

Collection Systems

Dry recyclables
Biowaste

RDF

RDF sort

MRF sort

Sec. Materials

Energy

MSW Compost

Putrescibles

Residue

RDF

Presort

Biogas-ification

Compost-ing

Compost

Mass-Burn

RDF burn

Fuel burn

Mass burn

Compost residue

Air Emissions

Ash

Filter dust

Water Emissions

Landfill

Pretreat

Landfill

Haz. Landfill

Final Inert Waste

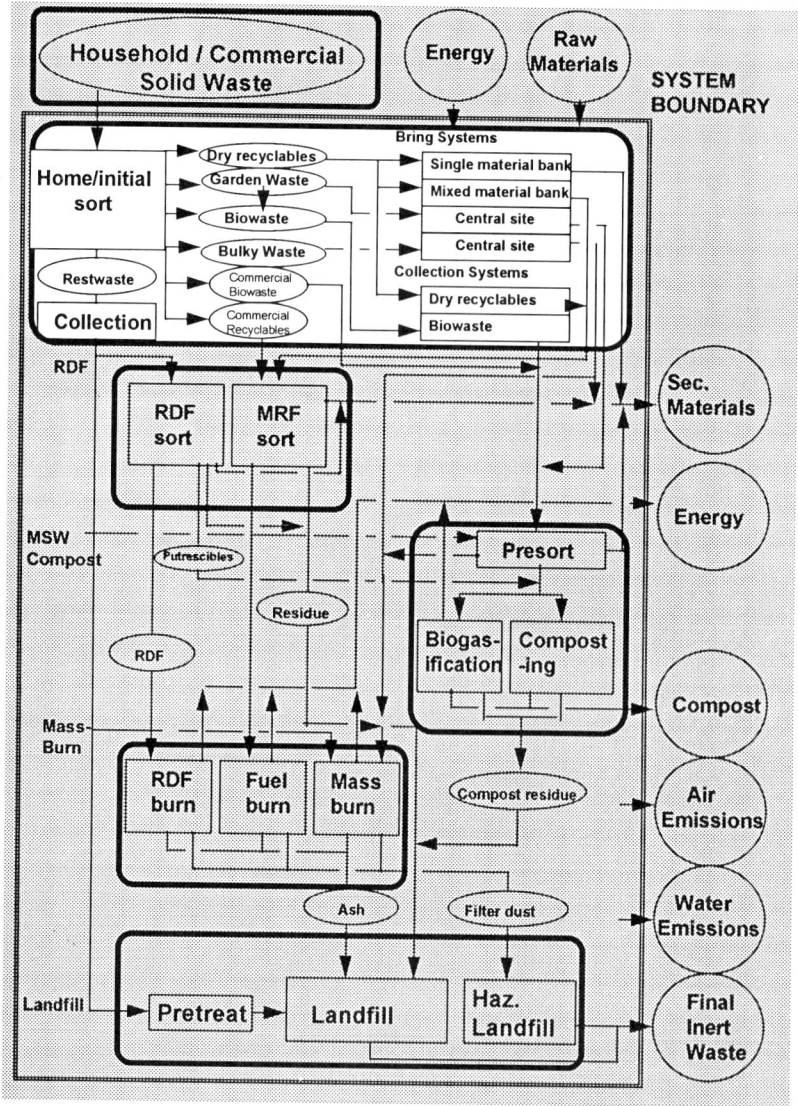

Figure 6.1 Pre-sorting and waste collection within integrated waste management.

6 Pre-sorting and waste collection

Summary

This chapter emphasises the importance of the collection operation in integrated waste management. It looks at the processes of home sorting and waste collection, from the creation of waste up to its delivery at a central sorting or treatment site. The characteristics and effectiveness of different collection methods are discussed, including both collection of separated fractions and the collection of commingled materials. The limitations of the common division into 'bring' and 'kerbside' schemes when comparing systems are emphasised, as is the need for effective communication between waste collectors and waste generators. The main environmental impacts of pre-sorting and collection processes (due to the vehicle transport involved) are discussed, and available data presented to allow these to be calculated. Limited information is also given on likely economic costs of collection systems. Finally, the module of the computer spreadsheet which models waste collection systems is presented and explained.

6.1 Introduction

There are good reasons why collection lies at the very hub of an integrated waste management system (Figure 2.3). The way that waste materials are collected (and subsequently sorted) determines which waste management options can subsequently be used, and in particular whether valorisation methods like materials recycling, biological treatment or fuel burning are feasible in an economically and environmentally sustainable way. The collection method will significantly influence the quality of recovered materials, compost or fuel that can be produced, which will in turn determine whether markets can be found. This importance of markets for valorised materials cannot be over-emphasised. In the absence of suitable markets, useful products cannot be produced. Therefore, either the collection method defines the subsequent treatment options, or, taking the reverse case, the existing or potential markets will define how materials should be collected and sorted if they are to be valorised. In any event, there must be a match between market need and material collected and sorted (Figure 6.1).

Waste collection is also the contact point between the waste generators (in this case households and commercial establishments) and the waste management system, and this relationship needs to be carefully managed to have an effective system. This householder-waste-collector link needs to be a customer-supplier relationship (in the total quality sense (Oakland, 1989)). The householder needs to have his/her solid waste collected with a minimum of inconvenience, whilst the collector needs to receive the waste in a form compatible with planned treatment methods. There is clearly a balance to be struck between these competing needs; waste management systems that fail to achieve balance in this relationship are unlikely to succeed.

Collection operations are often not independent of sorting operations, since the type of collection will determine the amount of subsequent sorting needed, and some collection methods (e.g. 'Blue Box' recovery schemes) themselves involve a level of sorting. This chapter focuses on the collection process, including any household, kerbside or bank sorting, i.e. pre-sorting, but leaves centralised sorting until Chapter 7.

6.2   Home sorting

From the householder's viewpoint, commingled collection of all solid waste probably represents the most convenient method, in terms of both time and space taken up. This collection method will limit, however, the subsequent options for valorisation. Most valorisation methods will require some form of separation of the waste into different fractions at source, i.e. in the home, prior to collection. At its simplest this might involve removing recyclable materials, e.g. glass bottles for delivery to a bottle bank; more extensive sorts involve separation of household waste into several different material streams. The degree of home sorting achieved in any scheme will be a function of both the ability and, especially, the motivation of householders.

Sorting ability.   Several schemes and pilot tests have demonstrated that, given clear guidance, householders are able to accurately sort their solid waste into different categories. A study carried out in Leeds, UK, for example, showed that householders could sort their waste into six different categories with a 96.5% success rate (Forrest *et al.*, 1990). A US study (Beyea *et al.*, 1992) showed a similar result. Clear instructions to the householder are essential for success, to which end many schemes run extensive communications programmes and publish frequent newsletters.

Sorting motivation.   The above sorting experiment in Leeds showed that accurate sorting of household waste is possible, but participants in the trial were volunteers, and therefore likely to be highly motivated. Is it realistic

Box 6.1 ATTRIBUTES OF 'BRING' AND 'KERBSIDE' COLLECTION SYSTEMS.

	Bring	Kerbside
Definition	Materials taken from property to collection point by householder.	Materials collected from property/home.
Sorting	Sorted by householder. May or may not be centrally sorted.	Sorted by householder. May also be sorted at kerbside and/or centrally.
Materials collected:	Separated materials or mixed materials	Separated materials or mixed materials
Containers	Communal	Individual (may be communal for apartments)
Consumer Transport needed:	High <----------------------------> None	
Collection Transport needed:	Low <----------------------------> High	
Amount collected	Low <------> High (depends on bank density)	High (assuming effective motivation)
Contamination level	Low: (separate collection) to High: (mixed collection)	Low: Kerbside sorted e.g. Blue Box) to High: (mixed collection)

to expect that the majority of householders will be similarly motivated? Whilst the degree of sorting in the Leeds study may be excessive, participation rates in other schemes with more limited sorting demands suggests that householders can be motivated (see Boxes 6.2–6.7). Participation rates are very difficult to measure, since what people claim that they do, and what they actually do, are not the same. ERRA define the participation rate by the number of waste generators putting out their recyclable materials at least once in a 4-week period (ERRA, 1992). A range of ERRA schemes report participation rates in voluntary home sorting and collection schemes between 60 and 90% (ERRA, 1993b); in one scheme (Adur, UK), the rate was actually measured and found to be 75% (Papworth, 1993). Similar high levels of voluntary participation are reported from North America; the Blue Box scheme covering Quinte, Ontario, had a measured participation rate of 85–91% in 1992 (Quinte, 1993). Where consumer research has been carried out to seek the views of householders, the most frequent comments voiced were that recycling was seen as a good idea, and that it 'helps the environment' (IGD, 1992). Participation rates for voluntary schemes will also depend on the economics. If householders have to pay for an

---

Box 6.2   DRY RECYCLABLES COLLECTION SCHEME, BARCELONA, SPAIN.

Scheme outline:	Collection of commingled packaging materials (dry recyclables), paper/board and glass via 'bring' streetside containers.
Date started:	May 1991 (Phase 1). Relaunched (Phase 2) Dec. 1992.
Area covered:	78 000 inhabitants of Sagrada Familia District A total of 27 956 generators (households and commercial premises)
Participation rate:	Not known
Housing type:	High-rise apartment buildings, plus commercial premises (restaurants, bars etc.)
Materials collected:	Paper/board Glass Plastic Beverage cartons
Collection method: (Phase 2)	Refuse in this area is collected in large (2200 litre) communal streetside containers (green bins). Mixed packaging materials are similarly collected in large streetside containers (blue bins). Both containers are emptied daily. Paper and board are collected separately in another large container and glass is collected in igloos. Collected mixed recyclables are transported to a Materials Recovery Facility (MRF) for sorting.
Quantities collected:	144 kg/generator/yr
Total waste stream:	982 kg/generator/yr

Source: ERRA, 1993b. Barcelona Waste Stream Analysis. (pers. comm); ERRA, pers. comm. 1994

---

additional recylables container, participation rates will be lower; if households get a cost reduction for having less non-recoverable material in their restwaste bin, participation rates are likely to be higher. Of course, in some schemes such as Lemsterland (Box 6.6), participation is not voluntary, as no alternative waste collection is provided, and separation of certain fractions of the waste at source is required by law in some countries (e.g. separation of organic material in the Netherlands). In such cases, participation rates are likely to be higher still.

Overall recovery rates for waste materials depend not only on the number of households participating, however, but also on the householder's sorting efficiency. The actual amount of any material recovered from household waste by home sorting will be determined by:

amount of material in household waste stream × participation rate × separation efficiency
(ORCA, 1992)

Even if participation is compulsory, motivation is still required to ensure a high level of sorting efficiency (Box 6.8).

Box 6.3 DRY RECYCLABLES COLLECTION SCHEME, PRATO, ITALY.

Scheme outline:	A dry recyclables collection scheme using large streetside collection containers serving many households. Households are provided with re-usable 'Blue Bags' to store recyclables prior to transport to large blue containers located in the street near existing refuse containers. Glass collected in igloo containers.
Date started:	November 1992
Area covered:	11 750 households. Total of 14 500 generators.
Participation rate:	70% (initial estimate)
Housing type:	High rise urban housing
Materials collected:	Newspaper and cardboard Plastics Metals Glass
Collection method:	Streetside containers emptied twice per week using existing collection vehicles, and materials taken to Materials Recovery Facility (MRF) for sorting.
Quantities collected:	103 kg/generator/yr
Total waste stream:	917 kg/generator/yr

Source: ERRA, 1993b.; ERRA pers. comm. 1994

Box 6.4 DRY RECYCLABLES COLLECTION SCHEME, DUNKIRK, FRANCE.

Scheme outline:	Collection of commingled recyclables from each household in a separate blue wheeled bin. Residual refuse collected in another wheeled bin.
Date started:	December, 1989
Area covered:	40 000 generators
Participation rate:	90% (estimated)
Housing type:	Suburban, horizontal housing
Materials collected:	Paper Plastic Metal Glass
Collection method:	Blue wheeled bins (120 litre) placed at kerbside, and emptied once per week by existing refuse vehicles, using lower compaction pressures than normal. Recyclables sorted at Materials Recovery Facility (MRF).
Quantities collected:	234 kg/generator/yr.
Total waste stream:	870 kg/generator/yr

Source: ERRA, 1993b., ERRA pers. comm. 1994

Box 6.5 DRY RECYCLABLES KERBSIDE COLLECTION SCHEME, ADUR DISTRICT, WEST SUSSEX, U.K.

Scheme outline:	A typical 'Blue Box' scheme. Dry recyclables collected weekly at the kerbside in a separate blue box container, using a purpose built multi-compartment vehicle. Close-to-home drop-off centres for households not covered by Blue Box scheme. Residual waste collected weekly by normal refuse collection vehicle.
Date started:	May 1991
Area covered:	26 000 households (19 500 households covered by Blue Box scheme, 6500 households using close-to-home drop-off centres)
Participation rate:	75% among Blue Box households (measured) (1)
Housing type:	Low-rise suburban housing
Materials collected:	Blue Box: Glass, 3 colours
	Metal cans (Fe and Al)
	Newspapers and magazines
	Plastic containers and film
	Drop-off: As for Blue Box
Collection method:	Blue Boxes (volume 44 litres) containing recyclables are set out by householder at kerbside when full. Boxes are collected once a week by special Labrie multi-compartment vehicle. Box contents are sorted by operator at vehicle into 5 categories (green glass, brown glass, clear glass, paper and plastics/cans). Residual waste collected weekly by normal refuse vehicle.
Quantities collected:	129 kg recyclables/generator/yr
Total waste stream:	460 kg/generator/yr
Collection costs:	Collection of recyclables £17.60/hhld/yr (2)
	Collection of residual £35.00/hhld/yr (3)

Source: General – ERRA, 1993b; ERRA, pers. comm., 1994.
(1) Papworth, 1993. (2) IGD, 1992. (3) Estimated, Aug, 1993.

Motivation, and hence both participation rates and separation efficiency, will be influenced by factors such as the level of convenience or inconvenience to the householder. Schemes with extensive home sorting may require too much time or too much space to store the separate waste streams before collection. Any loss in comfort level, for example if odour becomes a problem when organic material is not collected regularly, will lower motivation levels. Housing type also has an effect; data from the Netherlands suggest that occupants of high-rise buildings are less likely to participate in source separation programmes than those in suburban areas (Table 6.1). This may reflect a lack of storage space, but is also likely to be due to lack of social peer pressure in such buildings, as it is not possible for neighbours to see who is participating, i.e. who is environmentally responsible.

Box 6.6 DUAL BIN COLLECTION SCHEME, LEMSTERLAND, THE NETHERLANDS.

Scheme outline:	All households in the area are supplied with 2 wheeled bins, each divided into 2 compartments. Households sort their waste to be collected into 4 streams. The contents of each bin are collected every 2 weeks, on an alternating schedule. Glass is collected separately in a drop-off system, using glass igloos (bring system). There is one battery of 3 igloos (for clear, brown and green glass) per 500 inhabitants.
Date started:	May 1991
Area covered:	4526 households around town of Lemmer.
Participation rate:	95% of households take part in recycling programme (no other waste collection service exists)
Housing type:	Low rise.
Materials collected:	Bin 1, compartment 1: Mixed recyclables (plastics, metals, beverage cartons)
	Bin 1, comp. 2: Mixed paper (newspaper, cardboard etc)
	Bin 2, comp. 1: Organic material
	Bin 2, comp. 2: Restwaste
	Glass – Colour separated in drop-off containers.
Collection method:	Bins emptied every 2 weeks using single vehicle with horizontally split compartment. Vehicle visits each property once per week, emptying bins alternately.
Quantities collected:	Glass: 54 kg/hhld/yr
	Paper/board: 138 kg/hhld/yr
	Plastic/metal/cartons 200 kg/hhld/yr
	Organic material 288 kg/hhld/yr (estimated)
Total waste stream:	796 kg/generator/yr

Source: ERRA, 1993b; ERRA, pers. comm., 1994

Levels of environmental awareness vary geographically across Europe, but, overall, there appears to be a willingness on the part of householders to participate in some level of home sorting. Assessing the prevailing level of motivation in any particular area and gearing the collection scheme to this will achieve the best level of home sorting obtainable from a given area.

6.3 Bring versus kerbside collection systems

Collection methods are often divided into 'bring' and 'kerbside' collection schemes. ERRA (1993a) define bring collection systems as those where 'householders are required to take recoverables to one of a number of (communal) collection points'. In kerbside collection schemes, the 'householder places recoverables in a container/bag which s/he positions, on a

---

Box 6.7 DRY RECYCLABLES AND BIOWASTE COLLECTION SCHEME, ISMANING (NR MUNICH) GERMANY.

Scheme outline:	Collection of glass, paper and textiles by bring system. Metals, garden waste and bulky waste delivered to central drop-off containers. Biowaste and hazardous household waste collected from kerbside.
Date started:	Bring and drop-off systems 1987/88 Biowaste collection October 1990
Area covered:	12 900 inhabitants, area of around 40 km^2. (Within this area 2 specific test areas were analysed, containing 1344 residents (599 hhlds) and 1150 residents (592 hhlds) respectively.)
Participation rate:	n/a
Housing type:	Surburban (low rise) area. Approx. 323 inhabitants/km^2
Materials collected:	Glass Paper Textiles Metals Biowaste Hazardous waste Restwaste
Collection method:	Biowaste collected every 14 days. Households use 5 litre bins inside the home and then transfer to larger containers (240 litres) for collection. Other materials (except restwaste and hazardous waste) collected via bring systems.
Quantities collected:	Glass 21 kg/person/yr (1988 data) Paper 37 kg/person/yr (1988 data) Metals 3.5 kg/person/yr (1988 data) Restwaste 222 kg/person/yr (1988 data, so includes biowaste) Biowaste 47–54 kg/person/yr (1991 data)
Collection costs:	n/a

Source: Tidden and Oetjen-Dehne, 1992.

---

specific day, in the immediate vicinity of the property for collection'. Note that collection need not be, literally, from the kerbside. The key distinguishing point being that in bring systems, the householders transport the materials from their home, whereas in kerbside collection they are collected from the home. In reality, however, bring and kerbside are just the two ends of a spectrum of collection methods (Figure 6.2). The extreme form of bring system is the central collection site, variously called a civic amenity site (UK), déchétterie (France), or recyclinghof (Germany), to which householders transport materials such as bulky and garden waste. Such sites often also have collection containers for recyclable materials such as glass bottles and cans. Next in the spectrum of bring systems come material

---

Box 6.8 INFLUENCES ON MATERIAL RECOVERY.

Amount of material recovered = Amount of material in waste stream ×
Participation rate × Separation Efficiency

Amount of Material in Waste Stream:
See Chapter 5.

Participation Rate:
% of householders providing sorted material at least once per month

Separation Efficiency:
% of material correctly sorted and separated

Both Participation Rate and Separation Efficiency will be influenced by:

Level of convenience:
- Amount of sorting
- Difficulty of sorting
- Extra storage space required
- Distance of collection point
- Hygiene problems

Level of motivation:
- Quality and frequency of communications
- General environmental awareness/concern
- Peer pressure
- Legal requirements
- Availability of alternative disposal routes

Table 6.1 Participation in source-separation programmes in the Netherlands

Suburban housing		High-rise apartment buildings	
Location	Participation rate (%)	Location	Participation rate (%)
Amsterdam	89	Amsterdam	65
Apeldoorn	91		
Purmerend	90	Purmerend	62
Medenblik	97		
Ede	87	Ede	72
Nuenen	90		

Source: Kreuzberg and Reijenga (1989).

banks at low density (i.e. a high number of connected inhabitants), often situated locally at supermarkets. As the density of bring material containers increases, they become 'close-to-home' drop off containers (ERRA, 1993a), to which householders can walk rather than drive. This applies particularly to high rise housing, where residents of apartment blocks usually take their refuse (and recyclables) to large communal containers positioned outside the apartment blocks or at the side of the street (see details of Barcelona scheme, Box 6.2). This is essentially having a refuse container in the street, just outside, rather than inside, the householder's property; the

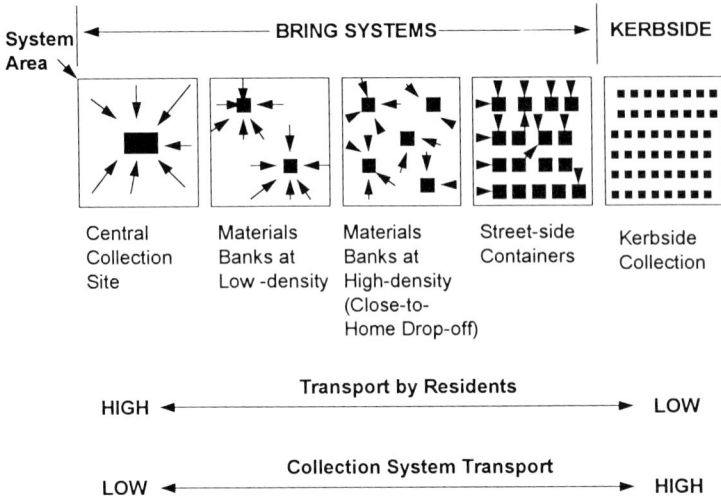

Figure 6.2 The spectrum of collection methods from 'bring' to 'kerbside systems'. (Arrow lengths indicate distances travelled by residents to collection points.)

only difference between this 'bring' system and a kerbside collection from individual properties is that the containers are communal, rather than for individual households.

The term 'bring system' clearly includes a range of different schemes. Kerbside collection is more narrowly defined, but there too collection can be of separated fractions or of commingled waste. As a result, blanket comparisons of 'bring' versus 'kerbside' approaches must be made with caution. Box 6.1 lists some of the common attributes of these categories. It can be seen that some attributes, particularly contamination, depend more on whether the material is collected separated or mixed, than on whether a bring or kerbside approach is used. Collection systems will therefore be discussed in the following section according to the materials collected, rather than whether bring or kerbside systems are involved.

Household waste has traditionally been collected in a mixed state, but where household sorting has occurred, the different waste streams are collected separately, whether by the same or different collection vehicles. The categories collected separately vary by geography. In Germany, for example, the Duales System Deutchland (DSD) collects packaging material as a separate stream, whereas in Japan householders separate out combustible material for separate collection. In Europe, separate collections are most commonly used for dry recyclables (paper, metal, glass, plastic), biowaste (kitchen and garden waste, with or without paper) and in some countries, household hazardous waste (batteries, paint, etc.). A collection for remaining residual waste is also needed. Garden waste and

Table 6.2 Glass recycling in Europe

Country	Collected volumes (tonnes)				Recycled % 1992
	1989	1990	1991	1992	
Austria	115 000	135 000	156 000	175 000	64
Belgium	208 000	204 000	223 000	216 000	54
Denmark	58 000	61 000	60 000	75 000	48
Finland	18 000	24 000	15 000	23 000	44
France	760 000	906 000	987 000	1 100 000	44
Germany	1 538 000	1 791 000	2 295 000	2 459 000	65
Greece	4000	18 000	26 000	30 000	20
Ireland	11 000	13 000	16 000	20 000	27
Italy	670 000	732 000	763 000	786 000	53
Netherlands	279 000	310 000	360 000	378 000	73
Norway	11 000	13 000	10 000	24 000	44
Portugal	34 000	46 000	50 000	62 000	30
Spain	287 000	304 000	310 000	312 000	27
Sweden	42 000	50 000	57 000	76 000	58
Switzerland	164 000	189 000	199 000	212 000	72
Turkey	47 000	58 000	54 000	52 000	25
UK	310 000	372 000	385 000	459 000	26
Total	4 566 000	5 226 000	5 966 000	6 459 000	

Source: FEVE (Fédération Européenne du Verre d'Emballage, Brussels).

bulky waste may be handled as separate streams, or, alternatively, included within the biowaste and residual waste streams, respectively.

6.4 Collection systems

6.4.1 Dry recyclable materials

This category currently employs the greatest range of collection methods, from central or low density materials banks, to kerbside collection of recyclable materials in specially designed trucks.

Single (mono) material banks. Materials banks ('drop-off') which collect a single material per container, represent one of the best known forms of material recovery, mainly due to the success of 'bottle banks' for glass. High levels of glass recovery have been achieved across Europe using this method, although there is considerable variation between countries (Table 6.2). Other industries (in particular steel and aluminium) have tried to emulate this success.

The success of material banks at low density (high number of connected inhabitants) for materials, recovery is hard to assess, however. At the national level, the recovery rate can be calculated by dividing the total amount of material recovered by the national consumption of that material, At the local level, in contrast, it is not clear what area a bring container covers, and, therefore, the base amount of material from which the collected material has been recovered. Keen recyclers, for example, may travel some distance to a bring container, so importing waste from outside the area considered to be covered by the bring scheme. The best estimates of success rates will come from relatively isolated communities, which have a saturation density of bring containers, and the data are best expressed in terms of amounts collected per person, or per household, rather than % recovery rates, since the latter depends on what figure is used for the base amount of waste.

In a bring system, the amount collected will depend on the density of banks or containers, since this will determine how far individuals will have to transport their recyclables to the bank, and thus their motivation. Differences in bank density probably account for a large amount of the geographical variation in glass recovery rates in Table 6.2. Lemsterland has one bank of three containers for every 500 inhabitants; the density for the Netherlands as a whole is around one bank per 890 inhabitants (ORCA, 1992), whilst in the UK it is around one bank site per 9600 people (Landbank, 1992). Increasing bank density will increase the amount collected, but with diminishing returns, i.e. the extra amount collected will decrease with every extra bank added. At a certain point, the additional economic cost and environmental impacts of emptying and servicing banks will outweigh the environmental gains from the collection of material, although at present there is insufficient evidence to identify this optimal bank density.

A more immediate problem with increasing bank density is finding suitable sites. At high densities, small containers on street corners may be suitable, and inhabitants should be able to walk to these with their recyclable materials. At lower densities with necessarily longer transport distances, car transport is likely to be used. These banks need to be placed in strategic sites that are already regularly visited (petrol stations, supermarkets, etc.) so that specific car journeys to bank sites are not necessary. Unfortunately, consumer behaviour in using bring systems is another area in which reliable data are lacking, although of prime importance. Given that the energy saving possible from the process of recycling glass (transport not included) is around 4 MJ/kg (Porter and Roberts, 1985; and see Chapter 8), and the fuel consumption of a car is around 2.5 MJ per kilometre, the need to minimise specific car journeys to materials collection banks is clear.

Mixed recyclables banks. Communal 'bring' systems have also been used for collection of mixed (commingled) recyclables from high rise housing areas, which present special collection problems. This represents the highest density of bring containers, with the density equalling that of the regular refuse containers. The dry recyclables collection schemes set up in some such areas (e.g. Prato (Italy) and Barcelona (Spain)) have tried to match the normal refuse collection, with large wheeled bins located next to the refuse containers.

Kerbside collection. A range of collection methods has been used to collect recyclable material, varying in the degree of sorting involved, and including boxes, bags and wheeled bins (see Boxes 6.2–6.7). In its simplest form, recyclables are separated by the household and stored together in a bag (e.g. DSD systems in Germany; Cardiff, UK) or wheeled bin (e.g. Dunkirk, France; DSD systems and for paper and glass, Germany) ready for collection. As with the collection of mixed recyclables from streetside containers, collectors can use existing collection vehicles, in some cases (e.g. Dunkirk) even with compaction. Commingled collection of recyclables, whether from communal kerbside containers or household bags or bins, requires extensive subsequent sorting at a materials recovery facility (MRF).

The collection method involving the highest level of sorting is probably the 'Blue Box' system, that has been imported into Europe from North America. Blue Box schemes supported by the European Recovery and Recycling Association (ERRA) have been operating in Sheffield and Adur, UK, since 1989 and 1991, respectively. Householders sort out the targeted materials and store them in the box, which is put out at the kerbside for collection in a specially adapted vehicle. At the kerbside, the box contents can be sorted by the vehicle operator into several different compartments on the vehicle, and the empty box left at the kerbside. Since this is a positive sort, any unwanted materials can be left in the box and subsequently returned to the residual waste collection by the householder. The material collected in this way has already been sorted, and therefore limited further sorting is required at the central MRF.

Amount of material collected. Reliable data on the performance of single material bank systems across Europe is not widely available. Lemsterland (NL), using igloo containers, collects 54 kg of glass per household per year (Box 6.6), whilst in Germany, banks at a density of 1 per 800–1000 individuals are reported to collect 18–25 kg of glass and 50–60 kg of paper per inhabitant (ORCA, 1992). Plastics banks in Hamburg were reported to collect 0.5–1.5 kg per person per year in 1986 (i.e. prior to the establishment of the DSD system (Härdtle *et al.*, 1986)).

Amounts of recyclables collected in material banks and kerbside collection schemes in the UK and Germany are given in Tables 6.3 and 6.4, as well as in the details of individual schemes (see Boxes 6.2–6.7).

Table 6.3 Comparison of UK bring and kerbside collection systems

Area and type of scheme (kerbside unless otherwise stated)	Programme recovery[a] rates (%)					Overall diversion rate[c]
	Paper	Glass	Plastic[b]	Cans	Textiles	
Separate wheeled bins						
Leeds (biweekly)	70–80	n/c	70–80	d	40–50	50
Blue box (weekly)						
Stocksbridge,						
Sheffield	28	45	21	17	n/a	6.6
SE Sheffield	52	66	28	14	32	15.3
Milton Keynes	57	44	57	24	n/c	18.7
Adur	67	71	60	54	n/c	27
No container (biweekly)						
Chudleigh, Devon	36	55	n/c	21	3	6.9
Green bag (biweekly)						
Cardiff	52	52	13	6	21	17.7
Bring systems						
Ryedale	13	40	n/c	n/c	3	4.1
Richmond-upon-Thames	18	61	n/c	3	8	8.2

Source: Atkinson and New (1993a, b).
[a] Recovery rate gives % of each material recovered after both collection and sorting, compared to amount of that material in the household waste stream.
[b] Programme recovery rates for plastics are for beverage bottles and food containers only.
[c] Diversion rates are calculated differently for bring and kerbside collection schemes and this will tend to flatter kerbside collection schemes. Bring recovery rates are calculated as a proportion of total household waste (collected and delivered). Kerbside recovery rates are usually calculated as a proportion of collected household waste only.
[d] 30–40% for aluminium cans, 50–60% for steel cans.
n/c, not collected; n/a, not available.

Contamination levels. The contamination level can be defined as the percentage of non-targeted material that is collected by a given method. This non-targeted material may be:

(a) the wrong material type for that part of the system, e.g. paper in a glass bank;
(b) the right material but in the wrong form, e.g. plastic film in a plastic bottle bank;
(c) dirty material, e.g. containers with contents still inside;
(d) non-recyclable material.

The contents of a single material bank may be bulked and sorted, but normally the collected material is shipped direct to the processors, and thus effectively leaves the waste management system as defined in this book. Any contaminants in the bring containers will also, therefore, leave the

Table 6.4 Comparison of German bring and kerbside collection schemes

	Paper		Glass		Metals	Plastics	Mixed dry recyclables	Biowaste
	kg/person per year	Recovery rate (%)	kg/person per year	Recovery rate (%)	kg/person per year	kg/person per year	kg/person per year	kg/person per year
Bring systems								
2000 persons/bank	5–15	8–25	5–15	13–38	0.5–2.5	1–2	15–50	5–30
1000 persons/bank	10–25	17–42	10–20	26–51				
500 persons/bank	15–50	25–50	15–25	38–64				
Kerbside collection								
Paper (collected in bundles)								
Every week	20–35	33–58	–	–	–	–	–	
Every 2 weeks	15–25	25–42	–	–	–	–	–	
Every 4 weeks	10–20	17–33	–	–	–	–	–	
Paper (in containers)	35–55	58–92	–	–	–	–	–	
Glass (in containers)	–	–	15–30	38–77	–	–	–	
Multi-material containers (glass, metal, plastics)	30–50	50–83	12–30	31–77	5–10	5–10	–	
Bag collection	5–25	8–42	5–20	13–51	1–2	5	30–60	
Biobin	–	–	–	–	–	–	50–140	

Source: Schweiger (1992).

waste management system, although they may re-enter it when they are screened out at a material reprocessing facility. Levels of contamination will vary with the material collected. In the case of glass for example, it is necessary in many cases to collect it colour-separated to achieve the highest market prices (particularly in the case of clear glass); so any failure to separate clear, brown and green glass where this is requested will constitute contamination. Additional contamination is likely to come in the form of organics (original container contents), ceramics and plastics (labels, closures) and metals (caps). Typical levels of contamination in recovered glass cullet that reaches the reprocessors (mainly via glass bank collection schemes) is around 5–6% (Ogilvie, 1992). In the case of paper and plastic collection, where it is likely that only certain types of the material (e.g. newspapers only) are required, contamination may arise from the public depositing materials not requested. Since the containers are left unattended in the open, there is also the possibility of contamination by people using the containers to dispose of litter or other refuse. As with other collection schemes, clear instructions from the collectors, and a reasonable level of motivation from the public are paramount.

Levels of contamination in mixed material banks and kerbside collection schemes for dry recyclables show a clear pattern (Figure 6.3). Kerbside box schemes generally have the lowest contamination levels (5–8%). The open nature of the box, and in some cases a kerbside sort allows inspection of the contents. Any unwanted materials can thus be left in the box, so do not enter the recyclables stream. Kerbside inspection and sorting is not possible where commingled recyclables are collected in a bin. Such schemes generally have a higher contamination level (27–32%) due to the inclusion of non-targeted materials. The highest contamination levels (35–56%) have been recorded from commingled material banks (i.e. bring system). These are communal bins; lack of 'bin ownership' and perhaps some contamination from litter probably explain why contamination is higher than in collection of mixed recyclables from kerbside bins serving individual households. The composition of the contaminants in the packaging collection bins in the Barcelona scheme is shown in Table 6.5. The wide range of possible contaminating materials is clearly shown, underlining the need for effective communication with households as to what materials are required in which container.

6.4.2 Biowaste and garden waste

Garden waste, if not dealt with at source (e.g. home composting), can be handled by a bring system at a central collection site. If kept separate, this material can be used as the feedstock for composting plants, to produce so-called 'greenwaste' or 'yardwaste' compost. Alternatively, it may be collected along with other biowaste, or restwaste, via a kerbside collection.

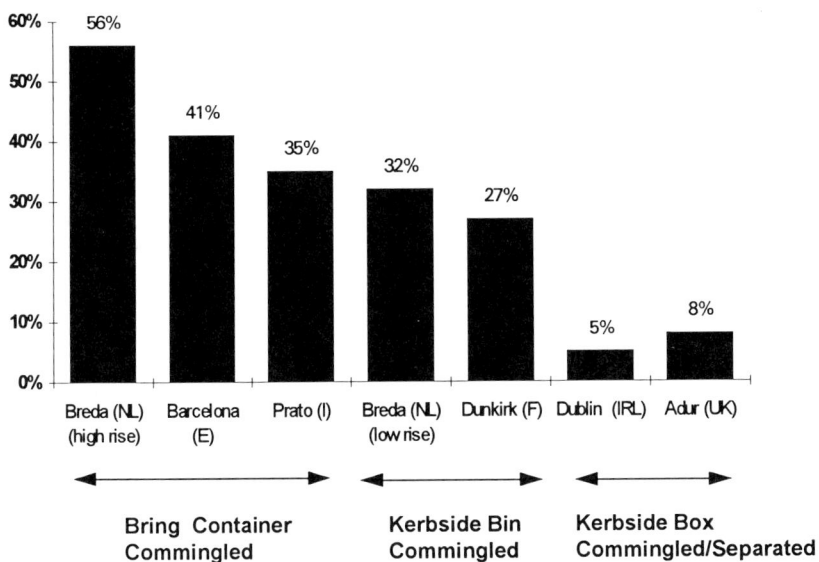

Figure 6.3 Contamination levels for dry recyclables collection schemes (figures are for packaging fraction only). These figures are for the residue left after collection and sorting, so include both non-targeted contaminants and some targeted materials that are not selected. See Section 7.2 for further discussion. *Source*: ERRA, personal communication (1993).

For biowaste, there is one main collection method available, kerbside or close-to-home collection. There has been a strong trend recently in Europe towards the separate collection of the organic fraction of waste for biological treatment. This trend has been particularly strong in German speaking countries, the Netherlands and in some parts of Scandinavia, where schemes have been in operation since the late 1970s and early 1980s (ORCA, 1991b). In the Netherlands, for example, legislation has already been passed that requires municipalities to introduce source-separated collection of biowaste. In the Flemish region of Belgium, plans have been made to ban compostable material from entering landfills or incinerators, and municipalities will be asked to implement separate collection of this material by 1996 (Ovam, 1991). In Germany, by 1991, some 88 projects were in operation for the collection of source-separated biowaste (Fricke and Vogtmann, 1992).

Biowaste definition. Whilst there has been a clear trend towards the separate collection of biowaste, there has been less agreement as to what should be included in the biowaste definition. Narrow definitions include only vegetable, fruit and garden waste (VFG). Collection of garden waste only can occur, but this is usually done through a central bring system

Table 6.5 Analysis of contaminants in Barcelona mixed recyclables (packaging) bins

Contaminant	% of bin contents
Clothes	5.66
Wood	1.47
Stones/sand	9.92
Miscellaneous	5.09
Plastic film	3.83
Tubs/cups	0.78
Other plastics	1.65
Other metals	3.26
Total contaminants	31.66

Source: ERRA, Barcelona Waste Stream Analysis Report (1993) (personal communication).

(e.g. at a civic amenity site in the UK); collection schemes usually involve at least some household organic waste. At the broader end of the spectrum, the biowaste definition can encompass the VFG material, plus part or all of the non-recyclable paper fraction. A range of biowaste definitions used in different schemes is given in Table 6.6.

A survey of collection schemes in Germany showed that 40% involved kitchen and garden waste only, 55% collected kitchen and garden waste, plus soiled paper (e.g. tissues, etc.); the remaining 5% utilised kitchen and garden waste plus the entire paper fraction (Fricke and Vogtmann, 1992). The collection of paper along with the organic fraction as biowaste for biological treatment has many advantages, including providing structure to the biowaste, without which the wet biowaste tends to become anaerobic with the production of offensive odours. Including paper in the biowaste also reduces seepage water from the bin, seasonal variability in the amounts of biowaste collected, production of leachate during composting and improves the final compost quality (see Box 6.9). It will also ensure that a larger proportion of the waste is valorised (if the paper would not otherwise be valorised by materials recycling or burning as fuel), provided that the inclusion of paper does not adversely affect the quality of the final compost produced.

Amounts collected. The actual amount of biowaste collected in any separate collection scheme will clearly depend on the amount of organic waste generated (i.e. the potential maximum amount) and the definition of biowaste applied. Organic waste generation rates vary both between urban and rural areas (Table 6.7) and seasonally. In Germany, for example, about three times more biowaste is collected in the spring and autumn, compared to mid-winter (Selle *et al.*, 1988). Given such variability, however, a recent German Government Report (1993) suggests that, on average, separate

Table 6.6 Definitions of biowaste

Biowaste elements	Netherlands	Copenhagen (DK)	Frederiksund (DK)	Mainz-Bingen (D)	Witzenhausen (D)	Diepenbeek (B)
Food waste (meat, fish, bones, cheese, fruit, vegetables, etc.)	×	×	×	×	×	×
Egg-shells	×	×	×	×	×	×
Coffee-filters and grounds, tea-bags	×	×	×	×	×	×
Nutshells	×	×	×	×	×	×
Flowers and house-plants	×	×	×	×	×	×
Pet litter	×	×	×	×	×	×
Grass, straw and leaves	×		×	×	×	×
Hedge cuttings, garden plants, small branches	×		×	×	×	×
Paper diapers			×			
Kitchen paper			×	×		
Sanitary towels			×			
Newspaper				×		
Newspaper with potato peelings	×	×		×	×	×
Wet newspaper			×			
Cardboard				×		

Source: ORCA (1991b). Items marked with '×' are included.

Box 6.9a INCLUSION OF WASTE PAPER IN THE BIOWASTE DEFINITION.

Advantages

Collection:

–Reduction of seepage water during storage and transport.
Biowaste without paper has a high moisture content, especially in inner city areas where garden waste is sparse. In the German inner city of Soln, for example, the biowaste had a total solids content of only 23% (Doh, 1990). The high water content leads to leakage during storage collection and transport.

–Reduction of malodours.
Odours are linked to the high moisture content of biowaste. The biowaste is highly putrescible, and with the high water content will rapidly become anaerobic especially in summer months, producing offensive odours. Addition of the paper fraction will absorb this moisture and so reduce odour generation

–Reduction in seasonal variability in amounts collected.
Where only kitchen and garden waste are collected, large seasonal variation will occur in the amount collected. In Germany, about three times more biowaste is collected in the spring and fall when compared to the winter (Selle *et al.*, 1988). The quality also varies, being limited to very moist kitchen waste in winter, but including drier garden waste at other times. Inclusion of paper will dampen the variability in both quantity and quality of biowaste collected.

Biological treatment (composting)

–Reduced production of leachate.
Biowaste including 20% or more paper can be composted in windrows without production of leachate (Fricke, 1990; see Chapter 9).

–Reduced requirement for bulking agents.
Biowaste without paper, i.e. with a high moisture content, requires bulking agents to absorb water and ensure free circulation of air. Otherwise, anaerobic conditions will occur. Some composting processes require up to 250 kg of wood chips to be added to every tonne of biowaste (Haskoning, 1991)

–Improved Carbon/Nitrogen (C/N) ratio.
Addition of waste paper corrects the C/N ratio from about 15–20 for biowaste, to 25 or more, which is optimal for biodegradation. At the lower C/N ratio the composting process is slowed down (see Chapter 9), and more odours such as ammonia are released (Jespersen, 1991).

–Increased organic content of final compost.
Biowaste compost in Germany has an organic content of around 26% (Selle *et al.*, 1988), whereas in some countries a minimum of 30–40% may be required (ORCA, 1991b). Adding paper to the biowaste can increase the organic content to this level.

–Reduced salt level of final compost
High salt levels (above 2 g of NaCl/litre) found in biowaste composts can limit their potential usage. Due to a dilution effect, adding paper will reduce the salt concentration below this critical level (Fricke, 1990).

Overall: –Increased valorisation rate.
Cities focusing on food waste alone in their biowaste will collect only 15–25% of their waste in this fraction. A broader definition of biowaste to include paper, paper products and some garden waste can valorise 40–50% of household solid waste in this way (ORCA, 1991b).

Box 6.9b INCLUSION OF WASTE PAPER IN THE BIOWASTE DEFINITION.

Possible Disadvantages

Collection:
  −Increased contamination level
  A wider biowaste definition could cause confusion in households as to what
  materials are required. This should be overcome by a well organised and
  frequent education and communications programme.

Subsequent compost quality:
  −Heavy metal levels from inks and other contaminants.
  The general heavy metal content of waste paper and paper products is low,
  but inks used in magazines and wrapping paper often use metallic
  pigments (Rousseaux, 1988), which will contribute to the heavy metal
  content of the finished compost.

Table 6.7 Quantity of biowaste

Town		Area	% of total household waste	
Amsterdam	(NL)	Inner city	23.5	(1)[a]
Amsterdam	(NL)	High-rise buildings	16	(1)
Apeldoorn	(NL)	Suburban	43	(1)
Apeldoorn	(NL)	High-rise buildings	27	(1)
Nuenen	(NL)	Suburban	40	(1)
Purmerend	(NL)	High-rise buildings	20	(1)
Frederikssund	(DK)	Suburban	40	(2)
Copenhagen	(DK)	Inner city	20	(2)
Mainz-Bingen	(D)	Suburban	50	(3)
Witzenhausen	(D)	Suburban	29	(3)
Diepenbeek	(B)	Suburban	60	(4)

[a] References: (1) Kreuzberg and Reijenga (1989); (2) Jespersen (1991); (3) Selle
et al. (1988); (4) Rutten (1991).

collection of biowaste is likely to recover 90 kg of organic material, per
inhabitant, per year. For the ERRA scheme in Lemsterland, an estimated
288 kg is collected per household per year (see Box 6.6). The wider the
definition of biowaste used, however, the greater the amount of biowaste
likely to be collected. The range of amounts collected for different biowaste
definitions and collection areas in Germany are presented in Table 6.8.
Similar overall recovery rates have been achieved in source separated collection schemes for organics in the United States. In a scheme run by the
Audubon Society in Fairfield and Greenwich, Connecticut, an average
of 6.4 kg of organics and soiled paper waste was collected per household
per week (equivalent to 333 kg/household per year) (Beyea *et al.*, 1992).
Given the higher household waste generation rate for the United States
(in this project 1110 kg/household per year), this meant that biowaste
collection was able to divert 30% of the total household waste from landfill.

Table 6.8 Comparison of collected biowaste in Germany

	Quantity (kg/person/yr)	Rate of recovery (%)[a]	Level of contamination (%)
Urban districts	73	69	2.24
Inner city areas	46	49	4.02
Rural districts[b]	102	73	1.77
National average	92	70	2.02
Biowaste including paper	184	85	7.50

Source: Fricke and Vogtmann (1992).
[a] Recovery rate = amount collected/amount available in waste.
[b] Excluding projects which include the entire waste paper fraction in the biowaste definition.

Contamination levels. As with other separate collection systems, contamination of the biowaste with unrequested materials will occur. The evidence suggests, however, that the contamination level is low (Table 6.8). In Germany, contamination levels are around 5% (Selle *et al.*, 1988; Fricke, 1990), consisting mainly of plastic. Less has been reported elsewhere. Results from the Netherlands show that the sum of glass, metal and plastic contaminants account for less than 1% of the biowaste (Kreuzberg and Reijenga, 1989), and a similar level has also been reported for the Diepenbeek scheme in Belgium (Rutten, 1991). Much of the contamination will consist of refuse bags, used to collect the waste and transport it to the refuse container.

Like the overall amounts collected, the levels of contamination also vary between city and rural areas and with biowaste definition. Contamination in inner city areas can rise to 10–15% (Fricke, 1990; ORCA, 1992), apparently due to lack of household motivation and effective peer pressure in high-rise accommodation. Similarly, with a broader definition for the biowaste, there is scope for more confusion as to what should be included, leading to an increased level of nuisance materials (Table 6.8). It should be possible to counter this trend, however, by clear instructions and an active communications programme to householders.

Along with such 'nuisance materials', biowaste will also be contaminated with heavy metals. This is of particular importance if the biowaste is to be processed into marketable compost, since heavy metal levels may determine whether the resulting compost is of acceptable quality (see Chapter 9). Typical heavy metal levels for German biowaste, and recommended maximum levels, are given in Table 6.9.

Collection methods. Biowaste is generally collected in bins or bags. Bins, either split into compartments or not, have the advantage that they do not add to the level of plastic contamination that must be removed

Table 6.9 Heavy metal levels in German biowaste (in mg/kg dry weight)

Parameter	Range	Average	German (UBA) recommended limits
Zinc	29.8–178.0	117.8	300
Lead	3.5–94.4	37.4	100
Copper	13.5–44.6	29.0	75
Chromium	13.0–20.8	17.1	100
Nickel	6.8–16.0	11.3	50
Cadmium	0.14–0.25	0.19	1
Mercury	0.07–0.18	0.11	1
Moisture content (% of wet weight)	66.6–72.7	69.7	
Organic content (% of dry weight)	68.5–82.7	78.1	

Source: Tidden and Oetjen-Dehne (1992).

at the biological treatment plant (unlike bags), but they may need washing out, especially in hot weather. A possible disadvantage of using bins, especially large wheeled bins, is that householders are tempted to add their garden waste too, rather than composting it at home. This can result in an increase in the waste entering the system (Selle *et al.*, 1988). Another method has been to use paper bags, with a moisture barrier, which are then biodegradable in the subsequent composting process. The Audubon scheme in Connecticut cited above successfully used such 'wet bags' (Beyea *et al.*, 1992), and there are also trials underway using bags made from biodegradable polymers in other parts of the United States (Goldstein, 1993). A survey of biowaste collection schemes in Germany reports that most schemes (81%) use 120- or 240-l wheeled bins, with 2% using 35-l bins and 9% multi-chambered bins. Only 7% used bags, evenly split between paper and plastic (Fricke and Vogtmann, 1992).

Collection of biowaste can also require specially modified vehicles, involving rotating drums, pneumatic presses or multi-chambered bodies. One of the major problems is the leachate leaking from the trucks, so in many cases they need to be specially sealed to prevent this. Alternatively, collection of some paper with the biowaste can reduce leachate production, as discussed above.

6.4.3 Hazardous materials in household waste

Household waste contains hazardous materials such as used motor oil, pesticides and solvent and paint residues in used cans and bottles. Contaminants such as heavy metals also occur; they are found in small quantities in a range of household waste items but are mainly concentrated into a few items such as used batteries, discarded light bulbs and tubes and mercury thermometers (see Chapter 5). Normally such materials are included in the residual or restwaste collection, but their presence can limit the options available for treating this waste stream. High heavy metals levels, for example, may exclude the possibility of mixed MSW composting, since the resulting compost would exceed the acceptable heavy metal concentrations. Similarly, the presence of organic contaminants such as persistent pesticides could result in groundwater contamination if the mixed waste were landfilled.

One solution is to separate out the hazardous materials at source, and deal with them separately. This approach has been taken in the Flanders region of Belgium, where 2 million households have been supplied with a 'KVA box' (Klein Gevaarlyk Afval: small, dangerous waste) for small hazardous waste (ORCA, 1992). As the amount of hazardous waste generated will be small (around 5–10 kg/household per year), this material can be collected separately on an infrequent basis. Alternatively, it can be taken to a central collection site (i.e. a bring system). In Germany, for example, there is typically a small container for such waste at each multi-material collection point.

Clearly there can be advantages to separate collection, so long as the extra collection/bring system can be integrated into the normal collection system, and provided that there are effective ways of dealing with the hazardous waste once collected. Kerbside collection in various US cities, for example, has collected 7–18% of the available batteries (Warmer, 1993). Many current schemes collect small batteries which have been sorted out of the waste by householders, but do not have access to appropriate recycling or disposal technology to deal with them, so are forced to stockpile them. Under existing legislation, at least in the UK, once these elements of waste are concentrated in this way, they are classified as a special waste, which limits the ways they can be stored, handled and treated.

6.4.4 Bulky waste

This solid waste can make a significant contribution, but is generally not included in estimates of household waste generation (see Chapter 5). In the UK, for example, bulky waste plus garden waste probably represents around 30% (by weight) of household waste generation (Atkinson and New, 1993a). Bulky wastes can either be delivered to a central collection site (i.e. a bring system) or picked up from households using a separate and infrequent collection. Once delivered to a central site, some bulky objects such as furniture or appliances can be recovered intact for re-use. Other bulky wastes can be recovered for metal recycling (appliances) and the residue, if sorted appropriately can be either incinerated or landfilled. There is also the possibility of recovering other materials (e.g. CFCs from refrigerators).

6.4.5 Restwaste

In most traditional systems, this category would contain all of the household waste, collected in a completely mixed state. Some restwaste can be handled in a bring system in city centre or high rise areas, such as in the Sagrada Familia area of Barcelona (Box 6.2), where large (2200-l) streetside restwaste containers are used. This system is common in southern Europe, since it allows daily removal of refuse. In most areas of northern Europe, however, collection from each property or kerbside is the norm. Where bring or kerbside collection schemes have been introduced for dry recyclables, biowaste and/or hazardous household waste, both the amount and composition of restwaste will be altered. Whatever methods are used for separate collection of parts of the waste stream, however, there will always be residual waste for collection.

Although the amount collected may be reduced in weight there may be little saving in either environmental or economic terms compared to the traditional collection of mixed waste. This is because the same number of properties have to be visited, and so the same distances need to be driven. The decreased amount collected per household can lead to longer collection rounds before the collection vehicle needs to be emptied. Such efficiency improvements may lead to some cost saving; a report in the UK suggested marginal savings of around £9 per tonne in restwaste collection (about 25% of current costs) when recyclables are collected in a separate round (DoE/DTI, 1992). An alternative is to reduce the frequency of collection, for example to every other week where current collections are weekly. This may not be possible, however, in regions where the frequency of waste collection is fixed by legislation, or where odour problems from the restwaste make regular collections necessary.

6.5 Integrated collection schemes

Collection is at the hub of an integrated waste management system. An integrated approach is the key to an effective collection system. The pre-sorting and collection stage needs to collect all of the waste, separated into suitable streams for subsequent valorisation methods. To be efficient in both economic and environmental terms, it must do this with the minimal use of transport, including both collection trucks and householders' private cars.

An integrated collection system could include any combination of bring systems (materials banks, close-to-home drop-off centres, central collection sites for garden/bulky waste) and/or kerbside collections (for recyclables, biowaste and/or restwaste). The key is that all methods form part of one system with the objective of collecting all the waste materials in suitable streams with minimum environmental and economic impact.

Key lessons on the importance of an integrated collection have come from a variety of collection schemes. Collection of dry recyclables in Blue Box schemes (e.g. Adur, Sheffield, UK; Ontario, Canada) in many ways represents an additional, rather than integrated collection system. A second truck travels the same route as the residual waste collection vehicle, collecting a dry recyclables fraction. This additional truck is likely to result in increased collection costs as well as increased environmental impacts due to the vehicle's emissions. In most cases, an improvement would be to collect both the recyclables and the residual waste on the same visit. This has been introduced in Worthing (a neighbouring district to Adur, UK) using a specially designed truck that has two compartments for recyclables, plus a normal compaction compartment for the restwaste. This allows both recyclables and restwaste to be collected on a single visit to each household. An alternative that has been developed in the United States (Omaha, Nebraska) is to collect commingled recyclables in blue bags, which are loaded along with the bagged rest waste into the same vehicle compartment, for separation at the sorting facility (Biocycle, 1992). This form of 'co-collection' has not been developed to any extent in Europe, however.

With more than three waste fractions, it is difficult to collect all fractions efficiently on a single visit, so an alternating collection schedule is often employed. Lemsterland in the Netherlands (see Box 6.6) for example, collects four different waste streams: mixed paper, mixed dry recyclables (excluding glass), organics and restwaste. The materials are stored in two separate wheeled bins, each with interior partitions (giving four compartments) and then collected by split-compartment vehicles on an alternating basis. Leeds (UK) similarly collects three different waste fractions, with one split-compartment truck and one normal compactor truck visiting properties in alternate weeks. A range of collection methods and schedules is used for waste collection in Germany (Table 6.10).

Alternating collections can be effective in collecting several separate fractions of household waste, without increasing the overall number of visits to each property, so long as it does not affect the comfort level of the participants. Biowaste collection is one area where this comfort level is likely to be compromised, since it is necessary to retain frequent collection to prevent severe odour nuisance during summer or year round in warmer climates. This factor needs to be considered in the design of effective collection systems. In the Netherlands, for example, source separation of biowaste is mandatory and biowaste collection is often in alternate weeks as a result. During hot weather, households are instructed to put their organic waste into the residual waste, rather than into the biobin, if

Table 6.10 Survey of biowaste collection methods in Germany

Schedule	Proportion (%)
Weekly	18.8
Weekly with multiple-chamber bins	9.4
Every 2 weeks in addition to the weekly collection of non-recyclables	20.0
Every 2 weeks, alternating with the collection of non-recyclables	51.8

Source: Fricke and Vogtmann (1992).

this is the next bin due for collection. This results in a loss of organic material from the biowaste stream (hence less compost produced), and a corresponding increase in the organic material going to landfill or incineration (hence a decreased diversion rate). Alternation of biowaste and restwaste collection is also the most frequently used method in Germany (Table 6.10). There are ways round this problem, however. Including paper and paper products in the biowaste definition can result in less odour nuisance from the bin (see Box 6.9), or alternatively the schedule can be devised so that biowaste collection is kept at the previous frequency (e.g. weekly) whilst the dry recyclable and the restwaste collections are alternated.

There is a general conflict between the needs of sorting and the ease of collection. Valorisation methods generally, and materials recycling in particular, require effective separation of the waste into several streams. To reduce any cross contamination, especially by organic material, it is best to separate materials as early as possible in the waste management system, i.e. at source. The number of categories for home sorting is limited, however, by householder ability and motivation, available storage space and the ability to collect many different waste streams without increasing the number of collection visits to the property. To keep the total number of fractions to a manageable level, therefore, fractions that can easily be separated by subsequent sorting (e.g. steel cans, aluminium cans and plastics) should be collected mixed.

The net result of the above conflicts is that there is no one best collection system for all areas. The best and most integrated system for any area will depend on the way the waste needs to be collected for the local treatment and disposal methods, the composition of the waste, the type of housing and population density, and the motivation of the residents. Bring and kerbside collection systems can both be appropriate, for different materials or fractions, within an integrated collection system.

6.6   Environmental impacts

The major environmental impacts associated with collection systems will be due to the transport used, which consumes energy and produces atmospheric emissions. The function of the collection system, after all, is to transport the waste from the household or commercial property to the sorting or treatment site. There may be other impacts, however, such as from the production of plastic bags used in the collection, or the cleaning of bins.

6.6.1   Transport impacts

The transport involved in collection systems involves a mixture of house-holders' cars and municipal waste collection vehicles, the exact combination varying with the collection method used (Box 6.1). At one extreme, in the use of central collection sites, most if not all of the transport will involve the use of householders' vehicles. Low density materials banks will often involve householders driving to the bank sites, and then special collection vehicles emptying the banks and transporting the materials to bulking depots, prior to sale and transport on to the materials processors. At the other end of the spectrum, close-to-home bring schemes (within walking distance of each property), and kerbside collections involve only municipal (or contractors') collection vehicles.

Calculation of the energy consumption and emissions resulting from each type of transport requires data on the distances driven and the average fuel consumption of the vehicles used. Whilst the latter information is available (Table 6.11), details of distances driven will vary widely between different areas, so cannot easily be generalised. As a result, these are included as variable data in this analysis, and have to be inserted by the user for the geographical area under study.

For bring systems (both central collection sites and low density materials banks) the number of special journeys made by car, each year, to the collec-tion points needs to be estimated, as well as the average distance. Note that depositing recyclables in a material bank in a supermarket car park during a shopping trip would not count, as the transport there could be allo-cated to the shopping trip. (Theoretically, the impacts should be divided between the two functions, but this is beyond the level of detail of the present study. Data of this type are not readily available, but could be acquired for any given region via a consumer survey.) Fuel consumption levels for private cars (petrol and diesel) are given in Table 6.11. Both urban and non-urban figures are given, but the urban figures are pro-bably more typical for driving to bring systems so will be used in this model.

For transport on from central collection sites to treatment plants, and

---

Box 6.10

RECYCLABLES' COLLECTION, RECOVERY AND RECYCLING
- TERMS AND DEFINITIONS USED IN THIS BOOK

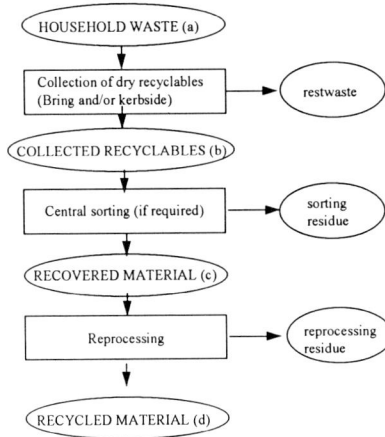

```
                    ( HOUSEHOLD WASTE (a) )
                              |
                              v
              +----------------------------+
              | Collection of dry recyclables | -----> ( restwaste )
              | (Bring and/or kerbside)      |
              +----------------------------+
                              |
                              v
                  ( COLLECTED RECYCLABLES (b) )
                              |
                              v
              +----------------------------+
              | Central sorting (if required) | -----> ( sorting
              +----------------------------+            residue )
                              |
                              v
                    ( RECOVERED MATERIAL (c) )
                              |
                              v
              +----------------------------+
              |       Reprocessing          | -----> ( reprocessing
              +----------------------------+            residue )
                              |
                              v
                    ( RECYCLED MATERIAL (d) )
```

Definitions:

Overall recyclables collection rate	$= 100 \times b/a$ %
Material-specific collection rate	$= 100 \times b_{m1}/a_{m1}$ % (where a_{m1} = amount of material m1 in a, b_{m1} = amount of material m1 in b)
Overall recovery rate	$= 100 \times c/a$ %
Material-specific recovery rate	$= 100 \times c_{m1}/a_{m1}$ % (where a_{m1} = amount of material m1 in a, c_{m1} = amount of material m1 in c)
Overall recycling rate	$= 100 \times d/a$ %
Material-specific recycling rate	$= 100 \times d_{m1}/a_{m1}$ % (where a_{m1} = amount of material m1 in a, d_{m1} = amount of material m1 in d)

---

for general waste transport elsewhere in this model, Table 6.11 also gives the fuel consumption figures from a heavy goods vehicle. Although the fuel consumption (33 l/100 km) is for non-urban conditions, it agrees well with the figure used by Boustead (1993) as average fuel (diesel) consumption for a 20 tonne payload vehicle (32.1 l/100 km).

Calculation of the energy consumption and emissions from kerbside collection is less straightforward. Clearly the stop-start nature of most kerbside collection makes the use of standard heavy goods vehicle (HGV)

Table 6.11 Fuel consumption data for different road vehicles

	Fuel consumption (1/100 km)		
Source	EC (1992)	Boustead (1993)	Data used in LCI of waste
Private car			
Petrol urban	11.6		
non-urban	5.3		
average[a]		14.6	11.6
Diesel urban	9.4		
non-urban	5.8		
average[a]		9.2	9.4
Heavy Goods vehicle			
Diesel non-urban	33.0		
average (10-tonne load)		26.7	26.7
average (20-tonne load)		32.1	32.1

[a] Data for 1.05 tonne unladen weight vehicles.

data inappropriate. Data for stop-start collections of dry recyclables from schemes in Adur and Milton Keynes (UK) report fuel consumption of 44 and 20 1/100 km, respectively (Adur District Council, personal communication; Porteous, 1992). What is needed for this LCI study, however, are the impacts per household serviced, or per tonne of material collected. Calculating this from fuel consumption data requires the average distance travelled by the collection vehicle per household visited. Again this will vary with housing density, and with the distance that the collection vehicle travels from the collection area to the sorting or treatment plant for emptying. This problem can be avoided by using data that should be available to every collection system operator, the average fuel consumption per round. Knowledge of the average number of households served then allows calculation of the average fuel used per household visit. Data of waste collected allows conversion into fuel used per tonne collected. The above reported results from Adur give values of 32 1 per 1000 households visited or 14.3 1/tonne collected (calculated from data in IGD, 1992), and 7.2 1/tonne for Milton Keynes (Porteous, 1992). Kerbside Dublin, a similar scheme collecting dry recyclables, has reported a fuel consumption of 17 1 per 1000 households visited (5.8 1 per tonne collected) over the first 6 months of 1993 (ERRA, personal communication). Fuel consumption per household visited is likely to be the more reliable measure, however, since the same distance must be covered by the vehicle, no matter how much waste is picked up from each property. Consumption of diesel fuel can then be converted into primary (thermal) energy consumption, emissions to air and water, and solid waste, using the generic data given earlier in Chapter 4.

6.6.2 Other impacts

The input boundary for this LCI study has been defined as waste at the point that it leaves the waste generator, i.e. the household or commercial property (Chapter 4). Any materials needed to get the waste from this point to a collection site or vehicle (e.g. refuse bags, blue boxes, refuse bins) will therefore fall within the defined boundaries, and the relevant lifecycle inputs and outputs, from cradle to grave, should be included. It is important to consider these additional impacts, since the range of collection systems discussed above differs in the number and composition of different waste fractions collected, and hence in the relative need for different bags or bins for the collection process.

Collection bags. Although they are normally included in analyses of household waste (Table 5.9), collection sacks are not strictly 'waste' during the collection stage as they are performing a useful function, i.e. containing the waste; they do become waste when the waste is subsequently delivered to a treatment site (e.g. a materials recovery facility or composting plant). Collection bags in common use are either paper or plastic, so the lifecycle inputs and outputs in terms of energy, emissions and solid waste can be calculated from generic production data for these materials, given the average weight per bag and the average number of bags used per household per year. The number of bags used will vary with the quantity of waste generated, size of bags and the degree of waste sorting and separation required in the home. Sorting the waste into many fractions could lead to the collection of numerous half-filled collection bags. The generic data for production of both paper and plastic bags used in this study are given in Table 6.12.

Note that impacts due to small refuse bags that are used to convey waste from the house to a dustbin or large refuse sack will not be included, since these are part of the waste before it leaves the property. Similarly, some schemes (e.g. Chudleigh, Devon, UK; Table 6.3) use ordinary carrier bags for collecting materials such as dry recyclables. Since these would have been included in the waste stream in any case, the upstream impacts of producing these bags should not be included.

Collection bins. Inclusion of the production impacts for collection bins in this LCI study is debatable, since they can be considered to represent capital equipment, rather than operating consumables. (Under the goal definition section in Chapter 4, capital equipment was excluded from the study.) This would artificially bias any comparisons between collection systems using different collection container types. There is fundamentally little difference between the use of bins and bags in a lifecycle study. Bins consume a large amount of material (most often plastic) initially, rather

Table 6.12 Lifecycle inventory data for the production of paper and low density polyethylene bags and polypropylene bins used in waste collection systems

	Virgin LDPE (per tonne) PWMI (1993)	Bag production (per tonne) Habersatter (1991)	Total for LDPE bags (per tonne)	Virgin PP (per tonne) PWMI (1993)	Injection moulding of bins (per tonne) Habersatter (1991)	Total for PP bins (per tonne)	Paper bags (unbleached sulphate pulp) (per tonne) Habersatter (1991)
Energy consumption (GJ)	88.55	9.5	98.05	80.03	9.5	89.53	53.0
Air emissions (g)							
Particulates	3000	197	3197	2000	197	2197	4346
CO	900	349	1249	700	349	1049	3165
CO_2	1 250 000	441 657	1 691 657	1 100 000	441 657	1 541 657	
CH_4							
NO_x	12 000	1236	13 236	10 000	1 236	11 236	5114
N_2O		70	70		70	70	345
SO_x	9000	2502	11 502	11 000	2 502	13 502	10 868
HCl	70		70	40		40	4
HF	5	0.01	5.01	1	0.01	1.01	0.01
H_2S				10		10	
Total HC	21 000	2112	23 112	13 000	2 112	15 112	6258
Chlorinated HC							
Dioxins/furans (TEQ)							
Ammonia	0.49	0.49	0.49	0.49	0.49	0.49	3.4
Arsenic							
Cadmium							
Chromium							
Copper							
Lead							
Mercury							
Nickel							0.004
Zinc							

Water emissions (g)							
BOD	200	0.15	200.15	60	0.15	60.15	2921
COD	1500	0.44	1500.44	400	0.44	400.44	25 423
Suspended solids	500	0.15	500.15	200	0.15	200.15	1
Total organic compounds	320	4.7	324.7	620	4.7	624.7	30
AOX							3
Chlorinated HCs							
Dioxins/furans (TEQ)							
Phenol		0	0		0	0	0
Ammonium	5	0.62	5.62	10	0.62	10.62	0.876
Total metals	250		250	300		300	
Arsenic							
Cadmium							
Chromium							
Copper							
Iron		0.003	0.003		0.003	0.003	
Lead							
Mercury							0
Nickel							
Zinc							
Chloride	130	0.02	130.02	800	0.02	800.02	22
Fluoride		1.335	1.335		1.335	1.335	1.89
Nitrate	5	1.32	6.32	20	1.32	21.32	7
Sulphide							
Solid waste (kg)	39.4	49.1	88.5	31.0	49.1	80.1	150.2

than on a weekly basis, but like bags, bins will also enter the solid waste stream eventually at the end of their useful life. The impact of bin use is therefore included; once calculated, it will be possible to determine whether the impact of this 'capital equipment' is insignificant, as originally predicted.

It is possible to calculate the relevant impacts from the use of bins given generic data on the materials used (Table 6.12), the weight of material per bin/container, and the expected useful life-span. For example, the Blue Boxes used for dry recyclables used in the Adur scheme in the UK are made from 1.6 kg of injection moulded polypropylene. After 3 years of operation, the boxes are starting to be replaced (D. Gaskell, ERRA, personal communication). This gives a requirement of around 0.5 kg of material, per household per year for this part of the collection. Wheeled bins (capacity 240 l) have a weight of around 15 kg; a 10-year life expectancy would give a material requirement of 1.5 kg of polypropylene per household per year.

Use of bins can lead to a further source of impacts, due to the need to wash the bins. This is likely to be particularly relevant where unlined bins are used to collect biowaste, since this can cause severe odour and/or fly nuisance in hot weather. This source of impacts should also be included, especially in comparison of bin versus bag collections for biowaste, but data on the level of bin cleaning, and on the typical amounts of water, etc. used are not readily available. Assuming that around 25 l of warm water (heated 20°C above ambient) are used per bin, 2.14 MJ would be needed. For simplicity, it will be assumed that water is heated by electricity with 100% efficiency, so around 0.6 kW-h would be consumed per wash. Any impacts due to the use of cleaning agents should also be included. A recent lifecycle study by Procter & Gamble on hard surface cleaners has shown, however, that the heating of the water represents the major source of both energy consumption and solid waste generation from such cleaning operations (P&G internal report, to be published), so the impacts of bin cleaning included here will be restricted to this element.

Pretreatment of waste. There will also be some environmental impacts due to waste sorting and treatment within the household. Some collection schemes for dry recyclables request that food cans, for example, are rinsed out prior to collection. These impacts occur prior to waste leaving the property, so are not included in this study, but could be included in other LCI studies with wider defined boundaries.

6.7 Economic costs

Care is needed when extracting data on the actual costs of individual collection systems. Although costs are often quoted for various collection and sorting schemes, it is important to understand exactly what is included in these costs, and equally importantly, what is not. Often, quoted figures for materials recovery schemes include not only the collection system, but also the sorting system and the revenues from the subsequent sale of material. Some costs also include disposal savings for any material that is diverted from landfill. Another problem where the collection system is run by the municipality is that the costs of waste collection are not separated from other areas of expenditure, so the actual collection cost is not known. This seems to be especially true of restwaste collection in some countries (notably the UK).

As with materials recovery systems, comparisons between the costs of different collection systems require that standard accounting methods are used. In parallel with their work on standardising data collection and terminology in materials recovery, the European Recovery and Recycling Association is developing a standard cost reporting structure. This will ensure comparable data evaluation, something which to date has been impossible.

To calculate the economic costs of different collection systems, even more so than with environmental impacts, it is necessary to have local data. Salaries, a major component of collection costs, will vary geographically, so no general figures are applicable. This section presents collected data to demonstrate typical figures for different countries, and the ranges thereof.

6.7.1 Material bank systems

Data for the collection costs of bring systems are often presented inclusive of the sale of the collected material. This revenue is likely to be significant compared to the collection cost. For example, Landbank (1992) estimate a UK collection cost for glass (via bottle bank) at 39 ecu per tonne, and an average sales price of 33 ecu per tonne, giving a net collection cost of 6 ecu per tonne. This inclusive figure will therefore vary with market price fluctuations of the materials. For modelling purposes, it is more useful to have the collection cost separated from any subsequent revenue. Table 6.13 gives a range of collection costs for material banks systems, indicating whether the costs are inclusive or exclusive of material sales.

No reliable data were found for the cost of running a central collection site for garden and bulky wastes.

6.7.2 Kerbside collection systems

Collection costs for kerbside collection systems are also quoted in a variety of ways: collection only, collection plus sorting, collection plus sorting

Table 6.13 Economic costs of collection using material banks (bring systems) (in ecu/tonne)

	Glass	Paper	Metal	Plastic	Beverage cartons	Textiles	Source
Europe	29[a]						ORCA (1992)
UK							
average	40	29	59	105			Landbank (1992)
Richmond-upon-Thames	(16)[a,b]	(16)[a,b]	(16)[a,b]			(16)[a,b]	Atkinson and New (1993a, b)
Rydale	(8)[a,b]	(8)[a,b]				(8)[b]	Atkinson and New (1993a, b)
Germany							
average	31	57					ORCA (1992)
(DSD system):							
Container costs	89–92	89–90	92–138 (Al) 45–68 (Fe)	132–198	160–240		Berndt and Thiele (1993)
Collection costs	26–31	46–61	138–229 (Al) 68–113 (Fe)	198–329	240–400		

Parentheses indicate profit.
[a] Cost net of sale of materials.
[b] Averaged over all materials collected.

plus sale of recovered materials, collection plus disposal, etc. Where full accounts have been published (e.g. IGD, 1992), it is possible to calculate the contributions of these various components. In the Adur Blue Box Scheme (W. Sussex, UK) for example, collection of dry recyclables alone costs 23.16 ecu per household per year (equivalent to 166.51 ecu per tonne collected and sold on). Including subsequent sorting of the material increases this figure, but only slightly, to 26.45 ecu per household per year (190.17 ecu per tonne). When the revenue from sale of materials is included, this falls again to 22.37 ecu (160.83 ecu per tonne) (IGD, 1992). Therefore, in contrast to the situation with bring systems, income from materials is relatively insignificant in the inclusive cost; the key factor is the actual cost of the collection.

Reported costs for a variety of kerbside collection schemes, for dry recyclables, biowaste, restwaste or fully integrated integrated systems are presented in Table 6.14.

6.8   Operation of the collection module of the LCI spreadsheet

Details of the inputs needed for the collection module of the computer model are shown in LCI Boxes 2a and 2b. The module is divided into collection of household and commercial waste, and the former is separated into bring and kerbside systems.

At the outset, it is necessary to stress the need for compatibility between the waste generation data (covered in Chapter 5 and inserted into the model in LCI Box 1) and the waste collection data to be inserted here. The collection module attempts to cover all of the possible ways that waste can be collected, for all possible wastes, including bulky household wastes, garden waste and commercial waste. Clearly all of these must have been included in the waste generation data used. If, for example, the waste generation data inserted does not include garden waste or commercial waste, the collection system should not include their collection.

2a COLLECTION OF HOUSEHOLD WASTE

Bring system
  Residents' vehicle distribution (%):
The split between petrol and diesel cars is required for calculation of fuel impacts due to journeys to and from bring systems.

(a) Central collection sites for garden and bulky wastes
  Residents' transport to central site:
    Average no. of special trips to site per household/year:
    Average car journey length (each way):

Table 6.14 Economic costs of kerbside collection (ecu/tonne)

	Dry recyclables	Biowaste	Restwaste	Comments	Source[a]
UK Adur	167			Blue box	1
Leeds	89[b,c]	89[b,c]	89[b,c]	Wheeled bins, bi-weekly collection	2
Sheffield, Stocksbridge	171		46	Blue box	2,3
SE Sheffield	145[b]			Blue box	2
Milton Keynes	86[b]			Blue box	2
Cardiff	83–108[b]			Green bag (bi-weekly)	2
Chudleigh, Devon	78[b]			No container (bi-weekly)	2
Germany Average	171[c]	171[c]	132	Single collection/week	4
Average				Dual bins	
Rural/suburban			66	wet/dry waste	4
Cities			132	Based on 120-litre containers	5

[a] 1, IGD (1992); 2, Atkinson and New (1993a, b); 3, Birley (1993); 4, ORCA (1991b); 5, Bongartz and Naumann (1991)
[b] Includes processing and sale of recovered materials
[c] Averaged over different collected fractions

The model calculates the total consumption of petrol and diesel for the whole area on a round trip distance basis and adds this to the fuel consumption columns. When the total fuel consumption over the whole lifecycle has been calculated, this is multiplied by the impacts of petrol or diesel use respectively, as given in Chapter 4, to give the overall primary energy consumption and emissions.

Garden waste delivered (kg/household per year):
The amount of garden waste (kg/household per year) inserted is converted into the total amount of organic material collected for the area per year, which is added to the biological treatment input stream.

Bulky waste delivered (kg/household per year):
Recovery of materials (as % of delivered):
For the amount of bulky waste delivered, the % recovered is subtracted from the total and added to the recovered materials stream. This allows for the recovery, mainly of metals, which occurs at such sites due to removal of used domestic appliances from the waste stream. The remaining bulky waste will enter the streams for landfilling or thermal treatment by mass incineration as defined in the following lines.

Transport distance to biological treatment plant (km):
Transport distances to the biological treatment plant, incinerator and landfill are used to calculate overall fuel (diesel) consumption, using the data in Table 6.11 for a 20 tonne truck load, based on a round-trip distance (i.e. assuming no return load is carried).

Cost
Cost of central collection site and transport to treatment plants (net of material income):
The cost of the central collection site, net of any revenues from the sale of recovered materials (mainly metals), needs to be inserted in terms of cost per tonne of material handled per year.

(b) Materials collection banks
Residents' transport to materials bank sites:
Average no. of special trips to site per household/year:
Average car journey length (each way):
As above for central collection sites, numbers of journeys and average distances allow total fuel consumption of petrol and diesel for the area to be calculated. Again only special journeys to materials banks should be included, visits to materials banks in supermarket car parks as part of regular shopping trips should not be included. Similarly, where close-to-home materials banks are used and transport to the bank is on foot, zero special car journeys should be inserted.

LCI BOX 2a

2a. COLLECTION OF HOUSEHOLD WASTE

Bring Systems

Residents' vehicle distribution (%)

Petrol	Diesel

(a) Central Collection Sites for Garden and Bulky Waste

(N.B. These waste streams must have been included in waste generation data used in Box 1, if not leave section blank)

Residents' Transport to Central Site:

Average # of special trips to site per household/yr

Average car journey length (each way) [   ] km

Garden Waste delivered (kg/hhld/yr)

Organic

Bulky Waste delivered (kg/hhld/yr)
Recovery of materials (as % of delivered)

Glass	Metal (Fe)	Metal non Fe	Plastic film	Plastic rigid	Other

Transport distance to biological treatment plant (km) [   ] (each way)

Bulky Waste Residue

	Incineration	Landfill
Treatment %		
Transport distance (km)		(each way)

Cost Cost of Central Collection Site and transport to treatment plants [   ] ecu per tonne handled
(net of material income)

(b) Materials Collection Banks.

Residents' Transport to Materials Bank Sites

Average # of special trips to site per household/yr

Average car journey length (each way) [   ] km

Amounts collected (kg/hhld/yr):

	Paper	Glass	Metal-Fe	Metal non-Fe	Plastic film	Plastic-rigid	Textiles
In single material containers							
In mixed material containers							
Market Prices (ecu)							

Collection and Transport to Bulking Depot or MRF

Average fuel (diesel) consumption per tonne collected [   ] litres

Cost Average collection and transport cost / tonne collected [   ] ecu

Kerbside Collection Systems

Waste Fractions Collected

Biowaste bin/bag

Paper	Organic

Contents [   ] kg/household/yr

Dry Recyclables

Contents

Paper (PA)	Glass (GL)	Metal (ME)	Plastic (PL)	Textiles (TE)
					kg/hhld/yr

Kerbside sort of dry recyclables? [   ] Insert "0" for NO, "1" for YES

RestWaste Collection [   ] kg/h hld/yr

Collection Containers:

Bags

	Bag wt. (g)	bags/hhld/yr	Type of bag used (%)	
			Plastic(LDPE)	Paper
Biowaste				
Dry Recyclables				
Restwaste				

Bins

Total weight of bins used [   ] kg/hhld

Average lifespan of bins [   ] years [   ] equivalent to [   ] kg/hhld/yr
(bins assumed to be polypropylene)

bin washes/hhld/yr [   ]

Collection Vehicles

Total number of collection vehicle visits/property per year

Average collection truck fuel consumption per 1000 property visits [   ] litres

(includes transport on to MRF, RDF plant, Biological treatment plant, incinerator, or transfer station/landfill site)

Costs

Total Kerbside Collection cost per property per year [   ] ecu

Go to Box 2b	(all collections included)

LCI BOX 2b

| 2b COLLECTION OF COMMERCIAL WASTE. | | (material not collected by household waste collection vehicles) | | | | | | |

	Paper	Glass	Metal Fe	Metal non-Fe	Plastic-film	Plastic-rigid	Textiles
Dry Recyclables Collection tonnes/yr							

	Paper	Organic		
Biowaste collection			tonnes/yr	

Restwaste collection			tonnes/yr

Transport:

Average Distance to		
MRF		km (each way)
RDF plant		km
Biol. Treatment		km
Incinerator		km
Landfill		km

Costs (assume collection costs borne by commercial waste generator)

	Recyclables	Biowaste	Restwaste	
Cost charged to waste generator (per tonne) for waste management				ecu per tonne

Go to Box 3.

Amounts collected (kg/household per year):
In single material containers:

This accounts for materials such as colour separated glass or aluminium cans, which can be shipped on to material processors without further sorting. This material is added directly to the secondary materials stream.

In mixed material containers:

This allows for collection of material that needs a subsequent sort, e.g. at a MRF, prior to sale. This could apply to the collection of mixed plastics, for example, or to a system such as Barcelona where mixed recyclables from high-rise housing are collected in communal containers. This material is added to the sorting stream, which forms the input to the MRF.

Market prices:

The market prices that need to be inserted are per tonne of this material, ex-collection bank.

Collection and transport to bulking depot or MRF:
Average fuel (diesel) consumption per tonne collected:

This accounts for the transport from materials banks to either a central bulking site (for separated materials prior to sale and onward shipment to materials reprocessors) or to a regional MRF. Since a variety of transport types is likely to be used, the best form of data to use here is the average fuel (diesel) consumption, per tonne of material collected. This value should be available to system operators.

Cost
Average collection and transport cost per tonne collected:

This needs to be exclusive of revenue from the sale of the recovered materials, which has already been accounted for above. If only net data inclusive of revenues are available, these can be used if zero sales values are inserted for the collected materials.

Kerbside collection systems

(a) Waste fractions collected
Biowaste bin/bag contents:

If biowaste is separated, the amount of organic material and paper (if a broad biowaste definition is used) collected in this stream needs to be inserted. The model will remove 5% of the total collected of both paper and organic and add an equivalent weight of plastic as contamination. As discussed above in this chapter, even where biowaste is collected in bins rather than bags, there is typically this level of plastic contamination.

Dry recyclables bin/bag contents:

If separated, the amount of each material collected as dry recyclables is inserted here in the form of amount per household per year.

Kerbside sort of dry recyclables?:

This asks whether kerbside sorting occurs, e.g. as in Blue Box schemes. A '1' is inserted if it occurs, '0' if it does not. Kerbside sorting helps prevent contaminants in the form of non-requested materials (e.g. organic or other waste) or non-targeted items of requested materials (e.g. films, foils) from entering the recyclables collection system. In the absence of a kerbside sort, a 5% level of contamination by other waste materials is added at this stage, and a lower sorting efficiency is assumed in the subsequent MRF sorting stage (LCI Box 3a; see Chapter 7).

Rest waste collection:

The model calculates the amount of household waste remaining to be collected as restwaste, per household per year, and displays it here.

(b) Collection containers:
Bags:

This section accounts for the upstream impacts of the use of bags (i.e. during raw material acquisition, processing, bag manufacture and transport). The user is required to insert the average weight of the bags, the bag material (paper or plastic), and the number of bags used per household per year, for each fraction collected in this way. The model calculates the total amount of paper and plastic required, and converts this to energy consumption and emissions by multiplying by the generic data for production of paper and plastic given in the materials impacts section of the spreadsheet. These data are shown in Table 6.12. It is assumed that the paper bags are made from unbleached paper, and that the plastic bags are made from low density polyethylene (LDPE). The data used in Table 6.12 are for production of virgin LDPE; although many refuse sacks are made at least partly from recycled LDPE, reliable data for such processes are scarce.

The amounts of paper and plastic used for bags are also added to the biowaste, dry recyclables and restwaste streams, respectively, since they will need to be dealt with in any subsequent treatment. Whilst paper bags for biowaste could be used as feed stock in subsequent biological treatment, plastic bags would represent a contaminant and therefore would need to be removed. Adding the bag material to the waste at this stage does, however, assume that the amounts of waste defined previously in LCI Box 1 (Chapter 5) did not already include refuse sacks. This will not always be the case. This modelling approach does allow the operator to compare the effects of using bags versus bins; if it is assumed that all bags were already in the waste stream, this would not be possible.

Bins:

For comparison with bag systems, the effects of bin use are calculated. The user inserts the (average) total weight of bins used per household, and the average lifespan of the bins used. This is converted into an equivalent total usage of material per year for the system area, and then into energy consumption and emissions using the generic data in the materials impacts section (data shown in Table 6.12). It is assumed that bins are made from injection moulded polypropylene.

The amount of plastic used in bin construction (on a per year basis) is added to the waste stream defined in LCI Box 1, since this material will also enter the solid waste management system eventually. (Again, this assumes that the waste input data defined for any given area is exclusive of waste bins or containers. Since this material is likely to end up in a bulky waste stream, it may well not be already accounted for.)

No. bin washes/household per year:

Bin washing accounts for additional energy consumption and emissions due to the heating of the water used. The user inserts the average number of bin washes carried out per household per year. The model calculates the total number of bin washes per year in the area, and converts this to electricity consumption using the estimated figure of 0.6 kW-h per bin washed. Electricity usage is totalled throughout the solid waste lifecycle and converted to primary energy consumption and emissions using the generic data given in Chapter 4.

(c)   Collection vehicles

Total no. of collection visits/property per year:

This includes collections for all of the different fractions (biowaste, dry recyclables, restwaste). For example, a weekly collection of all waste together would result in 52 visits per year. Weekly collection of restwaste and an additional weekly collection of dry recyclables would result in 104 visits to each property per year.

Average collection truck fuel consumption per 1000 property visits:
This needs to be calculated from local collection-round information. Given the fuel consumption over a given time (e.g. per year) for all of the trucks serving a given area and the number of households covered, the fuel consumption per household served can be calculated. From this, the fuel consumption per single property visit can be calculated; the model asks for the consumption per 1000 property visits as this is a more convenient figure to use. The model then calculates total fuel consumption per year for the area, and adds this to the fuel consumption column.

Costs
Total kerbside collection cost per property per year:
This figure should include all of the collections made, but not include any element for subsequent sorting or other treatments.

2b COLLECTION OF COMMERCIAL WASTE
This box defines any addition collection systems that deal with commercial waste that is not handled by the normal household waste collection vehicles.

Dry recyclables collection (tonnes/yr):
Biowaste collection:
The amounts of each material collected in these separate fractions need to be inserted here. It is assumed that dry recyclables are shipped to a MRF for sorting prior to sale, although this need not always be the case. The model adds the specified amounts of material to the sorting stream (dry recyclables) or biological treatment input stream (biowaste). Since materials from commercial sources are likely to be less mixed than from household sources, no contamination is added to the streams at this point.

Restwaste collection:
This is calculated by the model by subtracting the amounts specified for recovery from the total commercial waste entering the system.

Transport:
Average distance to: MRF
RDF Plant
Biological treatment site
Incinerator
Landfill
The model uses these inputs to calculate the respective fuel consumption,

assuming a 20 tonne load and a round-trip distance (no return load). This fuel consumption is then added to the fuel consumption column.

Costs

The model assumes that transport costs to a treatment site are borne by the waste generator.

Cost charged to waste generator (per tonne) for waste management:
The treatment/disposal charges levied are inserted here. These represent an income for the waste management system, so will be subtracted from overall costs.

The model then goes on to consider sorting processes, which are the subject of the next chapter.

References

Atkinson, W. and New, R. (1993a) An overview of the impact of source separation schemes on the domestic waste stream in the UK and their relevance to the Government's recycling targets. Warren Spring Laboratory Report LR 943, Warren Spring Laboratory, Gunnels Wood Road, Stevenage, Herts, UK.

Atkinson, W. and New, R. (1993b) Kerbside collection of recyclables from household waste in the UK – a position study. Warren Spring Laboratory Report LR 946, Warren Spring Laboratory, Gunnels Wood Road, Stevenage, Herts, UK.

Berndt, D. and Thiele, M. (1993) Status des Dualen Systems und seine Kosten. Verpackungs-Rundschau, pp. 84–88.

Beyea, J., DeChant, L., Jones, B. and Conditt, M. (1992) Composting plus recycling equals 70 percent diversion. *Biocycle*, May.

Biocycle (1992) Cocollection at curbside. *Biocycle*, September, 56–57.

Birley, D. (1993) Does the Blue Box have a future? *Warmer Bull.* **35**, 10–12.

Bongartz, T. and von Dörte Naumann, M. (1991) Aktuelle Entsorgungskosten in der Bundesrepublik Deutschland. *EntsorgungsPraxis*, March 1991 pp. 84–86.

Boustead, I. (1993) Resource Use and Liquid Food Packaging. EC Directive 85/339: UK Data 1986–1990. A report for INCPEN, May 1993.

Doh, W. (1990) *Biologische Verfahren der Abfallbehandlung*. EF-Verlag, Berlin.

DOE/DTI (1992) *Economic Instruments and Recovery of Resources from Waste.* Department of Trade and Industry and Department of the Environment, 75 pp.

EC (1992) *Green Paper on the Impact of Transport on the Environment – A Community Strategy for Sustainable Mobility.* 20th February, 1992. Brussels.

ERRA (1992) Programme ratios. Reference report of the ERRA Codification Programme. Available from European Recovery and Recycling Association, 83 Ave E. Mounier, Box 14, Brussels 1200, Belgium. Reproduced, with permission, as Appendix 3.

ERRA (1993a) Terms and definitions. Reference report of the ERRA Codification Programme. Available from European Recovery and Recycling Association, 83 Ave E. Mounier, Box 14, Brussels 1200, Belgium. Reproduced, with permission, as Appendix 4.

ERRA (1993b) Project summary sheets. European Recovery and Recycling Association, 83 Ave E. Mounier, Box 14, Brussels 1200, Belgium.

Forrest, P., Heaven, S. and Sandels, C. (1990) *Sorting at Source, Separation of Domestic Refuse.* Save Waste and Prosper, Leeds, UK.

Fricke, K. (1990) *Grundlagen der Kompostierung*. EF-Verlag, Berlin.

Fricke, K. and Vogtmann, H. (1992) *Biogenic Waste Compost: Experiences of Composting In Germany.* IGW, pp. 1–33.

German Government (Deutscher Bundestag) (1993) Anaerobe Vergärung als Baustein der Abfallverwertung. *Drucksache* **12**, 4905.

Goldstein, N. (1993) The curb and pile trial. *Biocycle*, August, 37–39.

Habersatter, K. (1991) Oekobilanz von Packstoffen Stand 1990. Bundesamt für Umwelt, Wald und Landschaft (BUWAL) Report No. 132, Bern, Switzerland.

Härdtle, G. Marek, K., Bilitewski, B. and Kijewski, K. (1986) Recycling von Kunstoffabfällen. *Fachzeitschrift für Behandlung und Beseitigung von Abfällen*, Vol. 27.

Haskoning (1991) Conversietechnieken voor GFT-afval, NOH, 53430/0110.

Henstock, M. (1992). An analysis of the recycling of LDPE at Alida Recycling Limited. Report by Nottingham University Consultants, Ltd.

IGD (1992) Sustainable waste management: the Adur project. Report by the Institute for Grocery Distribution, Letchmore Heath, Watford, UK, 85 pp.

Jespersen, L. (1991) Source separation and treatment of biowaste in Denmark. Verwerkingsmogelijkheden en scheidingsregels van groente-, fruiten tuinafval, Koninklijke Vlaamse Ingenieursvereniging (KVIV). Syllabus studiedag 7 maart, 1991.

Kreuzberg, G., and Reijenga, F. (1989) *Handboek gescheiden inzameling groente-, fruit-end tuinafval.* Provinciale Waterstaat Noord-Holland.

Landbank (1992) WARM Report. A proposal for a model waste recovery and recycling system for Britain. A Gateway Foodmarkets report, prepared by the Landbank Consultancy.

Oakland. J.S. (1989) *Total Quality Management.* Heinemann, Oxford, UK, 316 pp.

Ogilvie, S.M. (1992) A review of the environmental impact of recycling. Warren Spring Laboratory report LR 911 (MR).

ORCA (1991a) The role of composting in the integrated waste management system. In *Solid Waste Management, An Integrated System Approach. Part 5.* Organic Reclamation and Composting Association, Brussels.

ORCA (1991b) Composting of biowaste – the important role of the waste paper fraction. In *Solid Waste Management, An Integrated Approach. Part 8.* The Organic Reclamation and Composting Association, Brussels, Belgium.

ORCA (1992) *Information on Composting and Anaerobic Digestion.* ORCA Technical Publication No. 1. Organic Reclamation and Composting Association, Brussels. 74 pp.

Ovam (1991) Ontwerp Afvalstoffenplan 1991–1995 van Vlaanderen (Waste Management Plan of the Flemish region for the period 1991 to 1995).

Papworth, R. (1993) An assessment of the Adur District Council kerbside collection scheme utilising a simplistic analytical approach based on the ERRA codification programme – April 1993. Warren Spring Laboratory Report CR 3815.

Porteous, A. (1992) LCA study of municipal solid waste components. Report prepared for Energy Technology Support Unit (ETSU), Harwell, Oxon UK.

Porter, R. and Roberts, T. (1985) *Energy Savings by Wastes Recycling*, Elsevier Applied Science, London, ISBN 0 85334 353 5.

Procter and Gamble (1992). Life cycle inventory for consumer goods packages. A copy of this spreadsheet can be obtained from Procter and Gamble European Technical Center, Temselaan 100, B-1853 Strombeek-Bever, Belgium.

PWMI (1993) Eco-profiles of the European plastics industry. Report 3: Polyethylene and polypropylene. Report by Dr. I. Boustead for The European Centre for plastics in the Environment (PWMI), Brussels.

Quinte (1993) The first year. Report of Quinte Blue Box 2000 project, Quinte, Ontario, Canada.

Rousseaux, P. (1988) Les métaux lourds dans les ordures ménagères origines, formes chimiques, teneurs. R and D programme on recycling and utilisation of waste EEC, DG XII.

Rutten, J. (1991) Gescheiden inzameling en verwerking van GFT-afval te Diepenbeek, syllabus studiedag 7 maart, 1991. Verwerkingsmogelijkheden en scheidingsregelsvan groente- fruiten tuinafval, Koninklijke Vlaamse Ingeneurs-vereniging (KVIV).

Schweiger, J.W. (1992) Planung, Genehmigung und Betrieb von Recyclinghöfen. presented at Integrierte Abfallwirtschaft, Forum Zukunft e.v. conference, Kloster Banz, Germany.

Selle, M., Kron, D. and Hangen, H.O. (1988) Die Biomüllsammlung und Kompostierung in der Bundesrepublik Deutschland. Situationsanalyse 1988. Schriftenreihe des Arbeitskreises für die Nutzbarmachung von Siedlungsabfällen (ANS) e.v., Heft 13.

Tidden, F. and Oetjen-Dehne, R. (1992) Modellversuch Bioabfallvergärung Ismaning-Sammlung und Vergärung von Bioabfällen aus dem Geschosswohnungsbau. *Abfallwirtschafts J.* **4**(10), 787.

Warmer (1993) Batteries. *Warmer Bull.* **39**, 7.

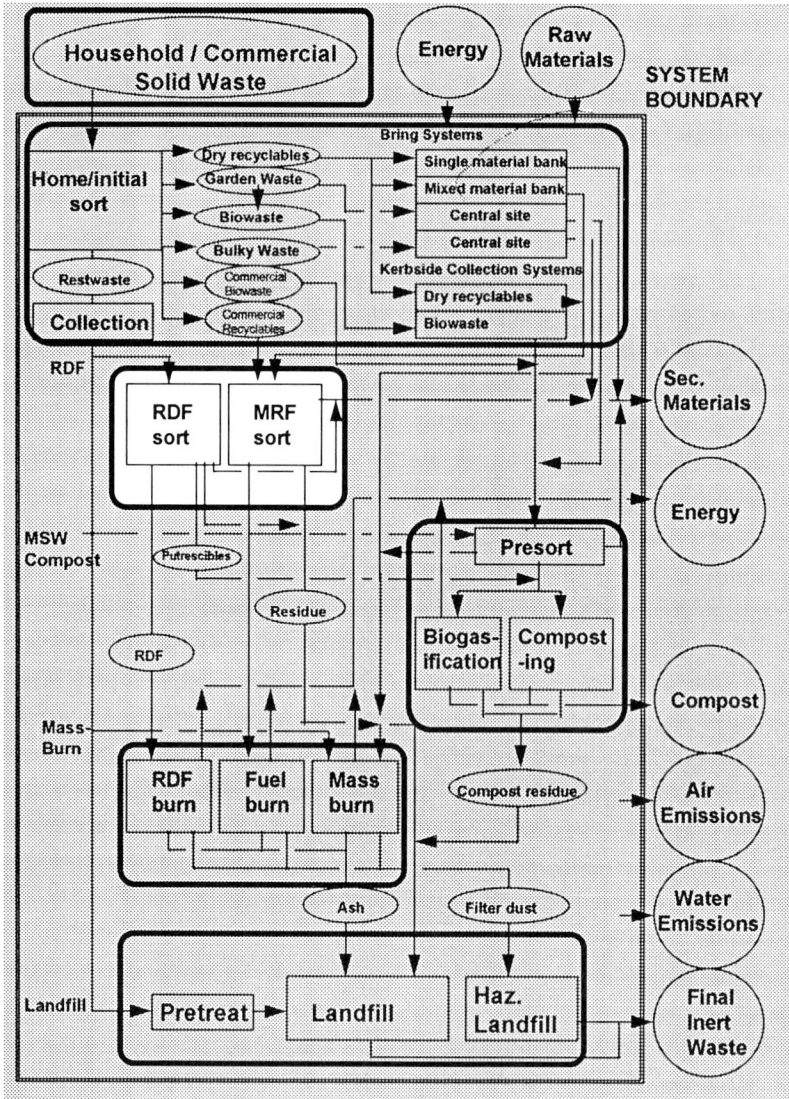

Figure 7.1 The role of central sorting in integrated waste management.

7 Central sorting

Summary

This chapter deals with two distinct types of central sorting: sorting of mixed recyclables at a materials recovery facility (MRF) and the sorting of mixed waste to produce refuse-derived fuel (RDF). The stages of each sorting process are described, as are the waste inputs and the outputs in terms of both products and residues. Available data are presented on the typical energy consumption of the two sorting operations. Economic data, both processing costs and revenues from the sale of recovered materials, are included where possible. The central sorting module of the lifecycle inventory model for solid waste is presented and explained.

7.1 Introduction

Sorting is an important part of any waste's lifecycle. Wastes are almost invariably mixed, and household wastes are amongst the most mixed in terms of material composition. Separation of the different materials in waste, to some extent, is an essential part of almost all methods of valorisation. This sorting can, and does, occur at any point in the lifecycle of waste; similarly, it can occur any number of times. The earliest sorting will occur in the home, when, for example, materials are separated from the residual waste stream, but the same materials can be sorted further during and/or after collection. Sorting of the input also represents the first stage of many waste treatment processes, such as composting, biogasification, and in some cases sorting of the outputs also occurs (e.g. removal of ferrous metal from incinerator ash residues). Sorting is thus ubiquitous in the lifecycle of waste, and is covered in each chapter of this book that deals with a particular waste management process. This chapter focuses on two particular central sorting operations that are not covered elsewhere: sorting of mixed recyclables at a materials recovery facility (MRF) and the sorting of mixed waste to produce refuse derived fuel (RDF).

These two processes are distinct, with different inputs and outputs (Figure 7.1), so are described and discussed separately.

7.2 Central sorting of recyclables at a materials recovery facility (MRF)

Materials recovery facilities (MRFs) are needed wherever recyclable materials are collected in a commingled fraction. The exact process required at a MRF will vary according both to which fractions are collected commingled, and to the markets that exist for either separated or mixed materials. Thus, as emphasised in the previous chapter, there is a need to start by considering the end markets along with the waste stream composition, and then to design collection and sorting systems together to produce the materials that these markets require.

Not surprisingly, therefore, there is no standard MRF operation. While some collection schemes (e.g. Dunkirk, see Box 6.4) deliver all recyclables to the MRF in a mixed state, other schemes will have already separated some materials out by this stage. Blue box schemes, such as Adur and Sheffield (UK), usually involve a kerbside sort that separates out paper and glass (colour separated). Although such materials may need to be bulked up before sale and onward transport, no further sorting is necessary. Where materials such as glass are collected via glass banks, there may be no need to process it at a MRF at all, since it can be shipped on to the materials processors directly.

Thus, MRFs may process all dry recyclable materials, or just a restricted range. Since the number of different fractions that can be collected separately is limited by practical considerations (see Section 6.5), it is likely that at least some recyclables, in particular plastics, aluminium cans and steel cans, will be collected in a commingled fraction, for subsequent central sorting.

MRF sorting techniques.   The simplest and most widespread separation technique is hand sorting from a raised picking belt. Operators remove the materials required from the belt and the remaining unselected materials are discarded as residue. In one US survey, the productivity of such sorters was found to average around 5 tonnes of sorted material per person, per day, but this varied widely between different materials handled (Table 7.1). This is clearly labour intensive, but some schemes (e.g. Milton Keynes, UK; Omaha, Nebraska, USA) use this as an opportunity for job creation, or for training disadvantaged societal groups. There have been some concerns over health and safety issues surrounding such waste sorting, and because of this (plus the need to increase efficiency), there is a trend towards increasing the mechanisation of the sorting process, to increase the possible throughput and sorting efficiency.

Electro-magnets located over the picking belt are used to separate out ferrous material in several MRFs (e.g. Dublin, Ireland). A further refinement has been the development of eddy current separators, as employed

Table 7.1 MRF employee productivity in the
United States (in tonnes per employee per day)

Material	Average	Range
Paper	7.54	4.36–13.05
Metals	6.06	1.21–17.53
Glass	4.28	1.02–16.11
Plastics	1.60	0.60– 3.21
Average	5.12	2.69– 8.44

Source: NSWMA (1992).

in the Adur (UK) MRF. This technique is capable of separately removing both ferrous and aluminium material from a material stream. This means that both metals and plastic can be collected commingled and the three streams separated mechanically at up to 5 tonnes per hour (Newell Engineering Ltd., 1993).

The area where hand sorting is generally still required is in the separation of plastic resin types. This is difficult due to the number of resins in common use and the difficulty in automated resin identification. Automated separation of plastic bottles is possible: PVC can be separated from PET bottles using an X-ray based sensor that detects the chlorine atoms in PVC; polypropylene can be separated from clear HDPE, and coloured HDPE can be sorted into different colours by means of colour sensors. Together, an automated plastic bottle sorter can be assembled capable of handling up to 2 tonnes per hour (Magnetic Separations Systems Inc.), but such systems have not been widely installed yet in Europe.

After materials have been adequately separated, they are either baled (plastic bottles, cans) or shipped on in bulk (paper, glass, flattened cans).

7.3 Sorting of mixed waste for refuse-derived fuel (RDF)

Refuse-derived fuel (RDF) is produced by mechanically separating the combustible fraction from the non-combustible fraction of solid waste. The combustible fraction is then shredded, and may also be pelletised. RDF production thus forms part of a thermal treatment system, which aims to valorise part of the waste stream by recovering its energy content. The second stage, RDF combustion, can either occur on the same site, or the RDF can be transported for combustion elsewhere. In this book, production and combustion of RDF, even if they occur on the same site, are treated separately. Since RDF production is a central sorting process, it is discussed in this chapter. RDF combustion is considered alongside other thermal treatment processes in Chapter 10.

A further reason for considering RDF sorting separately from thermal treatment is that the process need not only produce solid fuel; it can also produce an organic fraction which can form the feedstock for biological treatment. As a result, in some cases, the RDF sorting process occurs in combination with a biological treatment process (e.g. at Novaro in Italy (ETSU, 1993)). Again, although the RDF sort may occur on the same site as biological treatment, it is considered here as a separate process.

There are two basic RDF processes, each producing a distinctive product, known as densified RDF (dRDF) and coarse RDF (cRDF) (also referred to as fluff or floc), respectively. dRDF is produced as pellets, often similar in size and shape to wine corks. Prior to pelletising it is dried, so is relatively stable and can be transported, handled and stored like other solid fuels. It can either be burned alone, or co-fired with coal or other solid fuels.

dRDF requires considerable processing, including drying and pelletising, and so has a relatively high processing energy requirement. As a result, there has recently been interest in the alternative coarse RDF (cRDF). This comes in the form of a coarsely shredded product, that has been compared in appearance to 'the fluff from a vacuum cleaner' (Warmer, 1993). cRDF requires less processing, but as it has not been dried, cannot be stored for long periods. It is suitable for immediate use in on-site combustion for power generation and/or local heating. Depending on the level of processing, it can be suitable for combustion on conventional grates or in fluidised bed systems (ETSU, 1992).

Current status of RDF in Europe. The early development of RDF technology occurred mainly in the UK and to some extent in Italy, with plants built from the mid-1970s onward. Many of the early plants have since closed down, however, often due to difficulty finding markets for the dRDF fuel product. For example, of the nine dRDF plants built in the UK, only four (three full scale plants and a pilot plant) are still operating (ETSU, 1993). Lack of off-site markets has also led to the development of cRDF technology for on-site power generation. The current extent of RDF processing in Europe and elsewhere is shown in Table 7.2.

RDF sorting processes. Details of the processing line in different RDF plants vary, but the basic dRDF process can be broken down into five distinct stages (ETSU, 1993; Figure 7.2). As shown in Figure 7.2, the production of cRDF is a simpler process, which omits either one or two of these stages.

1. *Waste reception and storage*: Mixed waste is delivered by the collection vehicles and tipped into a hopper or onto a tipping floor, where any unwanted 'rogue' items (e.g. car engines, logs, etc.) can be removed. This initial short-term storage stage acts as a buffer to provide the RDF production process with a steady feedstock level.

Table 7.2 Locations of RDF sorting plants

Country	Number of RDF plants operating	Locations	Source[a]
Belgium	0		1
France	1 (dRDF)	Laval, Mayenne	1
Germany	1 (dRDF)	Herten	1,2,3
Italy	11 (5 dRDF)	Rome (2)	1
		Perugia	
		Milan	
		Modena	
		Novaro	
		Pieve di Corano	
		Ceresara	
		Tolmezzo	
		Udine	
		St. Georgio	
The Netherlands	1 (dRDF)	Amsterdam	1
Spain	1 (cRDF)	Madrid (under construction)	3
Sweden	5 (cRDF)		3
Switzerland	1 (dRDF)	Chatel St. Denis	3
UK	4 (dRDF)	Byker, Newcastle	1
		Polmadie, Glasgow	
		Hastings	
		Isle of Wight	
Canada	1		1
South Korea	1	Seoul	3
USA	28 (6 dRDF)	dRDF plants:	1,3
		Edin Prairie, MN	
		Thief River Falls, MI	
		Northern Tier, PA	
		Yankton, SD	
		Iowa Falls, IO	
		Cherokee, IO (under construction)	

[a] *Sources*: 1, ETSU, (1993); 2, Barton *et al.* (1985); 3, Warmer (1993).

2. *Waste liberation and screening*: The purpose of this stage is to free the waste from any refuse bags or containers, and to provide the main fuel/non-fuel sort. Bag opening can involve the use of flail mills, shredders, spikes or ripping devices, although experience has shown that non-shredding devices have the advantage of not shredding or mixing the waste excessively, which can make separation more difficult. Screening often involves a drum or rotary screen, which performs three functions. It completes the bag emptying process; it removes the undersize fractions (fines), and it separates the oversize (>500 mm) material from the fuel fraction. The fines fraction contains the high-moisture-content organic/putrescible material, as well as ash, dust and broken glass. The oversize fraction consists mainly of large pieces of paper, board and plastic film, and is usually landfilled along with other residues. The remaining fraction produced by this stage can be used as a crude cRDF (cRDF type A, ETSU,

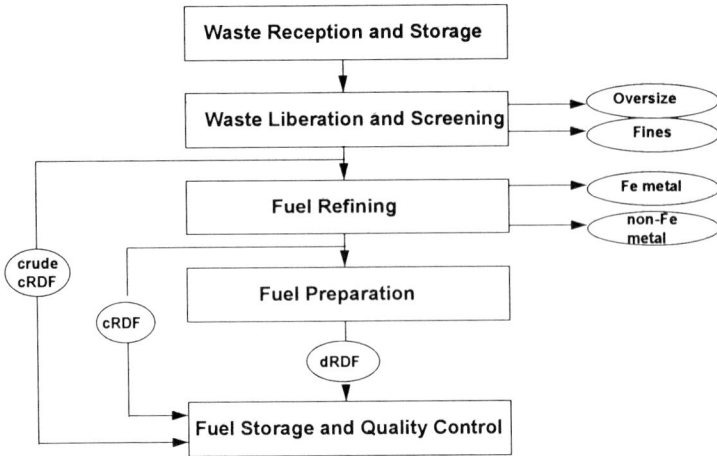

Figure 7.2 Stages in the production of refuse-derived fuel (RDF). *Source*: ETSU (1993).

1992) although it will still contain metals and other non-combustible materials.

3. *Fuel refining*: This stage involves size reduction, classification and magnetic separation. Size reduction using a shredder or hammer mill aids the separation into light and dense fractions and in fuel preparation. The density separation (classification) stage is necessary to separate the heavy fraction (metals, dense plastics) from the combustible light fraction (paper, plastic film) which will go on to form the dRDF product. Two main methods can be used to achieve this, air classification and ballistic. separation, which rely on the behaviour of the objects in an airstream and their 'bouncing behaviour', respectively. Magnetic separation can then be performed on the heavy fraction to remove both ferrous metal and in some plants also aluminium (by eddy current separation). The light fraction, together with the remains of the magnetically sorted heavy fraction, can be used as a more refined form of cRDF (cRDF Type B; ETSU, 1992).

4. *Fuel preparation*: This stage represents the main difference between the cRDF and the dRDF processes. It involves the conversion of the fuel rich fraction (floc) into a dry, dense pellet form by re-shredding, drying, and pelletising it. Secondary shredding is needed to reduce the particle size of the fuel fraction to the size needed for the pelletising operation, and drying reduces the moisture content from about 30% to around 12%. Low moisture levels are needed for good storage and combustion characteristics. dRDF can either take the form of pellets or briquettes, although most plants use a pellet mill to densify the

product. Pellet mills are very energy intensive, consuming over 35 kW-h per tonne of fuel produced, so the fuel needs to be cooled prior to storage, to remove the heat produced on compression. Pellet mills are also prone to damage from dense contaminants left in the fuel fraction, so that a further magnetic extraction stage and a ballistic separator are often used to remove both ferrous metal and other dense materials prior to the final pelletising stage (see Figure 7.4).

5. *Fuel storage and quality control*: Once dried and in pellet form, dRDF can be stored before use; cRDF, in contrast, needs to be burned soon after production.

7.4 Environmental impacts: input–output analysis

The major environmental impacts of both types of central sorting considered in this chapter are associated with their usage of energy in whatever form: electrical, gas or diesel. If consumption data can be obtained for the relevant processes, these can be converted into primary energy consumption and emission levels, using the generic conversion data for fuel and energy usage presented in Chapter 6. It is also necessary to consider the input and output material streams, which determine where the products and residues of the process are destined to go, as these will eventually cause environmental impacts in subsequent processes.

7.4.1 MRF sorting

Inputs. Average energy consumption figures for MRF sorting are likely to vary significantly between schemes, since there is no standard MRF process. Energy consumption is likely to be higher where more materials are separated in the MRF, as opposed to at the kerbside, and to increase with the level of mechanised sorting, in replacement of hand picking.

Recyclables streams that are collected commingled, such as mixed plastics and metals, need more sorting, and hence more energy, than streams which are collected in a pre-sorted fraction, such as glass or paper. To fully predict the likely energy consumption of any particular MRF process, it is necessary to have data on the sorting energy needed for each of these individual streams. Because of the recent introduction of MRFs into Europe, however, such data are not available. Indeed, there is a paucity of any data on MRF operations, that is only now being addressed by operators of recovery schemes. Table 7.3 gives information that has been collected for MRF processing. Electrical energy is used to power conveyor belts, ferrous/eddy current separators and other equipment. Diesel is consumed mainly by auxiliary vehicles such as fork-lift trucks, mechanical shovels etc., and gas,

Table 7.3 Energy and fuel consumption data (per input tonne) for materials recovery facilities

Scheme	Electricity (kW-h/tonne)	Diesel (l/tonne)	Natural gas (m³/tonne)	Source[a]
Adur, UK	24	0.87	–	1
Dublin, Ireland	22	1.35	2.3	2
Prato, Italy	27	n/a	–	3

[a] *Sources*: 1, R. Moore, Community Recycling, Sompting MRF, personal communication (1993); 2, ERRA/Kerbside Dublin, personal communication (1993); 3, ERRA, personal communication (1993).

where used, is normally for heating. Energy and fuel consumption have been averaged over the total input to the MRF, since the fuel consumption cannot be allocated to individual materials.

Outputs.   Residues from MRFs arise from two sources:

1. Material collected but not requested (i.e. the collection contaminants discussed in Section 6.4.1). The amount of such contamination can only be determined by a waste analysis of what is collected in the recyclables fraction. This material will not be selected in the MRF, so will become part of the residue.
2. Requested (i.e. targeted) material that is not separated out in the MRF (i.e. a sorting efficiency below 100%).

The total amount of residue from a MRF will represent the sum of these two contributions, and can vary from around 5% of the collected material, for Blue Box schemes with a kerbside sort, to over 50% for commingled recyclables collected in a communal 'bring' container (Figure 6.3). Analysis of the residue is necessary to determine the exact contribution of the two factors. Table 7.4 shows residue analysis results from the MRF at Prato, Italy, where on average the residue accounted for 35% of the input material (this level of contamination is now falling). During trials, the level of targeted material in the residue varied from 13% to 37%, depending on the rate of throughput, leaving 63–87% of the residue made up of non-requested contaminating materials. This represents a loss of 7–17% of the targeted material entering the MRF, or a sorting efficiency of 83–93%. Of the targeted material in the residue, some was damaged or contaminated, so not all would actually be recoverable. This would mean that the true sorting efficiency would be actually higher still.

Sorting efficiency will clearly vary with the level of contaminants present and the throughput of the MRF. The ERRA-supported collection scheme for dry recyclables in Barcelona provides a good example of this (see Box 6.2 above for scheme details). After 19 months of collecting a wide range of recyclable materials in communal bins (Phase 1), the scheme

Table 7.4 MRF residue composition, Prato, Italy[a]

Material	Analysis 1 (%)	Analysis 2 (%)
Paper/board	31.2	35.4
Plastic	45.8	40.9
Glass	2.6	1.1
Metal	4.0	3.9
Food/garden	4.6	0.5
Textiles	4.2	11.8
Other	7.4	6.4
Total	99.8	100
Level of targeted material in residue	36.7	13
Level of targeted material in residue that could be recovered	25.3	n/a[b]

Source: ERRA, personal communication (1993).
[a] The two analyses were conducted during tests to determine how sorting efficiency varied with throughput rate (analysis 1 high throughput rate, analysis 2 low throughput rate).
[b] = data not available.

was relaunched in December 1992. Under Phase 2, a narrower range of materials was collected, and an extensive communications programme was run to ensure that the participants understood which materials were targeted by the scheme. As a result of these two measures, the amount of materials collected fell by around 34%. However, the amount of material recovered from the scheme actually rose by 17% (Figure 7.3). The relaunch also coincided with the opening of a new MRF. The overall sorting efficiency of the new MRF (53%) was significantly higher than for the former MRF (41%) (ERRA, personal communication 1993). Whilst part of this increase probably reflects better working conditions and equipment, it is also due to the presence of a lower level of contaminants.

7.4.2 RDF sorting

RDF inputs

Waste material input. The input to the RDF process is normally mixed or residual waste which has been collected commingled and unsorted. Data are available on the inputs and outputs of the RDF process for wastes with typical MSW compositions (e.g. Figure 7.4). The introduction of separate collection systems for individual waste fractions such as the dry recyclables, whether using a bring or kerbside system, however, is likely to significantly alter the composition of the residual waste. Therefore, it is important to

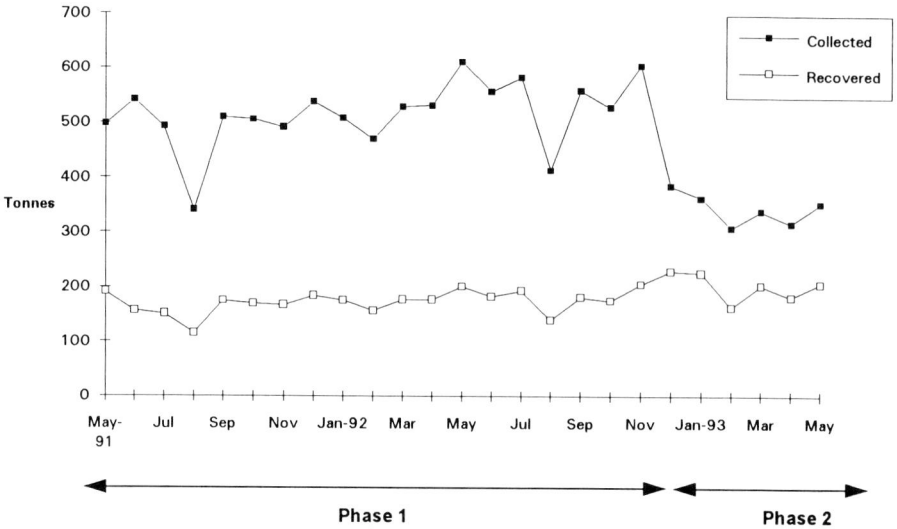

Figure 7.3 Material collected (■) and recovered (□) in Barcelona dry recyclables collection scheme. Note the fall in amount collected during Phase 2, due to narrower range of materials targeted and less contamination, but no fall in the amount of material actually recovered. Note also seasonal troughs in August each year due to holidays. *Source*: ERRA, Barcelona Waste Stream Analysis Report, personal communication (1993).

be able to predict the inputs and outputs of the RDF process for any input waste composition.

Energy consumption. Several operations in the RDF process have significant electrical energy consumptions, in particular primary shredding (12.5 kW-h per tonne of rated capacity), secondary shredding (8.5 kW-h/tonne) and pelletising (9.5 kW-h/tonne). The overall electrical energy consumption for the dRDF process has been estimated as 55.5 kW-h/tonne of rated capacity, i.e. per tonne of annual plant input. In addition, the drying process prior to pelleting requires around 400 MJ of heat energy per tonne of rated capacity (ETSU, 1993). In plants where on-site combustion of RDF occurs, this drying heat requirement can be met by burning some of the RDF produced, or by using waste heat from the power generation system. Where no on-site burning of RDF occurs, heating by gas or other fuels will be needed.

The cRDF process does not involve so many energy intensive stages, nor the drying process. As a result, the energy consumption is lower and has been estimated at 6 kW-h/tonne of plant input for the crude Type A cRDF (Figure 7.2) and 21.5 kW-h/tonne of input for the more refined Type B cRDF (ETSU, 1992).

INPUT   OUTPUTS

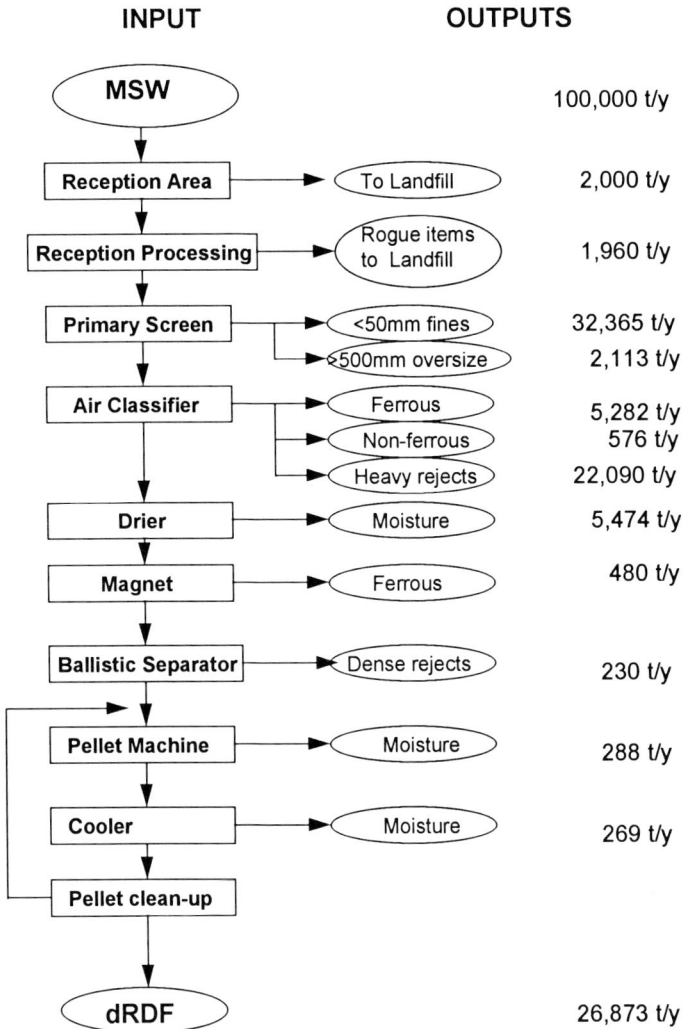

Figure 7.4 Detailed input and output analysis for production of densified refuse-derived fuel (dRDF). *Source*: ETSU (1993).

RDF outputs. The individual outputs from the processes in a typical dRDF flow line are shown in Figure 7.4. When aggregated, these outputs give the total amounts of RDF, recovered materials, putrescible fines, residue and air emissions from the dRDF process, as shown in Figure 7.5. This mass balance applies to a 'typical' input waste composition, and shows that on average around 27% of the input, by weight, is converted into dried, pelletised dRDF. To estimate the mass balance for any input waste

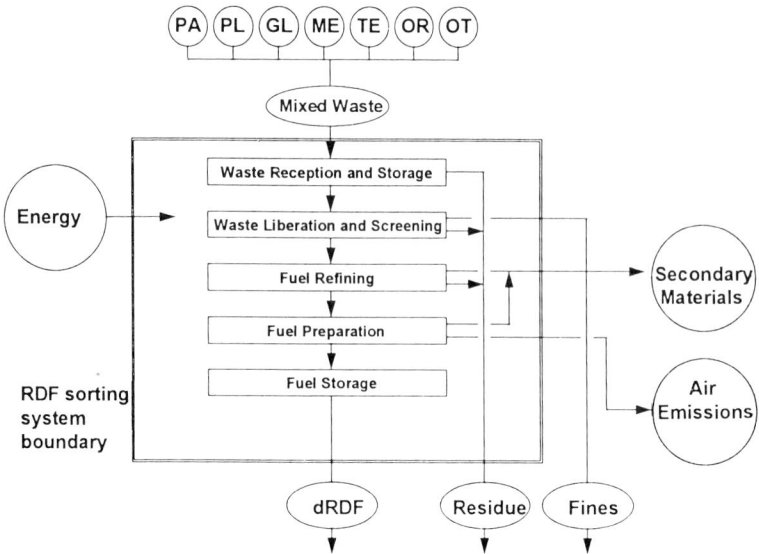

Input materials:
PA = paper/board   PL = plastics   GL = glass   ME = metals   TE = textiles
OR = organics   OT = other

Outputs (as % of input by weight):

dRDF	26.9%	Residue:	
		crude waste	2.0%
		rogue items	1.9%
Secondary materials:		+500 mm oversize	2.1%
ferrous	5.8%	heavy rejects	22.1%
non-ferrous	0.6%	ballistic rejects	0.2%
Fines	32.4%	Air emissions:	
		water vapour	6.0%

Figure 7.5 Mass balance for dRDF process with typical mixed waste input. *Source*: ETSU (1993).

composition, it is necessary to have information on the composition of the different output streams (Table 7.5) and the distribution of each material in the input waste stream that enters the plant between the various RDF sorting process outputs (Table 7.6). Using this information, it is possible to determine both the amount, and composition of the various process outputs (RDF, recovered materials, residue, etc.) for any input waste amount or composition.

Table 7.5 Refuse-derived fuel processing: typical composition (% by weight) of the process output streams

Stream	Output (%)	Composition of output streams (%)									Total
		Paper	Plastic film	Plastic rigid	Textile	Glass	Metal Fe	Metal non-Fe	Organic	Other[a]	
Fines	33.7	8.0	0.6	0.9	0.4	16.5	0.9	0.6	37.5	34.6	100
Gross oversize	2.2	23.9	24.5	2.4	21.6	1.0	5.6	0.7	4.7	15.6	100
Fe fraction	6.7	1.1	0.0	0.1	0.4	0.2	96.3	0.0	0.7	1.2	100
Heavy rejects	15.4	14.8	0.5	10.0	4.1	4.9	3.5	5.2	15.2	41.8	100
Scalp magnet fraction	0.1	16.6	9.3	0.2	14.2	0.0	19.8	0.5	0.6	38.8	100
Fuel fraction	41.9[b]	56.0	6.3	1.7	4.8	0.0	0.0	0.4	2.1	28.7	100
TOTAL	100.0										

Source: ETSU (1993).
[a] Includes miscellaneous combustibles, miscellaneous non-combustibles and <10 mm fines categories.
[b] Fuel fraction includes moisture content, which will be removed by drying before final fuel is produced.

Table 7.6 Refuse-derived fuel (dRDF) processing: distribution of incoming waste materials (% by weight) between process outputs streams[a]

Material	Process outputs (%)				Total (%)
	Fuel fraction	Fe fraction	Fines	Residue	
Paper/board	81.3	0	8.8	9.9	100
Plastic (film)	80.8	0	4.5	14.7	100
Plastic (rigid)	28.3	0.3	8.8	62.6	100
Glass	0	0.5	70.5	29.0	100
Metal (Fe)	0	86.7	4.0	9.3	100
Metal (non-Fe)	14.1	0	17.1	68.8[b]	100
Textile	61.0	0.8	4.1	34.1	100
Organic	11.6	0.6	56.2	31.6	100
Other[c]	14.2	0.2	61.5	24.1	100

Source: Calculated from data in ETSU (1993) (Appendix C).
[a] Table gives distribution of materials between fractions, not composition of fractions. Thus, for any individual category of material entering the RDF process, this table will show the distribution of this material between the different process outputs.
[b] Assuming no recovery of non-Fe material from heavy reject stream. If recovery occurs, residue contains 18.8% of non-Fe material; 50% of non-Fe material is recovered.
[c] Calculated from original data assuming that 'other' category contains 28% miscellaneous combustibles, 13% miscellaneous non-combustibles and 59% 10 mm fines (taken from core waste analysis used in ETSU, 1993).

7.5   Economic costs

7.5.1   MRF sorting

As with environmental impacts, it is impossible to predict an average MRF processing cost as there is no standard MRF process. Schemes that collect commingled recyclables, so simplifying and perhaps saving costs in collection, are likely to have higher MRF processing costs than schemes with kerbside sorting, where the MRF processing will be simpler. Thus, there is a trade-off of costs between the collection and sorting parts of the lifecycle.

Processing costs for MRFs, averaged over the total throughput, are given in Table 7.7. In most cases it is not possible to allocate costs to the processing of individual materials. One study in the United States has tackled this problem, and has published the MRF processing costs for different materials (NSWMA, 1992), and these are shown in Table 7.8. Ten privately owned MRFs were examined, and the results show two distinct features: firstly there are clear differences between individual materials, as might be expected due to the level of sorting needed. The second finding was that the costs vary widely between MRFs, generally by a factor of three, but sometimes by up to a factor of five for the same material. Data are also now becoming available for the sorting of packaging materials from the Dual System in Germany, and these are also included in Table 7.8.

Table 7.7 MRF processing costs and revenues (in ecu) from recovered materials

Scheme	Processing cost/tonne	Revenue from recovered materials/tonne						Source[e]
		Paper	Glass	Aluminium	Steel	Plastic	Average	
UK Adur	24						29	1
Sheffield							25	2
Ireland, Dublin		7[a]	20	526	26	132[c]	26	3
		13[b]				158[d]		
Germany	13–2327							4
USA (average)	45 (range 25–64)						22	5

[a] Board/magazines/cartons.

[b] Newspapers.

[c] Mixed plastics.

[d] Clear PET.

[e] *Source*: 1, IGD (1992); 2, Birley (1993); 3, Kerbside Dublin (1993), 4, Berndt and Thiele (1993); 5, NSWMA (1992) Recycling Times (8th September 1992).

Table 7.8 MRF sorting costs by material: USA and Germany

Material	USA		Germany (DSD system)
	Average cost ($/ton)	Cost Range ($/ton)	Cost (ecu/tonne)
Newspaper	33.59	19.94–55.33	
Corrugated board	42.99	20.29–56.26	44–57
Mixed paper	36.76	16.82–65.59	
Aluminium cans	143.41	72.88–362.59	412–1959
Steel cans	67.53	30.22–125.64	31–41
Clear glass	72.76	37.17–105.62	
Brown glass	111.52	69.70–148.92	13–18
Green glass	87.38	57.56–134.21	
Mixed glass	50.02	28.51–76.24	
PET plastic	183.84	64.43–295.35	665–2327
HDPE plastic	187.95	121.58–256.15	
Beverage cartons			399–1397
Other laminates			412–2062
Per average ton of recyclables	50.30 (44.51 ecu)	28.11–72.06 (24.88–63.77 ecu)	
Source	1	1	2

Sources: 1, NSWMA (1992); 2, Berndt and Thiele (1993).

The other economic factor that needs to be considered at this point is the revenue obtained from the sale of materials recovered at a MRF. Like any commodity, these are affected by market forces and their prices will fluctuate with supply and demand over time and geography. Some prices that have been reported are given in Table 7.7, but it should be stressed that these will not necessarily reflect current market conditions. At the time of writing, the market price for many grades of paper and plastic has been depressed, and in some cases negative values apply, due to the large amount of material flooding on to the European market from the Dual System in Germany.

7.5.2   RDF sorting

Two recent studies have attempted to predict the economic costs of processing Refuse Derived Fuel (ETSU, 1992, 1993). The resulting costs depend on the capacity of the plant used, due to economies of scale. The predicted break-even gate fee for a plant producing dRDF pellets for sale in the UK was 33.88 ecu per input tonne for a 100 000 tonnes/year plant, falling to 31.66 ecu for a plant with double this capacity. This gate fee is inclusive of revenues from the sale of the fuel pellets and recovered ferrous and non-ferrous metals, and the costs of transport and disposal of the residues. When these are split away, the processing cost for dRDF can be estimated at 24.39 ecu per input tonne, for a plant of 200 000 tonnes/year capacity (Table 7.9). For comparison, a plant of 10 000 tonnes/year capacity including a composting operation for further treatment of the putrescible fines fraction, is estimated to have a break-even gate fee of 36.28 ecu (ETSU, 1993). Similar estimates have been made for the production and, in this case, on-site combustion of cRDF (ETSU, 1992), giving a break-even gate fee of 47.79 ecu for a 200 000 tonnes/year capacity plant (Table 7.9).

7.6   Operation of the central sorting module of the LCI spreadsheet

The central sorting operations in the lifecycle of solid waste are handled in LCI Boxes 3a and 3b.

LCI BOX 3a. MATERIAL RECOVERY FACILITY (MRF) SORTING

Processing:
  Energy/fuel consumption:
    Electrical:
    Diesel:
    Natural gas:
Since there is no standard MRF process, the energy consumption needs to be inserted here for the process required. Some values, per input tonne are given in Table 7.3. The model adds the total energy/fuel requirement to the lifecycle energy/fuel consumption totals.

  Residue:
    Estimated % of input lost as residue:
The model automatically inserts a value here: 30% where commingled recyclables are collected and 8% where a kerbside sort is used (see

Table 7.9 Estimated processing costs (in £) for RDF plants

	dRDF plant	cRDF plant with on-site power generation via fluidised bed combustor	Notes
Break-even gate fee for 200 000 tonnes/year plant	24.06 (31.66 ecu) includes non-ferrous metal recovery, but not composting of residues	36.32[a] (47.79 ecu) includes building and operation of combustion plant	Discount rate 10% on capital costs
Inclusive of:			
Residue disposal costs (landfill)	7.59	5.79	@£12.50/tonne
Transport costs	4.25	2.44	@£7/tonne of residue
Revenue from dRDF	4.30	–	@£16/tonne
Revenue from non-Fe metal	1.73	1.78	@£300/tonne
Revenue from Fe metal	0.29	1.34	@£5/tonne
Revenue from sale of electricity (cRDF only)		10.93	@£0.025 per kW-h[a]
Calculated RDF processing cost	18.54 (24.39 ecu)	42.14 (55.45 ecu)	Break-even gate fee, plus revenues, minus other costs

Source: Calculated from data in ETSU (1992, 1993).
[a] Assumes no subsidy for power generation from waste/renewable resources applies. If the UK Non-Fossil Fuel Obligation (NFFO) subsidy is paid, electricity sales revenue will increase, and so the overall gate fee will decrease.

LCI Box 3

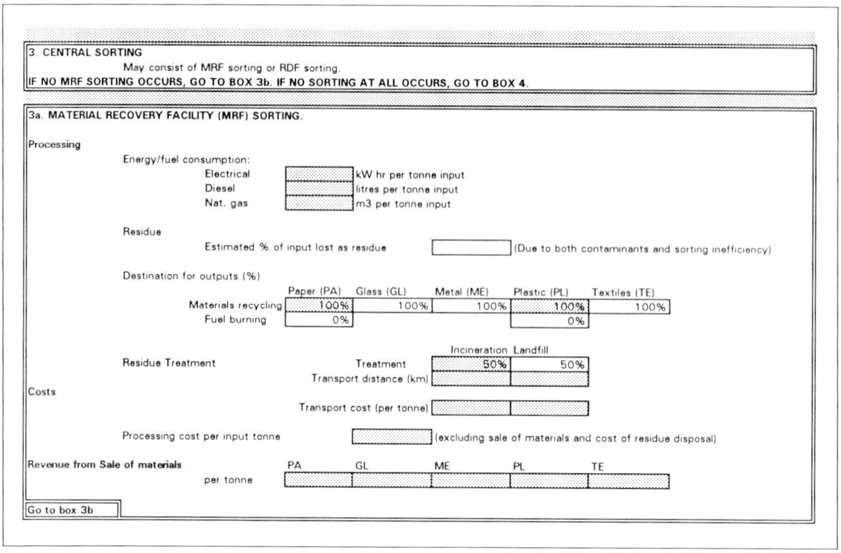

Figure 6.3). This loss includes both contaminants collected, and sorting inefficiency. The model removes this amount of each material collected, and adds it to the incineration or landfill streams, as appropriate.

Destination for outputs (%):
  Materials recycling:
  Fuel burning:

As it is possible to use paper and plastic as a fuel for energy recovery, as well as for materials recycling, it is necessary to insert the relative usages of paper and plastic at this point. It is assumed that glass, metal and textiles (if collected and sorted) are destined for materials recycling. The model calculates the amount of each material left after removal of the residue, and then adds the relative amounts to either the secondary material totals (materials recycling) or to the fuel burning inputs as appropriate.

Residue treatment: *Treatment: incineration/landfill:*
                         Transport distance (km):

The user defines the treatment method used for the residue and the average distance (one way) to each type of treatment plant. The model adds the appropriate amounts of each material to either the incinerator or landfill input streams, and calculates the fuel (diesel) consumed by transport (assuming a 20 tonne load on a round trip basis, i.e. no return load), which is added to the fuel consumption totals.

Costs:

Residue transport cost: incineration/landfill:
The user needs to insert transport cost per tonne of residue shipped.

Processing cost per input tonne:
This should exclude revenues from sale of recovered materials and costs of residue disposal. If data which include sales revenue are used, then zero revenues should be inserted in the next boxes.

Revenue from sale of materials:
Recovered materials destined for recycling leave the waste management

---

LCI Data Box 3 Fixed data used in LCI module on RDF sorting.

	cRDF	dRDF
Fuel consumption		
Electricity	21.5 kWh/input tonne	55.5 kW-h/input tonne
Natural gas		10.3 m^3/input tonne
Screening of input material		
Waste rejected due to unavailability of plant	2% of waste input	2% of waste input
Rogue items rejected	2% of plant input (1.96% of waste input)	2% of plant input (1.96% of waste input)
Process input	96.04% of waste input	96.04% of waste input

Distribution (%) of process input between outputs.

Process Outputs	Input materials to RDF process								
	Paper/ board	Plastic film	Plastic rigid	Glass	Metal Fe	Metal non-Fe	Textile	Organic	Other
cRDF									
Fuel	89.5	83.6	89.3	28.7	7.6	31.5	80.7	42.7	36.5
Fe	0	0	0.3	0.5	86.7	0	0.8	0.6	0.2
non-Fe	0	0	0	0	0	50.0	0	0	0
Fines	8.8	4.5	8.8	70.5	4.0	17.1	4.1	56.2	61.5
Residue	1.7	11.9	1.6	0.3	1.7	1.4	14.4	0.5	1.8
dRDF									
Fuel	81.3	80.8	28.3	0	0	14.1	61.0	11.6	14.2
Fe	0	0	0.3	0.5	86.7	0	0.8	0.6	0.2
non-Fe	0	0	0	0	0	50.0	0	0	0
Fines	8.8	4.5	8.8	70.5	4.0	17.1	4.1	56.2	61.5
Residue	9.9	14.7	62.6	29.0	9.3	18.8	34.1	31.6	24.1

Loss of moisture due to drying and pelletising of dRDF = 18.3% of fuel fraction. Final amount of dried dRDF produced = 81.7% of fuel fraction.

Notes: Non-ferrous separation is assumed in both cRDF and dRDF production.
       cRDF data calculated for a refined cRDF product, from dRDF data.

Sources: ETSU, 1992; 1993.

system defined in this study at this point. The revenue received, per tonne of material ex MRF, needs to be inserted in the relevant boxes.

LCI Box 3b. REFUSE DERIVED FUEL (RDF) SORTING

Is restwaste sorted to make RDF (1 = YES; 0 = NO):
This defines whether RDF is included in the waste management system. If the answer is no, there is no need to complete any more of this box, and the user should move on to Box 4. If the answer is yes, all of the remaining restwaste is transferred to the RDF sorting column of the model.

RDF Input:
The model automatically inserts both the amount of restwaste that will be processed for RDF, and its composition, in the relevant boxes.

RDF Process:
Define fuel type produced:
This box defines whether coarse (cRDF) or densified (dRDF) will be produced, and hence which fixed process data from Data Box 3 will be used. The model allows for the production of dRDF or of refined cRDF (Type B in ETSU, 1992).

Energy consumption per tonne of input:
This is inserted automatically using the appropriate data from Data Box 3. The fuel/electricity used per tonne is multiplied by the input tonnage

LCI Box 3b

to get total consumptions, which are added to the fuel consumption columns.

Fuel produced:
The model then allocates the input materials to the RDF process to the various outputs, according to the data in Data Box 3. The amount of RDF fuel produced is inserted automatically into this box, for the user's information. Note that the dRDF process includes drying, so there is a weight loss due to loss of moisture; the cRDF process does not include drying, so the weight is on an 'as received' basis. The amount of fuel is inserted into the RDF columns, which form the input for the thermal treatment section of the model.

Residues:

Is putrescible fines fraction sent for biological treatment? (1 = YES; 0 = NO):
If the answer is yes, the amount of putrescible fines is added to the input stream for biological treatment. If the answer is no, the fines are added to the residue stream, for landfilling.

Distance to biological treatment plant:
This allows calculation of fuel consumption impacts for the transport to the biological treatment plant. If the fines are treated on site, no transport is needed and a zero should be inserted.

Amount of residue to landfill:
The model automatically calculates the total amount of residue that needs to be landfilled, and displays it here.

Distance to landfill:
This allows calculation of fuel consumption impacts, and is calculated on a round trip basis, assuming no return load.

Costs:

Processing cost per tonne of input:
This should be exclusive of the costs of residue transport or disposal, and the revenues from sale of recovered materials or fuel.

Transport costs per tonne of residue:
Revenue from recovered materials:
These need to be accounted for here, but the revenue from the fuel should not be included as it does not leave the waste management system as defined in this model.

References

Barton, J.R., Poll, A.J., Webb, M. and Whalley, L. (1985) *Waste Sorting and RDF Production in Europe*. Elsevier Applied Science, London.

Berndt, D. and Thiele, M. (1993) Status des Dualen Systems und seine Kosten. Verpackungs-Rundschau, October 1993, pp. 84–88.

Birley, D. (1993) Does the Blue Box have a future? *Warmer Bull.* **35**, 10–12.

ETSU (1992) Production and combustion of c-RDF for on-site power generation. Energy Technology Support Unit report no. B 1374, by Aspinwall and Company Ltd. Published by the Department of Trade and Industry, 32 pp.

ETSU (1993) Assessment of d-RDF processing costs. Energy Technology Support Unit report no B 1314, by Aspinwall and Company Ltd. Published by the Department of Trade and Industry, 63 pp.

IGD (1992) Sustainable waste management: the Adur project. Report by the Institute for Grocery Distribution, Letchmore Heath, Watford, UK, 85 pp.

Kerbside Dublin (1993) Kerbside Dublin, one year on. Report of Kerbside Dublin, Unit 8, Cookstown Industrial Estate, Dublin 24, 16 pp.

Magnetic Separation Systems, Plastic bottle sorting. Literature from Magnetic Separation Systems Inc., 624 Grassmere Pk. Dr., Nashville TN 37211, USA.

Newell Engineering (1993) Black Magic non-ferrrous separator. Literature from Newell Engineering Ltd., Burnt Meadows Rd., North Moons Moat, Redditch, Worcs, UK.

NSWMA (1992) Processing costs for residential recyclables at materials recovery facilities. National Solid Wastes Management Association, Washington, DC.

Warmer (1993) Refuse-derived fuel. Warmer Information Sheet, Warmer Bulletin 39.

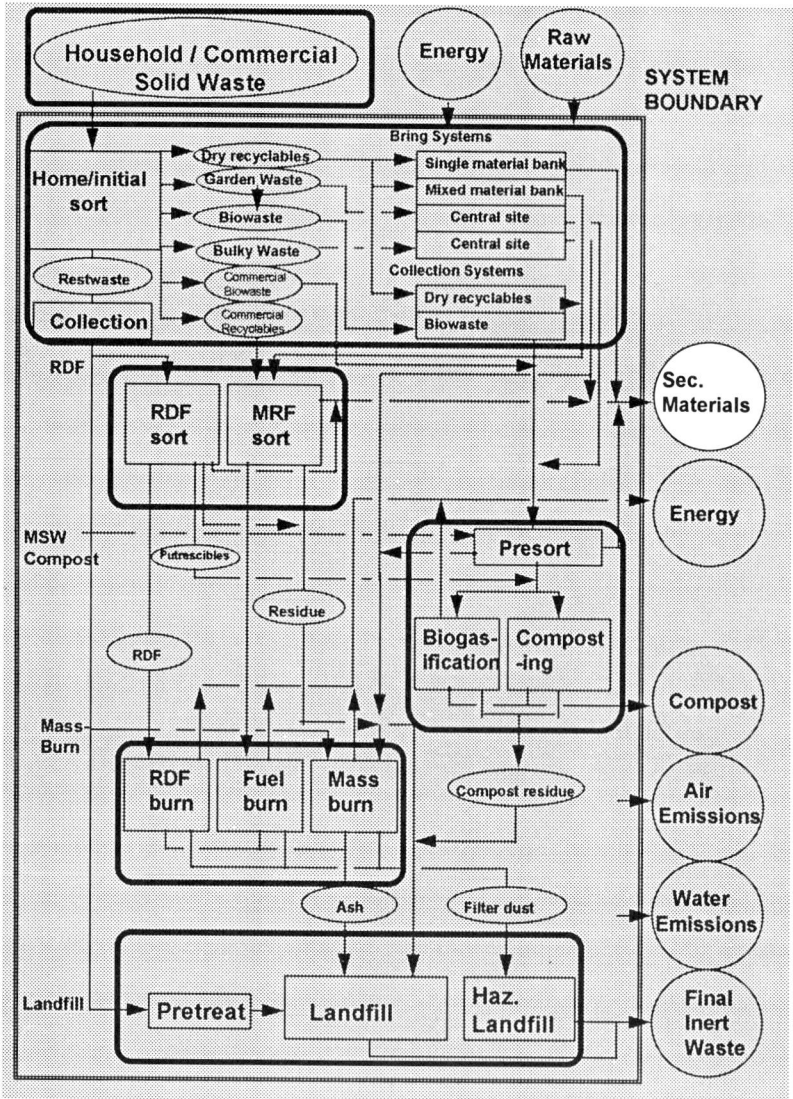

Figure 8.1 Materials recycling in relation to integrated waste management.

8 Materials recycling

Summary

The reprocessing of recovered materials into recycled materials is outside the boundary of the waste management system that is modelled in this book. Recovered material that is reprocessed can, however, be used to replace virgin materials, and this may result in overall savings in energy consumption and emissions. In this chapter, the recycling processes used for each material are briefly described and their energy consumption and emissions quantified where possible. These are then compared with the energy consumption and emissions associated with the production of an equivalent amount of the virgin material, so that overall savings or additional costs can be calculated. This is presented as an option within the computer LCI model, so that the savings associated with the production of recovered material can be considered in the overall balance.

8.1 Introduction

According to the boundaries defined in this book, materials recycling processes lie outside the waste management system (Figure 8.1). In this system, materials destined for recycling cross the system boundary as recovered secondary materials at the exits of materials recovery facilities, RDF sorting plants, biological treatment plants, mass-burn incinerators or transfer stations for mono-material bank-collected material. These materials then enter the industrial processing system for each particular material. This was chosen as the waste management system boundary because at this point the recovered and sorted material has generally re-acquired value (the secondary materials are sold on to the processors), and so has ceased to be 'waste' (as defined in Chapter 1). Recovered materials, rather than recycled materials, thus form one of the outputs of this waste management system.

One problem with this system definition is that it is based on economic criteria, and thus is not necessarily absolute. As with any commodities, the market value of recovered materials fluctuates with supply and demand; in the case of recovered materials it can actually fall to a negative value. This is currently (1994) particularly relevant to the market for recovered plastics. Due to the large amount of plastic material entering the European

Box 8.1   RELATIONSHIP OF MATERIALS RECYCLING PROCESSES TO AN IWM SYSTEM.

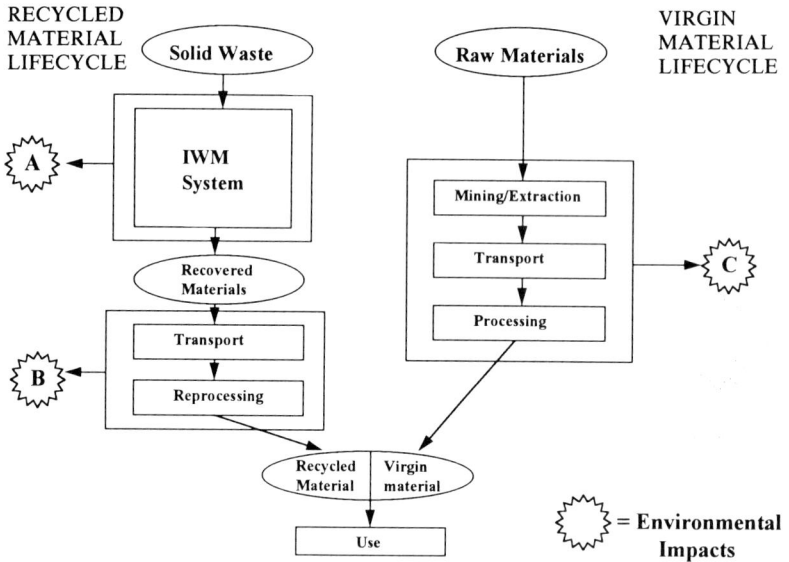

Scope of Chapter 8

The environmental impacts of collection and sorting of recovered materials are included within the environmental impacts of the IWM system (A). This is what the IWM LCI model developed in this book predicts.

Chapter 8 attempts to quantify the environmental impacts of converting the recovered material from the IWM system into recycled material for further use (environmental impact B in the above figure).

Thus: the environmental impact of an IWM system plus subsequent reprocessing of recovered material to produce recycled material = A + B

But using the recycled material leads to reduced virgin material usage and hence savings in environmental impact C.

Therefore, net overall environmental impact of waste management including reprocessing

$$= A + B - C$$

Tables 8.1 to 8.6 give values of B (excluding transport), C and (B − C) for individual materials.

market from the DSD collection system in Germany, compared to the limited reprocessing capacity, the price of recovered plastic has fallen to zero or even a negative value (i.e. requiring payment of a subsidy or 'gate-fee' to the reprocessors). A solution to this definition problem is to include the recycling of the recovered materials within the waste management system boundary. In such a case, this material would leave the system as recycled, rather than recovered material (i.e. as metal ingots, granules of recycled plastic resin, etc.). The energy consumption and emissions associated with transporting the material to the reprocessors, and the processing stages themselves would then also need to be allocated to the waste management system.

Including recycling industries within the waste management system will increase both the size and complexity of the model. Also, in some materials, e.g. glass, steel and aluminium, the recycling process is integrated into the normal production process for the virgin material, so may be difficult to separate out. The output of such an enlarged system would be recycled material. Provided that a market existed for this material (i.e. so long as it could be produced at a competitive cost compared to virgin material alternatives) it would then replace use of virgin material. The costs and environmental impacts of the virgin material production would thus be saved. However, to include these savings within the present model would require that the virgin material process, from cradle to grave, also be included within the system boundary. Clearly a model of this size and complexity is not feasible at present; it would contain a large proportion of all manufacturing industry, as well as the waste management industry (see Box 8.1).

To keep the present model to manageable dimensions, therefore, the system boundary for recovered materials will be kept at the point where they are sent on for reprocessing, i.e. at the exit of the sorting facility. This chapter provides data on the environmental impacts and economic costs of reprocessing the recovered materials, to give an indication of possible savings (or costs) compared to the use of virgin raw materials. This information is provided in the spreadsheet model as an additional optional feature, to allow calculation of the overall impacts and costs for an enlarged waste management system that includes the reprocessing stage and produces recycled, rather than recovered, materials.

8.2 Materials recycling processes

8.2.1 Transportation

The first stage in the conversion of all recovered materials into recycled materials is the transport from the sorting or collection facility, to the reprocessing facility. The distances involved clearly depend on the relative locations of the IWM scheme and the reprocessing plants, so there is a strategic need to locate such plants within easy reach of large potential sources of recovered materials. Brief details of the subsequent reprocessing stages are given below for each material.

8.2.2   Paper and board

Waste paper reprocessing varies according to the type of recycled paper product, which will in turn determine the type of waste paper that is used as the process feed stock. Waste paper is graded into numerous categories (11 in the UK; 5 main grades in Germany, with 41 sub-grades) according to quality (Cathie and Guest, 1991). The higher quality grades (UK grades 1–4) (paper mill production scrap, office and writing papers) which need little cleaning, are used to make printing and writing papers, tissues and wrapping papers, and are known as pulp substitute grades since they are used to replace virgin pulps. Newsprint (UK grade 5) and other papers needing de-inking are reprocessed for further use in the production of newspaper and hygiene papers. The lower (bulk) grades (UK grades 6–11) are mainly used for the production of packaging papers and board.

The details of the process stages will vary according to whether pulp substitute grades, newsprint or bulk grades are treated, but the basic steps are shown in Figure 8.2. After an initial soaking, the waste paper is pulped to separate the fibres, screened to remove contaminants (contraries), de-inked, thickened and washed. During these refining processes both contraries (nuisance materials) and some fibres are removed from the system; such losses have been estimated at 15% for newsprint reprocessing (Shotton, 1992). Therefore, the input of one tonne of recovered paper will result in the production of 850 kg of recycled paper.

8.2.3   Glass

Glass reprocessing has always been practised since the input of recovered cullet to the furnace lowers the temperature needed to melt the virgin raw materials, and thus leads to considerable energy savings (Ogilvie, 1992).

The first stage of glass reprocessing usually consists of a manual sort to remove gross contaminants (plastic bottles, ceramics, lead wine bottle collars) followed by automatic sorting to remove ferrous contaminants and low density materials (paper labels, aluminium bottle tops). The former is achieved by magnetic extraction, the latter by a combination of crushing, screening and density separation techniques. Around 5–6% of the recovered glass input is removed in this way (Ogilvie, 1992). The crushed cullet is then ready for mixing with virgin raw materials, prior to melting in the furnace and blowing or moulding of the final glass products.

Since the use of recovered glass cullet is integrated within the normal glass production process, the environmental impacts of glass reprocessing will be considered up to the production of finished glass containers.

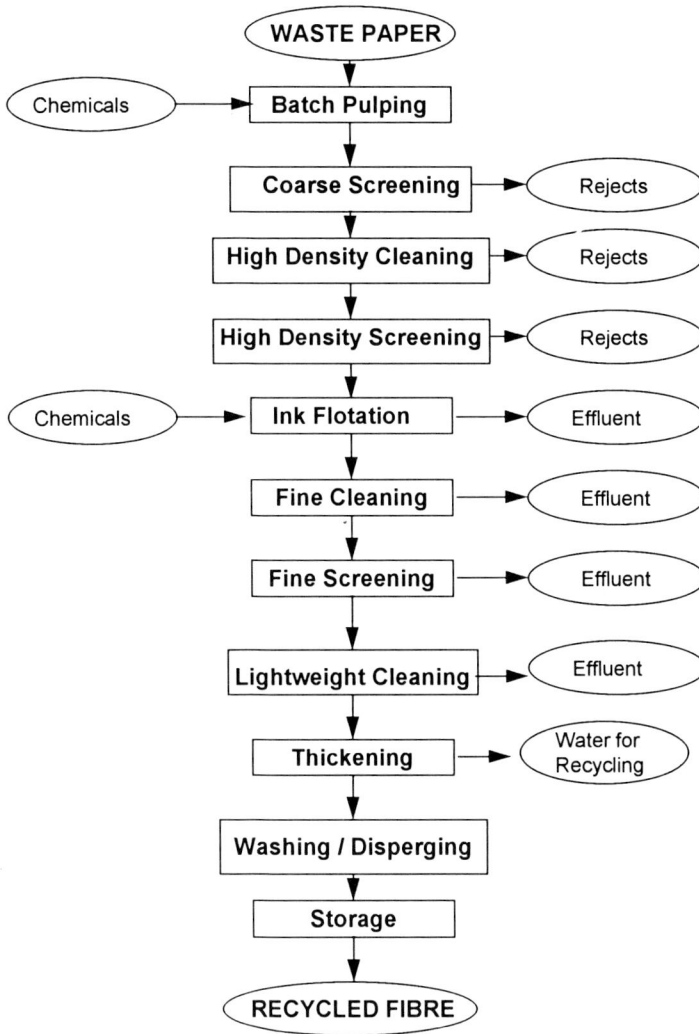

Figure 8.2 Stages in the production of recycled paper fibres. *Source*: Porteous (1992).

8.2.4 *Ferrous metal*

Ferrous metal within household and commercial waste is found in the form of iron and steel scrap, but the majority is in the form of tinplate in food and beverage cans. To reprocess steel from steel scrap merely involves a sort to remove contaminants, before the scrap is melted and recast.

To produce high grade steel from tinplate for further use, it must first be detinned. This process, shown in Figure 8.3, consists of shredding the

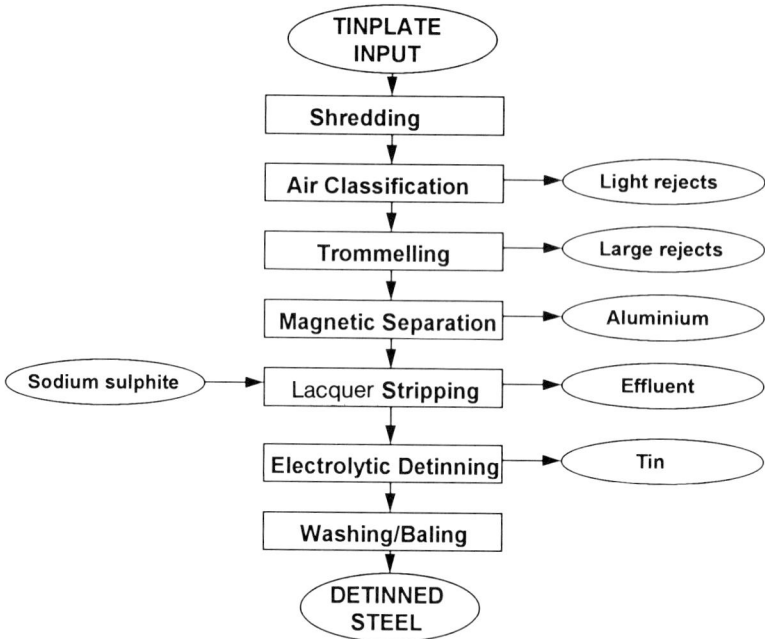

Figure 8.3 Stages in the reprocessing of tinplate steel. *Source*: AMG Resources (1992).

incoming tinplate and removing contaminants, before electrolytic removal of the tin plating. The tin only represents between 0.25% and 0.36% of the input material (Boustead, 1993b; Habersatter, 1991), but the value of this metal makes it worthwhile to recover, melt and cast the scrap tin for further use. The detinned steel scrap needs to be washed thoroughly to remove process chemicals, and then baled for delivery to the steel converting plant.

Such ferrous metal scrap is melted and recycled by two different methods. An electric arc furnace directly melts the metal, and can consequently use 100% scrap. Alternatively, the scrap can be melted by the excess heat generated in a basic oxygen furnace, where blast-furnace-smelted pig iron is converted into steel. In this latter case, the recycling of ferrous metal is integrated into the production of the virgin material.

8.2.5 Non-ferrous metal

The major non-ferrous metal to be recovered from waste is aluminium, so discussion of reprocessing will be restricted to this metal. The reprocessing of recovered aluminium is a much simpler and less energy intensive process than the production of virgin aluminium, which requires smelting of the metal from the bauxite ore. Reprocessing involves sorting of the recovered metal and then melting in a furnace. Since most aluminium is used in an alloy form with other metals or coatings, it is necessary to select an appropriate mix of recovered metal to give an ingot of the correct composition. Sometimes contaminants need to be diluted by the incorporation of high grade virgin material to meet the stricter specifications.

8.2.6   Plastic

Plastic is a generic term covering a host of different resins, but only thermoplastic resins are suitable for materials recycling. Thermoset plastics (duroplastics) cannot be reprocessed in this way. Within the thermoplastics, there are as many different production processes for the virgin materials as there are resin types. There is less variety, however, in the re-processing methods, with basically two categories. After separation of the resin types into individual, or at least compatible fractions, plastics can be either mechanically or chemically recycled. In mechanical recycling, the plastic is shredded or crumbed to a flake form, and contaminants such as paper labels are removed using cyclone separators. The flake is then generally washed (this stage may also be used to separate different resins on the basis of density), dried and then the flake extruded as pellets for sale into the plastics market.

   Chemical recycling involves a more complex process whereby the plastic polymer is broken down into the monomer form, and then re-polymerised. In this case, as with glass and steel, the recycled product is indistinguishable from the virgin material. This recycling method has been developed for certain resins, notably polyethyleneterephthalate (PET) where chemical recycling via a methanolysis process is commonly used. There have also been developments in the reprocessing of mixed plastics, which can be cracked to produce mainly ethylene.

8.2.7   Textiles

Textile recovery has a long history, and has mainly been practised for economic rather than environmental reasons. Of the textiles currently recovered, the majority are re-used as cloth, rather than recycled as fibres. In the UK, for example, some 26% of recovered textiles are re-used as second-hand clothing, 40% used as wiping cloths and 22% used as filling material. Only 7% is actually reprocessed to produce recycled fibres for cloth production (UK Textile Reclamation Association, cited in Ogilvie, 1992). The recycling process uses toothed drums, combs and pulling machines to tear the textiles apart and extract the fibres. As with paper recycling, the process leads to a shortening of the fibres. Consequently, textiles cannot be recycled indefinitely since at some stage the fibres become too short to be re-used.

8.3   Environmental impacts: input–output analysis

Several reports have recently been produced providing data on the energy consumption and emissions resulting from materials recycling (e.g. Habersatter, 1991; Henstock, 1992; Ogilvie, 1992; Porteous, 1992; Boustead, 1993a, b). Before using such data in the context of this chapter, however, it is necessary to determine their relevance.

In many cases, the aim of the studies (e.g. Henstock, 1992) has been to compare the environmental impacts of producing recycled versus virgin material. This comparison is made on a 'cradle-to-produced material' basis; for virgin material, this is from extraction of raw materials, whilst for recycled materials, the cradle is often defined as the start of the collection process for the waste materials. This is not the appropriate comparison to determine the environmental savings or costs that can be attributed to the recovered materials leaving an IWM system. The environmental impacts of collection and sorting are included in such reports, but in the IWM-1 model developed in this book, they have already been accounted for within the waste management system boundaries. Therefore, what the present study requires are data for just the transport to the reprocessors and for the reprocessing itself. These can be added to the impacts of the defined waste management system. If any relative savings/costs are to be calculated, the relevant comparison would be between the transport and reprocessing impacts for the recycled material, versus the total impacts for the virgin material, from raw material extraction to produced material (see Boxes 8.1 and 8.2).

A second note of caution is needed when interpreting the impacts of materials recycling. The impacts associated with material production (whether virgin or recycled) are usually presented on the basis of per kilogram (or per tonne) material produced, i.e. per unit of output (e.g. Habersatter, 1991). This is not the form of data relevant for an LCI of solid waste. The ultimate function of a waste management system is to manage a given amount of waste in an environmentally and economically sustainable way, not to produce recycled material (see Chapter 4). The functional unit of the whole system is therefore the amount of waste entering the system, not the amount of recovered or recycled material leaving the system. If the reprocessing stage is to be taken into account in the overall system, therefore, data on environmental impact need to be in the form of per tonne of recovered material sent for recycling, i.e. per unit of input, rather than output.

In the sections below covering the environmental impact of reprocessing each of the recovered materials, data will be presented for the following:

1. Energy consumption and emissions associated with the reprocessing of recovered materials into recycled material;
2. Energy consumption and emissions associated with production of the

Box 8.2 CALCULATING THE ENVIRONMENTAL BENEFITS OF RECYCLING VERSUS THE USE OF VIRGIN MATERIALS.

RECYCLED MATERIAL LIFECYCLE

VIRGIN MATERIAL LIFECYCLE

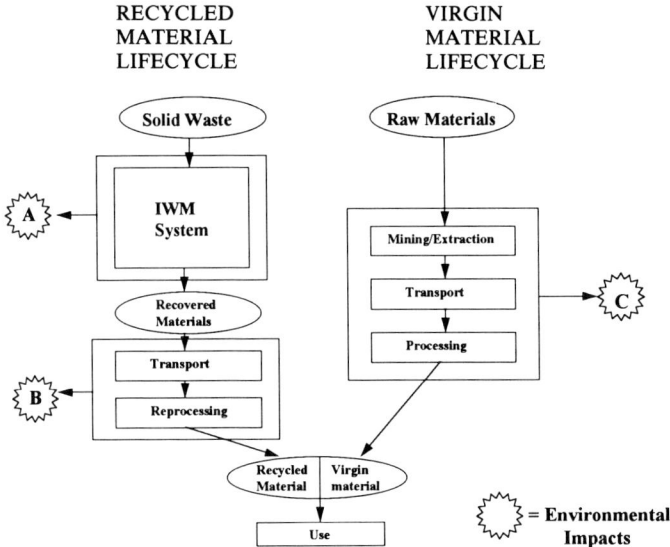

Solid Waste

Raw Materials

A

IWM System

Mining/Extraction

Transport

C

Recovered Materials

Processing

Transport

B

Reprocessing

Recycled Material | Virgin material

Use

= Environmental Impacts

Environmental Benefits (or costs) of including recycling in an IWM system.
Comparing the overall impacts of an IWM system with recovery plus reprocessing of materials ($A_r + B$), with a system without materials recovery, where all products use virgin materials ($A_{nr} + C$), the difference in the overall (system) environmental impact due to recycling is ($A_r + B$) − ($A_{nr} + C$) (where A_r is the impact of an IWM system that includes collection and sorting of recyclables, and A_{nr} is the impact of a waste system that does not include collection and sorting of recyclables).

Difference in overall impact due to recycling $= A_r + B - A_{nr} - C = \delta A + B - C$ where δA = difference between impact of IWM system with collection and sorting of recyclable, and system with only disposal of these materials i.e. ($A_r - A_{nr}$)

There will be an overall environmental benefit (reduced impact) so long as $\delta A + B < C$.

Choosing between recycled and virgin material.
In product or package manufacture, there will be an environmental reason for choosing recycled over virgin material if the impacts of producing the recycled material are less than those of producing an equivalent amount of the virgin material

i.e. $\delta A + B < C$ for that material,

where δA = the difference in impact between collecting and sorting the material and collecting and disposing of the material by incineration and/or landfilling,
B = the impact of transporting and reprocessing the material
C = the impact of production from virgin raw materials

virgin materials that this material could replace (starting from raw material extraction; hence:

3. The potential saving (or addition) of energy consumption and emissions for every tonne of recovered material sent for reprocessing;
4. Energy consumption and emissions associated with transporting the recovered material from the collection/sorting facility to the reprocessing facility. This needs to be subtracted from any potential savings to determine the actual savings (or additional costs) likely.

When considering the potential savings associated with materials recycling, the following notes of caution should be borne in mind. The data presented below tend to be generic (average data) or taken from specific studies of individual processes. Whilst their purpose is to give a broad indication of the available savings, they will not be universally applicable. It is assumed that recycled material performs equally well and can replace an equal quantity of virgin material. This is not always the case since some high grade materials, e.g. writing papers etc., cannot be replaced with recycled materials of equal quality.

8.3.1 Transport

The energy consumption and emissions (per tonne) associated with transporting the recovered materials from the collection or sorting site (at which point they leave the basic LCI model developed in this book) to the reprocessing site will obviously vary with the distance involved. The fuel consumption data in Chapter 6 and the fuel production and use data in Chapter 4 are used to calculate the energy consumption and emissions associated with this transport. The calculation assumes that 20 tonne (payload) trucks are used and that the trucks carry loads in both directions.

8.3.2 Paper

A wide range of data have been reported for the energy consumption and emissions associated with the production of paper, whether from recovered or virgin materials. The figures differ according to the type of pulp or paper that is being produced (e.g. newsprint or bleached sulphite paper), and the boundaries chosen for the calculations (e.g. whether the data are for pulp production only, or for complete paper production; whether de-inking is included for recycled paper or not, etc.). What is important when making comparisons, is that similar products are compared and that comparable boundaries are drawn.

For newsprint production using 100% recovered paper, for example, the primary energy consumption for pulping, de-inking, paper-making and

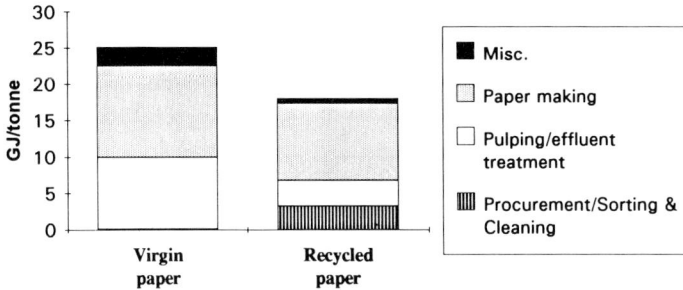

Figure 8.4 Process energy requirements for virgin and recycled paper production.
Sources: Porter and Roberts (1985); Porteous (1992).

effluent treatment has been calculated at 14.5 GJ per tonne produced. (This does not include collection, sorting and transport.) Equivalent production of newsprint from virgin wood consumes 21.0 GJ per tonne, giving a primary energy saving due to recycling of 6.5 GJ per tonne produced (Pulp and Paper, 1976). Similar savings have been calculated when energy consumption has been averaged across the different grades of paper and board produced (Figure 8.4). The average primary energy requirements for virgin and recycled paper processing were reported as 25.1 GJ (range 20–28) per tonne and 18.0 GJ per tonne, respectively in 1985, giving an average saving of 7.1 GJ per tonne (Porter and Roberts, 1985). It is likely that energy consumption has fallen since that time, due to more efficient processing techniques. The requirement for virgin paper making in the UK in 1989, for example, was around 21 GJ per tonne produced (Ogilvie, 1992).

Considerably higher energy consumption figures for both virgin and recycled paper have been quoted by some sources. Habersatter (1991), using data from Swedish and Swiss sources, gives energy consumptions of 53.0 GJ/tonne for production of virgin unbleached sulphite paper and 29.7 GJ/tonne for 100% recycled paper. The reason for this apparently higher energy consumption is that the inherent energy (i.e. calorific value) of the wood or recovered paper feed stock is also included in these totals. The inherent or feed stock energy of wood or waste paper is around 15 GJ/tonne and 2.02 tonnes of air dried wood or 1.02 tonnes of recovered paper are needed to produce 1 tonne of unbleached sulphite pulp or 1 tonne of recycled fibre pulp respectively (Habersatter, 1991). Subtracting the appropriate feed stock energies (30.3 GJ/tonne and 15.3 GJ/tonne respectively) leaves a processing energy requirement for unbleached sulphite paper of 22.7 GJ/tonne, and 14.4 GJ/tonne for recycled paper. These accord with the data quoted above, giving an energy

Box 8.3 SUMMARY OF ENVIRONMENTAL BENEFITS AND COSTS OF RECYCLING.
Table gives the calculated effect on energy consumption and emissions of reprocessing recovered materials, compared to the production of virgin material. Results are per tonne of recycled material produced.

Material	Process Energy saved by recycling (GJ/tonne)	Inherent Energy saved by recycling (GJ/tonne)	Air Emissions for recycling	Water Emissions for recycling	Solid Waste reduced (increased) by recycling (kg/tonne)	Comments
Paper	8.3	30.3	generally lower	generally lower	80	pulp and paper making included
Glass	3.8	–	generally lower	generally lower	(25)	process to finished container included. Data for 100% virgin extrapolated as all glass-making uses some used cullet
Metal -Fe (tinplate)	13.5	–	generally lower	generally lower	278	data for tinplate recycling up to production of new tinplate
Metal -Al	156	–	generally lower (except HCl)	generally lower	639	
Plastic (LDPE)	15.4	47.7	generally lower (except CO$_2$)	little data	(93)	Incomplete data for reprocessing of LDPE
Plastic (HDPE)	25.6	47.7	generally lower	poor data, but may be higher for recycled	(184)	Incomplete data for reprocessing of HDPE
Textiles	52–59	no data	no data	no data	no data	Energy range for woven and knitted wool only

Note: Figures are indicative only and will vary with processes and equipment used. The figures give the difference between the reprocessing burdens for producing 1 tonne of recycled material and the burdens of producing 1 tonne of virgin material. The burdens of collecting and sorting the recovered material, and transporting it to the reprocessors, are not included. Similarly, the diversion of the recovered material from landfill is not included in the solid waste savings.

saving due to the processing associated with recycling of 8.3 GJ/tonne of paper produced.

There has been considerable discussion as to whether feed stock energy should be included in comparisons between virgin and recycled materials. Habersatter (1991) includes the feed stock or inherent energy of both virgin and recycled materials, whereas Henstock (1992), in a comparison of recycled and virgin low density polyethylene, includes the inherent energy of the raw materials used to make the virgin resin, but not the inherent energy of the recovered plastic used to produce the recycled resin. This convention significantly increases the apparent energy savings due to recycling. The decision must reflect the aim of the study which is defined in the goal definition (see Chapters 3 and 4). The aim of the present study is to predict the energy consumption and emissions associated with managing the solid waste of an area in an environmentally and economically sustainable way. The inherent energy contained in the waste entering the system is not considered as contributing to the energy consumption. Therefore, the inherent energy of the recovered material will not be considered as part of the energy consumption of the recycling process. For consistency, therefore, the inherent energy of the virgin raw materials will also not be included as part of the energy consumption of virgin materials. It can be argued that the inherent energy of the material has not actually been consumed; it has merely been locked up and can be released at a later time, e.g by burning as a fuel. The net effect of this convention is to give conservative estimates of the energy savings due to recycling. The processing energy savings and inherent energy savings of recycling are both given in Box 8.3, however, so the higher value can be calculated if required.

Before leaving the subject of energy consumption in paper production, it is necessary to add a cautionary note on using primary energy totals alone to compare the environmental benefits of paper recycling. As well as the total energy consumption, it is necessary to know how the energy was produced, and in particular, whether it came from renewable or non-renewable sources. Many pulp and paper mills in Sweden, for example, generate their own electricity and steam on site using by-products from the pulping process (bark, process liquors, etc.). Some plants produce excess steam or electricity, which are exported from the site. Such use of biomass (a renewable energy resource) will result in no net production of some emissions, such as carbon dioxide, since these were absorbed during the growing of the trees in the first place. By contrast, recycling processes will tend to use power generated from fossil fuel with a net production of carbon dioxide and depletion of finite fossil fuel reserves (although, equally, power could be generated from on-site boilers fuelled by waste paper). The general point to be emphasised here is that energy consumption is not an environmental impact in itself; what is important are the environmental

Table 8.1 Energy consumption and emissions from recycled and virgin paper production

Source	Recycled paper/tonne produced Habersatter (1991)	Virgin paper/tonne produced (nature, unbleached) Habersatter (1991)	Savings/tonne produced	Savings/tonne recovered material used[a]
Energy consumption[b] (GJ)	14.4	22.7	8.3	6.8
Air emissions (g)				
Particulates	357	4346	3989	3270
CO	383	3165	2782	2280
CO_2				
CH_4				
NO_x	2295	5114	2819	2310
N_2O	280	345	65	53
SO_x	6054	10 868	4814	3947
HCl	0	4	4	3.3
HF	0.004	0.01	0.006	0.005
H_2S	0	15	15	12
HC	4195	6258	2063	1692
Chlorinated HC				
Dioxins/furans (TEQ)				
Ammonia	2.9	3.4	0.5	0.4
Arsenic				
Cadmium				
Chromium				
Copper				
Lead				
Mercury	0	0.004	0.004	0.003
Nickel				
Zinc				

Water emissions (g)

BOD	1	2921	2920	2390
COD	3	25 423	25 420	20 840
Suspended solids	1	1	0	0
Total organic compounds	25	30	5	4
AOX	0	3	3	2.5
Chlorinated HCs				
Dioxins/furans (TEQ)				
Phenol	0	0	0	0
Ammonium	0.331	0.876	0.545	0.447
Total metals				
Arsenic				
Cadmium				
Chromium				
Copper				
Iron				
Lead				
Mercury	0	0	0	0
Nickel				
Zinc				
Chloride	9	22	13	11
Fluoride	0.714	1.89	1.18	0.97
Nitrate	0	7	7	6
Sulphide				
Solid waste (kg)	70.6	150.2	79.6	65.3

a Assuming a material loss of 18%.
b Not including inherent energy of feedstock material.

Table 8.2 Energy consumption and emissions from recycled and virgin glass production

Source	Recycled glass (100%)/tonne produced Habersatter (1991)	Virgin glass/tonne produced[a] Habersatter (1991)	Savings/tonne recycled glass produced	Savings/tonne recovered glass used[b]
Energy consumption (GJ)	5.8	9.6	3.8	3.7
Air emissions (g)				
Particulates	428	17 780	17 352	16 831
CO	57	105	48	47
CO_2				
CH_4				
NO_x	1586	2270	684	663
N_2O	12	106	94	91
SO_x	2652	3627	975	946
HCl	6	75	69	67
HF	2.4	1	−23	−22
H_2S				
HC	1113	2300	1187	1151
Chlorinated HC				
Dioxins/furans				
Ammonia	2	4	2	1.9
Arsenic				
Cadmium				
Chromium				
Copper				
Lead	16	0	−16	−15.5
Mercury				
Nickel				
Zinc				

Water emissions (g)				
BOD	1	1	0	0
COD	2	4	2	1.9
Suspended solids	1	1	0	0
Total organic compounds	20	26	6	5.8
AOX				
Chlorinated HCs				
Dioxins/furans (TEQ)				
Phenol	0	0	0	0
Ammonium	0	0	0	0
Total metals				
Arsenic				
Cadmium				
Chromium				
Copper				
Iron				
Lead				
Mercury				
Nickel				
Zinc				
Chloride				
Fluoride				
Nitrate				
Sulphide				
Solid waste (kg)	29.3	4.0	−25.3	−24.5

[a] Calculated by extrapolation from data for 100% and 56% recycled glass.
[b] Assuming a material loss of 3%.

impacts resulting from the generation of the energy used, i.e. the emissions and the depletion of finite resources.

The emissions (to air, water and as solid waste) associated with production of recycled and virgin paper are given in Table 8.1. Note that these data relate to conditions in Switzerland where steam production for both virgin and recycled paper production uses fossil fuels, and electricity production is assumed to be according to the general UCPTE model introduced in Chapter 4. If virgin pulps were made in integrated pulp mills such as in Sweden, fuelled by biowaste by-products, the emission levels associated with virgin pulp production would be generally lower.

To calculate the energy consumption and emissions per tonne of recovered paper sent for reprocessing, rather than per tonne of recycled paper produced, requires data on material losses during the process. Habersatter (1991) assumes a loss of 2% only, but this would appear low. The actual process losses will depend on the quality of the input fibres, the level of filler in the recovered material, the type of recycling process employed and the quality of recycled product required. For recycled newsprint manufacture, a loss of 18% is typical (Claydon, 1991), and across different paper grades a general loss of 15-20% applies (Cathie and Guest, 1991). Taking the former figure and an energy saving of 8.3 GJ/tonne produced (above), the energy saving amounts to 6.8 GJ/tonne of recovered paper sent for reprocessing. There are also significant savings in most emissions and in the overall amount of solid waste generated (Table 8.1).

8.3.3   Glass

The use of recovered glass cullet in glass making has the advantage of lowering the furnace temperature needed to melt the other raw material ingredients. These energy savings can be estimated by the simple equation:

Energy savings (%) = 0.25 × % of scrap glass used

(Habersatter, 1991)

Considering the process of glass recycling up to the production of the hot 'gob' of glass, energy consumption and emissions data per tonne of 100% recycled glass produced are presented in Table 8.2. Note that in this case the data include the transport of the glass from the collection or sorting facilities to the reprocessing plant.

Corresponding data for the production of 100% virgin glass are not available, since all glass production processes incorporate some recovered glass, either collected post-consumer material or production scrap, because of the furnace energy savings available. Extrapolating linearly from the data for recycling rates of 100%, 75% and 56% given by Habersatter (1991), however, it is possible to estimate the energy consumption and associated emissions for glass production from virgin materials (Table 8.2).

Using this estimate, it can be seen that the recycling of glass saves around 3.8 GJ/tonne of glass produced. There are savings in the amounts of those emissions that are associated with this energy production, but there are slight increases in the amounts of some air emissions, notably hydrogen fluoride (HF) and lead, associated with recycling. The amount of process-derived solid waste also increases with the level of recovered material used, due to the removal of contaminants (metals, ceramics, paper) which are reported to constitute 2.7% of the feed stock (Habersatter, 1991).

Assuming contamination is around this 3% level, the energy savings and emissions can be calculated per tonne of recovered material reprocessed (Table 8.2); the energy savings will be of the order of 3.7 GJ/tonne.

8.3.4 Ferrous metal

Typical primary energy consumption and emissions associated with the full production process for tinplate from 100% recovered material are given in Table 8.3, along with data for the production of tinplate from virgin materials. These figures, from Habersatter (1991) suggest an energy saving of around 13.5 GJ/tonne of tinplate produced, when recovered tinplate is used. Ogilvie (1992) similarly cites an energy saving for the recycling of tinplate scrap recovered from mixed waste prior to incineration of 15.7 GJ/tonne. There are also savings in most, but not all emissions (Table 8.3).

The amount of recovered material input that is lost in the reprocessing of tinplate is given as 8.2% by Habersatter (1991); Porteous (1992) suggests a loss of 5% in the process up to the production of detinned washed steel. Taking the former value would mean that each tonne of recovered tinplate delivered to the reprocessors would produce 918 kg of recycled tinplate. The typical primary energy consumption associated with this production would therefore be 18.36 GJ/tonne of recovered tinplate used, or an energy saving (when compared to the use of virgin materials) of around 12.4 GJ/tonne (using the Habersatter data). Comparisons of the amounts of emissions (including solid waste) associated with the processing of one tonne of recovered tinplate scrap versus production of an equivalent amount of tinplate from virgin are also presented in Table 8.3.

For the large part of ferrous scrap not in the form of tinplate, reprocessing is simpler and consists only of removal of contaminants and then remelting. For iron, the energy needed to remelt is 1.8 GJ/tonne whilst iron production from ore requires 7.92 GJ/tonne. This would give an energy saving of 6.12 GJ/tonne of recycled iron produced. For steel, the energy saving of using electric arc melting of scrap versus virgin steel production using a blast furnace is around 15.8 GJ/tonne (Ogilvie, 1992). Assuming the same material loss as used above (8.2%), the possible savings

Table 8.3 Energy consumption and emissions from recycled and virgin tinplate production

Source	Recycled tinplate/ tonne produced Habersatter (1991)	Virgin tinplate/ tonne produced Habersatter (1991)	Savings/ tonne recycled tinplate produced	Savings/ tonne recovered tinplate used[a]
Energy consumption (GJ)	20.0	33.5	13.5	12.4
Air emissions (g)				
Particulates	864	26 955	26 091	24 004
CO	1909	1381	−528	−486
CO_2				
CH_4				
NO_x	2350	2733	383	352
N_2O	226	417	191	176
SO_x	5347	8450	3103	2855
HCl				
HF	0	0.5	0.5	0.46
H_2S				
Total HC	5262	16 527	11 265	10 364
Chlorinated HC				
Dioxins/furans (TEQ)				
Ammonia	1.6	73.5	71.9	66.1
Arsenic				
Cadmium				
Chromium				
Copper				
Lead				
Mercury				
Nickel				
Zinc				

Water emissions (g)				
BOD	4.3	4.7	5.2	0.5
COD	-0.09	-0.1	1.3	1.4
Suspended solids	16	17	318	301
Total organic compounds	-0.9	-1	514	515
AOX				
Chlorinated HCs				
Dioxins/furans (TEQ)				
Phenol	0.28	0.3	0.3	0
Ammonium	5.3	5.8	6.5	0.7
Total metals				
Arsenic				
Cadmium				
Chromium				
Copper				
Iron	0	0	100	100
Lead				
Mercury				
Nickel				
Zinc				
Chloride	0	0	0	0
Fluoride	10.9	11.8	33.4	21.6
Nitrate	-0.09	-0.1	0.3	0.4
Sulphide	0.18	0.2	0.2	0
Solid waste (kg)	255.0	277.5	398.6	121.1

[a] Assuming a material loss of 8%.

Table 8.4 Energy consumption and emissions from recycled and virgin aluminium production

Source	Recycled aluminium/tonne produced[a] Habersatter (1991)	Virgin aluminium/tonne produced[a] Habersatter (1991)	Savings/tonne recycled aluminium produced	Savings/tonne recovered aluminium used[b]
Energy consumption (GJ)	15.6	171.2	155.6	147.8
Air emissions (g)				
Particulates	1222	37 388	36 166	34 358
CO	474	17 713	17 239	16 377
CO_2				
CH_4				
NO_x	2527	27 711	25 184	23 925
N_2O	252	1673	1421	1350
SO_x	7090	75 793	68 703	65 268
HCl	760	50	−710	−675
HF	0	254	254	241
H_2S				
Total HC	4753	39 870	35 117	33 361
Chlorinated HC				
Dioxins/furans (TEQ)				
Ammonia	3	20	17	16
Arsenic				
Cadmium				
Chromium				
Copper				
Lead				
Mercury				
Nickel				
Zinc				

Water emissions (g)				
BOD	1	799	798	758
COD	3	19 020	19 017	18 066
Suspended solids	1	6	5	4.8
Total organic compounds	28	173	145	138
AOX				
Chlorinated HCs				
Dioxins/furans (TEQ)				
Phenol	0	0	0	0
Ammonium	0	1	1	0.95
Total metals				
Arsenic				
Cadmium				
Chromium				
Copper				
Iron				
Lead	0	1	1	0.95
Mercury				
Nickel				
Zinc				
Chloride	0	0	0	0
Fluoride	1	3	2	1.9
Nitrate	0	0	0	0
Sulphide				
Solid waste (kg)	237.6	876.5	638.9	607.0

a Using a 'Western World' scenario for electricity generation.
b Assuming a material loss of 5%.

Table 8.5 Energy consumption and emissions from recycled and virgin plastic production

	Recycled LDPE/ tonne produced	Virgin LDPE/ tonne produced	LDPE savings/ tonne recycled	LDPE savings/ tonne recovered LDPE used[a]	Recycled HDPE/ tonne produced	Virgin HDPE/ tonne produced	HDPE savings/ tonne recycled	HDPE savings/ tonne recovered HDPE used[b]	Virgin PP/ tonne produced	Virgin PS/ tonne produced
Source	Henstock (1992)[c]	PWMI (1993)			Deurloo (1990)	PWMI (1993)			PWMI (1993)	PWMI (1993)
Energy consumption (GJ)	25.4	40.82[d]	15.42	14.60	7.62	33.25[d]	25.63	21.79	32.30[d]	51.66[d]
Air emissions (g)										
Particulates		3000			158	2000	1842	1566	2000	3100
CO		900			280	600	320	272	700	1400
CO_2	1 299 900	1 250 000	−49 900	−47 405	353 325	940 000	586 675	498 674	1 100 000	1 600 000
CH_4										
NO_x	6390	12 000	5610	·5330	989	10 000	9011	7660	10 000	24 000
N_2O					56					
SO_x	13 870	9000	−4870	−4627	2002	6000	3998	3398	11 000	34 000
HCl		70				50			40	40
HF		5			0.01	1	0.99	0.8	1	1
H_2S									10	2
Total HC	21 000	21 000	21 000	19 950	1690	21 000	19 310	16 414	13 000	26 000
Chlorinated HC										
Dioxins/furans										
Ammonia										
Arsenic										
Cadmium										
Chromium										
Copper										
Lead										
Mercury										
Nickel										
Zinc										

Water emissions (g)										
BOD	200				2365	100	−2265	−1925	60	80
COD	1500				4620	200	−4420	−3757	400	1800
Suspended solids	500					200			200	1000
Total organic compounds	320					200			620	750
AOX					24.2					
Chlorinated HCs										
Dioxins/furans (TEQ)										
Phenol					0.55					
Ammonium	5					10			10	10
Total metals	250					300			300	1100
Arsenic					0.1					
Cadmium					0.055					
Chromium					0.33					
Copper					2.31					
Iron										
Lead					0.11					
Mercury					0.006					
Nickel					0.22					
Zinc										
Chloride	130				97.9	800	702	597	800	500
Fluoride						10			20	
Nitrate	5									
Sulphide					0.55					
Solid waste (kg)	**132.0[e]**	**39.4**	**−92.6**	**−88.0**	**216.0**	**32.0**	**−184.0**	**−156.4**	**31.0**	**67.0**

[a] Assuming a material loss of 5%.
[b] Assuming a material loss of 15%.
[c] Using best case data.
[d] Feedstock energy not included.
[e] Calculated from power consumption and material losses.

from recycling of iron and steel, per tonne of recovered material used, would be 5.0 GJ and 12.9 GJ, respectively.

8.3.5 Aluminium

There are clear energy advantages in the use of recovered material to produce aluminium, due to the large energy requirement of production from virgin materials (bauxite). Data from Switzerland for the energy consumption and emissions associated with the production, per tonne, of both virgin and recycled aluminium are given in Table 8.4. Since the production of virgin aluminium relies on electrolysis and consumes large amounts of electrical energy, the total primary energy consumption and emissions are heavily dependent on the method used for electricity genera-tion. The data here use a 'Western World' mode for electricity generation (Habersatter, 1991). The primary energy consumption and emissions would be less if hydroelectric power was used for the smelting process.

The data presented here show that energy savings associated with alumi-nium recycling can be in the order of 155 GJ/tonne of recycled aluminium produced. There are also large savings in most of the associated emissions to both air and water, and in the overall amount of solid waste produced.

If one assumes a further material loss of 5% during the recycling process, due to the removal of any contaminants in the recovered material feed stock, and that this material becomes an additional residue for landfilling, the savings in primary energy and emissions can be calculated per tonne of recovered aluminium sent for reprocessing (Table 8.4). From this it can be seen that there is an energy saving in the region of around 150 GJ/tonne of recovered material reprocessed, along with considerable savings in the generation of emissions and solid waste.

8.3.6 Plastics

There have been numerous reports calculating the energy consumption and emissions associated with the production of specific virgin plastic resins (e.g. Lundholm and Sundström, 1986; Kindler and Mosthaf, 1989; Habersatter, 1991; PWMI, 1993 (see review in Ogilvie, 1992)). Comparable detailed data are not available, however, for the process of plastics recycling, most likely due to its relatively recent introduction and rapid rate of development. Clark and New (1991) suggest that the energy savings from plastics recycling vary from 27 to 215 GJ/tonne, depending on the resin type, but there is no detail on how these figures are obtained. A more detailed study by Henstock (1992) reports the energy consumption and some emissions associated with the reprocessing of low density polyethylene film, collected from supermarkets, into recycled LDPE granules and then into recycled polyethylene bags. The primary energy consumption of the reprocessing from recovered LDPE film into recycled LDPE granules (excluding sorting at the stores and transport) is given as between 25.4 and 33.2 GJ/tonne of recycled LDPE produced. The air emissions that result from the electrical power and propane consumption during the process are given in Table 8.5, but no details of emissions to water are provided.

For reprocessing of rigid plastic bottles (HDPE), Deurloo (1990) gives a figure of 2.88 GJ of electricity per tonne of recycled HDPE produced (equivalent to 7.6 GJ thermal energy/tonne, using the UCPTE 88 generation efficiency of 37.8%; see Chapter 4). Again, the only air emissions included are those for the electricity generation, but data for water emissions were given (Table 8.5).

For the production of virgin LDPE and HDPE, the most recent data (averaged across Europe) give total primary energy consumptions of 88.55 and 80.98 GJ/tonne, respectively (PWMI, 1993). These include the inherent energy of the feed stock material used (47.73 GJ/tonne), and it can be argued whether this should be included when comparisons are made. As discussed above, using the recovered material does result in the saving of the raw material which contains this amount of energy, but using the convention stated above, the inherent energy of the plastic feed stock will not be included in the energy consumption of using virgin plastic. Taking the data given above for recycling, this results in a potential (processing) energy saving due to LDPE reprocessing of 7.6–15.4 GJ/tonne produced, and a saving of 25.7 GJ/tonne of recycled HDPE produced.

Table 8.5 also gives the energy consumption and emissions reported for the production of other virgin plastic resin types. Although no recycling data are available for these, if the processing consists of flaking, washing, drying and granulating, then they may well be similar to those reported for LDPE and HDPE. More data are clearly needed in this area, however.

The material loss during the recycling process is given as 5% for LDPE film (Henstock, 1993), and 15% for HDPE (Deurloo, 1990), although this will depend on how well the material has been sorted and is likely to vary between different resin types. Using these values, however, the potential energy saving associated with recycling will be in the order of 7.2–14.6 GJ/tonne of recovered LDPE, and 21.8 GJ/tonne of recovered HDPE, reprocessed.

One of the assumptions made when calculating possible savings due to materials recycling is that the recycled material performs in exactly the same way as the virgin material. This is not always the case, especially with regard to some plastics. For example, bags made from the recycled LDPE described above had to be 30 μm thick, compared to 20 μm for a bag made of virgin HDPE (Henstock, 1992). Similarly, use of recycled material is reported to result in higher wastage rates than virgin material (an increase of 3.5%) during bag production. Thus, while the figures calculated and used here will give a broad estimate of potential savings, their accuracy must be viewed with some caution.

On the other hand, it has been possible to use recycled HDPE material

LCI Data Box 4 Energy consumption and emissions savings due to recycling used in the LCI model. Data are per tonne of recovered material sent for reprocessing.

	Paper	Glass	Metal-ferrous	Metal-Aluminium	Plastic-film (LDPE)	Plastic-rigid (HDPE)	Textiles
Energy consumption (GJ)	6.8	3.7	12.4	147.8	14.60	21.79	52.0
Air emissions (g)							
Particulates	3270	16 831	24 004	34 358		1566	
CO	2280	47	−486	16 377		272	
CO_2					−47 405	498 674	
CH_4							
NO_x	2310	663	352	23 925	5330	7660	
N_2O	53	91	176	1350			
SO_x	3947	946	2855	65 268	−4627	3398	
HCl	3.3	67		−675			
HF	0.005	−22	0.46	241		0.8	
H_2S	12						
HC	1692	1151	10 364	33 361	19 950	16 414	
Chlorinated HC							
Dioxins/furans (TEQ)							
Ammonia	0.4	1.9	66.1				
Arsenic				16			
Cadmium							
Chromium							
Copper							
Lead		−15.5					
Mercury	0.003						
Nickel							
Zinc							

Water emissions (g)						
BOD	2390	0	4.3	758	−1925	
COD	20 840	1.9	−0.09	18 066	−3757	
Suspended Solids	0	0	16	4.8		
Total organic compounds	4	5.8	−0.9	138		
AOX	2.5					
Chlorinated HCs						
Dioxins/furans (TEQ)						
Phenol	0	0	0.28	0		
Ammonium	0.447	0	5.3	0.95		
Total metals						
Arsenic						
Cadmium						
Chromium						
Copper						
Iron			0			
Lead				0.95		
Mercury	0					
Nickel						
Zinc						
Chloride	11		0	0	597	
Fluoride	0.97		10.9	1.9		
Nitrate			−0.09	0		
Sulphide	6		0.18			
Solid waste (kg)	65.3	−24.5	255.0	607.0	−88.0	−156.4

successfully in laundry detergent bottles. By co-extruding a layer of recycled plastic between two outer layers of virgin material, over 25% of recycled material can be used, without the need to change the weight, performance or aesthetics of the bottle. In these cases, therefore, a straight comparison of the energy consumption and emissions of recycled versus virgin material production is valid.

8.3.7 Textiles

Information on the energy consumption and emissions of textile recycling processes is very limited. One study on the woollen industry reported that the energy consumption of producing woven cloth of virgin wool was 115.61 GJ/tonne, compared to 56.61 GJ/tonne for cloth with 100% recycled content, giving a saving of 59 GJ/tonne produced (Lowe, 1981). For knitted products, virgin wool use consumed 108.28 GJ/tonne compared with 56.61 GJ/tonne for recycled material (a saving of around 52 GJ/tonne). Note that these figures are per tonne produced, not per tonne reprocessed; no data are available on material losses during processing. The figures for virgin wool do not include the initial scouring process, so the actual savings associated with the use of recycled content are likely to be larger (Ogilvie, 1992).

8.4 Economic costs

Just as there can be savings or additional costs in energy consumption and emissions, there will be economic savings or costs associated with the production of recovered materials.

The additional costs will be the transport cost to the reprocessors, and the cost of the reprocessing operation. Since the price paid for the recovered material by the reprocessors has already been included in the income of the basic model, it must be included within the processing costs.

Additional income to the system when reprocessing is included comes from the sale of recycled material, values for which can be found in current commodity market prices.

Thus, if the transport and processing costs (including recovered material prices) exceed the value of the recycled material, there will be an extra cost incurred by reprocessing the recovered material. Alternatively, overall cost savings will occur if the value of the recycled material produced exceeds the transport and reprocessing costs.

In the final analysis, recycled material will only sell if it is priced competitively compared to virgin. How close the recycled material price comes to the virgin price will depend if it has equal performance for the intended use (e.g. in the case of glass) or whether there is a fall off in performance that requires a compensating discount. In either event, the price for recycled material is pegged to virgin material prices and is thus relatively fixed. For the recycling industry to expand, therefore, it is necessary for the transport and reprocessing costs to be competitively less than the virgin material price. For materials where there are large energy savings resulting from reprocessing compared to virgin production (e.g. aluminium, steel), this will hold. If there are high reprocessing costs and/or small associated energy savings, it may not be possible to produce the recycled material for a competitive price, as is often stated for plastics. In such cases, other options are still possible. Rattray (1993) conducted an exercise with different sectors in the plastics industry in the United States to determine where cost savings could be found in the recycling of HDPE, so that the cost of recycled resin could be reduced below that of virgin resin. By optimising the system rather than individual operations within it, the exercise generated ideas to reduce costs by 20 cents per pound. Likely savings would be more in the region of 6–8 cents per pound, but this would be enough to allow recycled material to compete well with virgin material for many end markets.

The alternative way to encourage materials recycling would be to lower the price the reprocessors pay for the recovered materials, either by improving the efficiency of the waste management system or by increasing the charges levied against users of the system.

8.5 Operation of the materials recycling module of the LCI spreadsheet

The materials recycling option of the LCI spreadsheet is shown in LCI Box 4.

MATERIALS RECYCLING (optional)

Transport from sorting facility/collection bank to reprocessing plant
User inserts the average distance (one way) in km to the reprocessing plant used for each material. The model calculates the energy consumption and emissions according to the data in Table 6.11. The model then calculates the savings in energy consumption associated with the recycling of the

LCI BOX 4

4. MATERIALS RECYCLING (optional)

This box calculates the potential savings (or additional costs) associated with the reprocessing of recovered materials.

(Results are presented with final output below.)

Transport from sorting facility/collection bank to reprocessing plant:

		Paper	Glass	Metal-Fe	Metal non-Fe	Plastic-film	Plastic-rigid	Textiles
	Distance (one way) km							

Costs:

		Paper	Glass	Metal-Fe	Metal non-Fe	Plastic-film	Plastic-rigid	Textiles
	Transport costs (ecu/ tonne)							
	Process costs (ecu /t. output)							
	Recycled Material price (ecu/t)							

Go to Box 5

amount of recovered material predicted in the rest of the model, using the data in LCI Data Box 4. The energy consumption and emissions from the transport to the reprocessing plant are subtracted to give the actual savings likely from the recycling process, and these are displayed in the final output box next to (and then included in) the overall energy consumption and emissions of the waste management system.

Costs

Transport costs (per tonne input):

Process costs (per tonne output) (including price paid for recovered material):

Recycled material price (per tonne):

User inserts these economic data and the model calculates the additional cost or saving attributable to the recycling of the recovered materials.

References

AMG (1992) Metal waste reclamation. *AMG Resources Brochure.* AMG Ltd., Harborne, Birmingham, UK.

Boustead, I. (1993a) Resource Use and Liquid Food Packaging. EC Directive 85/339: UK Data 1986-1990. A report for INCPEN, May 1993.

Boustead, I. (1993b) Aerosols and other containers. A Report for the British Aerosol Manufacturers' Association.

Cathie, K. and Guest, D. (1991) *Wastepaper.* PIRA International, UK, 134 pp.

Clark, H. and New, R. (1991) Current UK initiatives in plastics recycling. Paper presented to Plastics Recycling '91 European Conference, Copenhagen.

Claydon, P. (1991) Recycled fibre – the major raw material in quality newsprint. *Paper Technol.* **32**(5), 34–37.

Deurloo, T. (1990) Assessment of environmental impact of plastic recycling in P & G packaging. Procter & Gamble European Technical Center, internal report.

Habersatter, K. (1991) Oekobilanz von Packstoffen, Stand 1990. Bundesamt für Umwelt, Wald und Landschaft (BUWAL) Report No. 132, Bern, Switzerland.

Henstock, M. (1992) An analysis of the recycling of LDPE at Alida Recycling Limited. Report by Nottingham University Consultants, Ltd.

Kindler, H. and Mosthaf, H. (1989) Ökobilanz von Kunststoffverpackungen. BASF-Ludwigshafen, 1989.

Lowe, J. (1981) Energy usage and potential savings in the woollen industry. Wool Industry Research Association, Wira House, West Park Ring Road, Leeds.

Lundholm, M.P. and Sundström, G. (1986) Resource and environmental impact of two packaging systems for milk: tetrabrik cartons and refillable glass bottles. Mälmo.

Ogilvie, S.M. (1992) A review of the environmental impact of recycling. Warren Spring Laboratory report LR 911 (MR).

Porteous, A. (1992) LCA study of municipal solid waste components. Report prepared for Energy Technology Support Unit (ETSU), Harwell, Oxon UK.

Porter, R. and Roberts, T. (1985) *Energy Savings by Wastes Recycling.* Elsevier Applied Science, London, ISBN 0 85334 353 5.

Pulp and Paper (1976) Secondary vs virgin fibre newsprint. *Pulp and Paper* **50**(5).

PWMI (1993) Eco-profiles of the European plastics industry. Report 3: polyethylene and polypropylene. Report by Dr. I. Boustead for The European Centre for plastics in the Environment (PWMI), Brussels.

Rattray, T. (1993) Fixing plastic recycling. *Resource Recycling* May (suppl.) 65–71.

Russell, D., O'Neill, J. and Boustead, I. (1994) *Is HDPE Recycling the Best Deal for the Environment?* Dow Europe S.A., Bachtobelstrasse 3, 8810 Horgen, Switzerland.

Shotton Paper Company (1992) Cited in Porteous (1992).

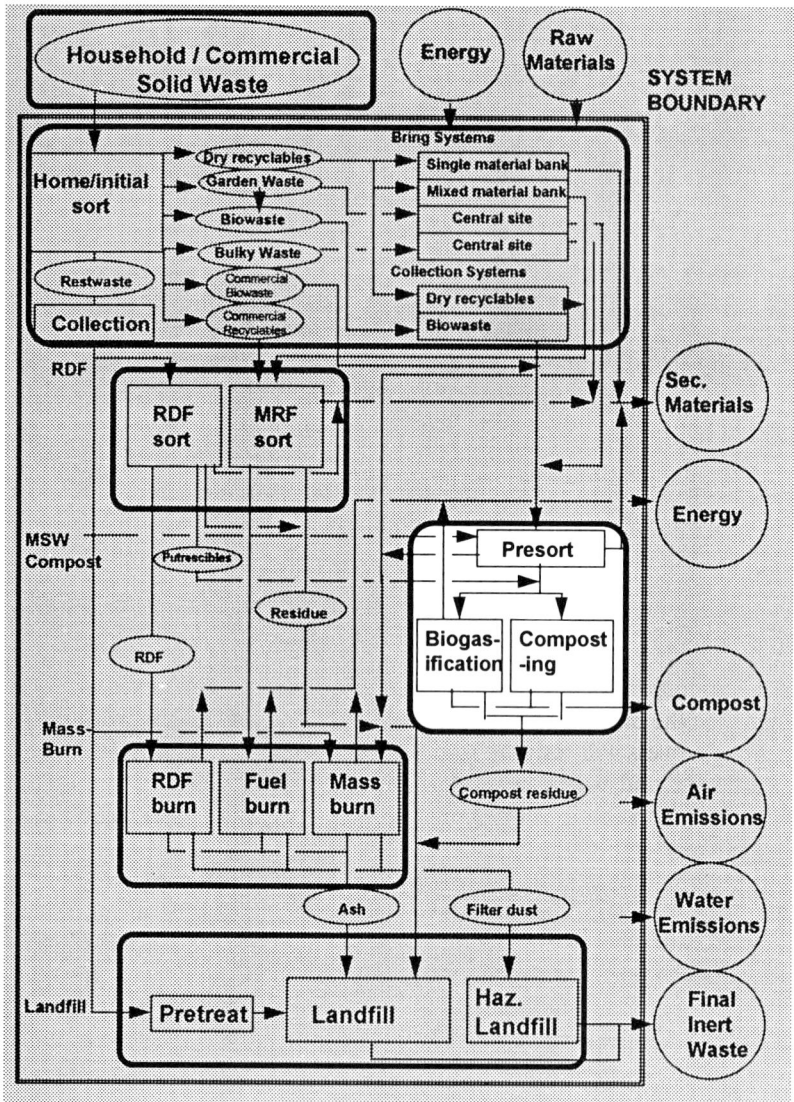

Figure 9.1 The role of biological treatment in an integrated waste management system.

9 Biological treatment

Summary

Biological treatment can be used to treat both the organic and paper fractions of solid waste. Two main treatment types exist: composting (aerobic) and biogasification (anaerobic). Either can be used as a pre-treatment to reduce the volume and stabilise material for disposal in landfills or as a way to produce valuable products, such as compost and (in biogasification) biogas plus compost, from the waste stream. The inputs and outputs of each process are discussed, using available data. Further development of biological valorisation depends on the further development of markets, and agreed standards, for the compost products.

9.1 Introduction

Biological treatment involves using naturally occurring micro-organisms to decompose the biodegradable components of waste. If left to go to completion, this process will result in the production of gases (mainly carbon dioxide, methane and water vapour) plus a mineralised residue. Normally the process is interrupted when the residue still contains organic material, although in a more stable form, comprising a compost-like material.

The garden compost heap is the simplest form of biological treatment. With some care and regular turning, this can transform vegetable scraps and garden refuse into a rich and useful garden compost. Garden compost heaps are a valuable method for valorising part of the household waste at source, but are limited to more rural and suburban areas where space and gardens are plentiful. The alternative method to treat organic waste not composted at source (in particular from urban areas), involves centralised biological treatment plants.

Almost any organic material can be treated by this method. It is particularly suitable for many industrial wastes from such sources as breweries, fruit and vegetable producers and processors, slaughter-houses and meat processors, dairy producers and processors, paper mills, sugar mills, and leather, wool and textile producers (Bundesamtes für Energiewirtschaft, 1991). At the local community level, it is widely used to treat sewage sludges and organic wastes from parks and gardens.

Household waste is also rich in organic material, consisting of kitchen

Table 9.1 The range of possible inputs to biological treatment plants

Category	Description
Mixed Wastes	
MSW	Municipal solid waste, commingled solid waste collected from households, commerce and institutions
HW	Household waste, commingled waste collected from households only
Centrally sorted waste	
RDF sort fines	Putrescible material sorted mechanically from mixed waste during the production of refuse derived fuel (RDF)
Separately collected waste	
Wet waste	Household waste from which dry recyclables have been removed
Biowaste	This term is widely misused, ranging in meaning from garden waste only, to VFG material or to VFG plus paper; here it refers to separately collected organic and non-recyclable paper waste only
VFG	Separately collected vegetable, fruit and garden waste only
Greenwaste (GW)	Separately collected garden waste only

and garden waste. According to geography, this accounts for between 25% and 60% of municipal solid waste by weight (see Chapter 5), with levels of organics particularly high in southern Europe. If one adds to this the paper fraction, which is also of organic origin and suitable for biological treatment, some 50–85% of the MSW can be treated by such methods. The suitability of biological treatment for wet organic material contrasts markedly with other treatment methods, such as incineration and landfilling, where the high water content and putrescible nature can be a source of major problems, by reducing overall calorific value and increasing the production of leachate and landfill gas. This potential of biological treatment is being exploited in some countries, but almost ignored in others.

Numerous variants of biological treatment exist, differing according to the feedstock used (Table 9.1) and the process employed. Feedstocks range from highly mixed wastes, e.g, MSW, which require extensive treatment to remove the non-organic fractions prior to, or occasionally after, biological processing, to the separately collected and more narrowly defined biowaste, VFG (vegetable, fruit and garden) and green wastes. Although there are many different types of plant available, there are two basic process types, aerobic and anaerobic. In aerobic treatment, usually known as composting, organic material decomposes in the presence of oxygen to produce mainly carbon dioxide, water and compost. Considerable energy is released in the process, which is generally lost to the surroundings. Anaerobic processes are variously described as anaerobic fermentation, anaerobic digestion or biogasification. Throughout this chapter, the term biogasification will be used. As the name implies this produces biogas, a useful product consisting mainly of methane and carbon dioxide, plus an organic residue which can be stabilised to produce compost, but differs somewhat from aerobically produced composts.

This chapter looks at the two basic processes, and their current use across Europe. It then considers the process inputs and outputs needed to contribute to the overall lifecycle inventory of solid waste management (Figure 9.1), and the economic costs involved.

9.2 Biological treatment objectives

Both composting and biogasification can fulfil several functions, and it is necessary to identify the key objective(s) required of the process. They can either be considered as pre-treatments for ultimate disposal of a stabilised material (normally in a landfill) or as a valorisation method.

9.2.1 Pre-treatment for disposal

Volume reduction. Breakdown into methane and/or carbon dioxide and water results in the decomposition of up to 75% (Bundesamtes für Energiewirtschaft 1991) of the organic material on a dry weight basis. From wet biowaste to normal compost the weight loss is generally around 50%. As organics and paper represent the two largest fractions of the household waste stream this is a significant reduction. Additionally there is considerable loss of water, either by evaporation (in composting) or by pressing of the residue (biogasification). The moisture content of the organic fraction of household waste is around 65% (Barton, 1986), whilst for compost made from biowaste, it is around 30–40% (Fricke and Vogtmann, 1992) and for material from biogasification 25–45% (De Baere, 1993; Six and De Baere, 1988). The breakdown and moisture loss together result in a marked volume reduction in material for further treatment and disposal. Removal of water will also reduce the formation of leachate if the residues are subsequently landfilled.

Stabilisation. As much of the decomposition has occurred during biological treatment, the resulting materials are more stable than the original organic inputs, and thus more suitable for final disposal in a landfill. The cumulative oxygen demand of the organic material, a measure of biological activity and thus inversely related to stability, can decrease by a factor of six during biological treatment (Table 9.2). Similarly, the carbon/nitrogen (C/N) ratio, which gives a measure of the maturity of a compost (high C/N ratio indicates fresh organic material, low C/N ratio indicates mature, stable material), falls markedly during biological treatment processes.

Table 9.2 Compost quality

	Fresh organic fraction	Windrow composting (after 6 weeks)	Biogasification (after 6 weeks)
C/N ratio	30	15	12
Cumulative oxygen demand (mg O_2/g organic matter over 10 days)	250–300	150–160	50–60
Pathogen destruction (colonies/g dry wt.)			
Faecal coliforms	3×10^3	2×10^2	0
Faecal streptococci	2×10^5	4×10^4	0

Source: OWS.

Sanitisation. Both composting and biogasification are effective in destroying the majority of pathogens present in the feedstock. Aerobic composting is a strongly exothermic process, and temperatures of 60–65°C are built up and maintained in composting piles or vessels over an extended period of time, sufficient to ensure the destruction of most pathogens and seeds (Table 9.2). Biogasification processes are only mildly exothermic, but may be run at temperatures of 55°C (thermophilic process) by the addition of heat. The combination of this temperature and anaerobic conditions is sufficient to destroy most pathogens (Table 9.2), though if lower process temperatures are used (mesophilic process), further heat treatment during the final aerobic stabilisation stage may be required to produce sanitary residues.

9.2.2   Valorisation

In contrast to the above, the main objective of most biological treatment is to produce useful products (biogas/energy, compost) from organic waste, i.e. to valorise part of the waste stream.

Biogas production. Biogasification produces a flammable gas with a calorific value of around 6–8 kW-h (21.6–28.8 MJ) per m³ (German Govt. report, 1993), which can be sold as gas, or burned on-site in gas engine generators to produce electricity. Some of the biogas will be burned to provide process heating on site, and some electricity will be consumed, but there can be a net export of either gas or electrical power from the plant. There will normally be a market for this product, at least for the electricity. As the gas can be stored between production and use for power generation, electricity export into the national grid can be timed to coincide with peak consumption times, and thus highest energy prices. This economic advantage is increased further where additional premium prices are paid for electricity generated from non-fossil fuel sources (e.g. under the UK Non-Fossil Fuel Obligation (NFFO) scheme).

Compost production. Both composting and biogasification produce a partly stabilised organic material that may be used as a compost, soil-improver, fertiliser, filler, filter material or for decontaminating polluted soils (Ernst, 1990). Alternatively the material can be considered as a residue and landfilled. The only point that determines whether the material is a useful product, and hence of value, or a residue to be disposed of at a cost, is the presence of a market.

Markets for compost will differ widely across Europe. In southern Europe, the lack of organic matter in the soil creates a large need for additional organic material. There is therefore a strong market for compost made from any feedstock, provided that the compost is safe for use, even though the level of contamination may be high. The same compost, if produced in Holland or Germany, however, would be considered only for landfill cover material or as a residue for disposal. As it would not meet the relevant quality guidelines, there would be no market for such a product, although markets do exist for higher quality composts. Since the main determinants of compost quality are the composition of the feedstock (Fricke and Vogtmann, 1992) and the process used, production of compost for sale in such countries may require the use of restricted feedstocks involving separate collection (biowaste, either with or without paper; VFG; green waste), or the use of more sophisticated processing techniques. There is a grey area, however, around the distinction between product and residue, as in many cases, waste-derived compost is freely distributed. Ernst (1990) reports that of the 225 290 tonnes/yr of compost produced in the former West Germany, 30% was given away, incurring no disposal costs, but equally supplying no revenue. In the Netherlands, on the other hand, agreements have been reached for agriculture to take 150 000 to 450 000 tonnes per year of compost.

The need to define the objective of any biological treatment process bears repeating. If the aim is to produce a quality compost for sale, then a restricted input (e.g. VFG, or biowaste) is preferable, and the necessary source separated collection schemes must be put in place. If, however, the objective is to maximise diversion from landfill, while still producing a quality compost, the biowaste definition can be widened to include paper as well as organic materials so long as there is a market for the resulting compost. If the objective is to pre-treat waste prior to final disposal, then treating a mixed waste stream will be effective.

In biogasification, where there are two possible products, biogas and compost, it is also necessary to decide which should be optimised. The German Federal Government has stated that whilst it considers the objective of composting is the production of good quality compost, the objective

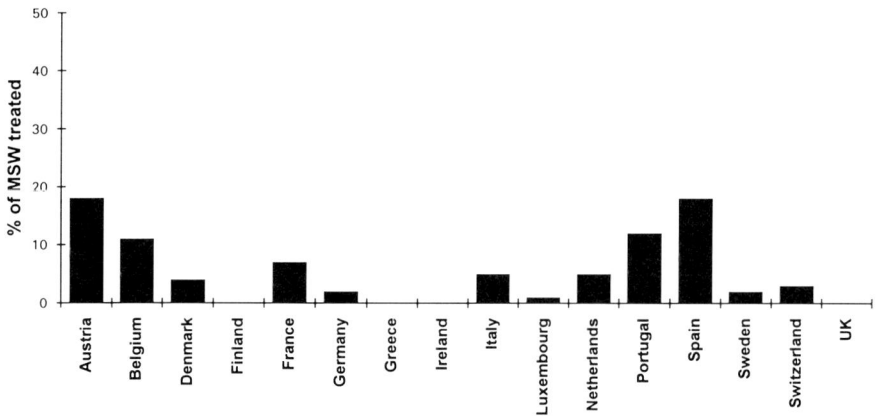

Figure 9.2 Biological treatment of MSW across Europe. *Source*: OECD (1991).

of biogasification is biogas, not compost production. In its opinion, the digested residue produced in Germany cannot be marketed as compost, as the quality is too low due to the presence of hazardous materials (German Government Report, 1993). In contrast, a biogas plant in Belgium, using a different process, produces a humus-like product for which there is a viable market (De Baere, 1993).

9.3   Overview of biological treatment in Europe

Use of biological treatment for dealing with MSW varies considerably across Europe (Figure 9.2). It is extensively used in Southern European countries, e.g. Spain, Portugal, France and Italy, which correlates with the generally high organic content of municipal wastes from these areas (see Chapter 5). The feedstock used for biological treatment also vary geographically: Southern Europe generally treats MSW, whilst countries such as Germany, Austria, Netherlands, Belgium and Luxembourg treat more narrowly defined feedstocks which are separately collected (Table 9.3).

By far the majority of the biological treatment in Europe involves aerobic processing, i.e. composting. Large numbers of often small plants exist in countries that exploit this treatment method. Switzerland alone has over 140 plants with a capacity of over 100 tonnes/year, of which 40 plants have capacities of over 1000 tonnes/year, and 15 plants over 4000 tonnes/ year (Schleiss, 1990). Anaerobic treatment is more limited, with less than 20 plants treating household derived waste across the whole of Europe (Table 9.4).

Table 9.3 Main feedstocks for biological treatment across Europe

Country	Main feedstocks
Austria	Biowaste, VFG
Belgium	MSW, greenwaste
Denmark	MSW, greenwaste
France	MSW, biowaste, greenwaste
Germany	Biowaste, greenwaste
Italy	MSW
Luxembourg	MSW, greenwaste
Netherlands	VFG
Spain	MSW
Sweden	MSW, biowaste
Switzerland	VFG, biowaste, MSW (plants > 1000 tonnes/year)
Turkey	MSW

Source: Thome-Kozmiensky (1991) with additions.

Table 9.4 Biogasification plants in Europe

Country	Location	Process	Feedstock	Capacity (tonnes/year)
Belgium	Ghent	DRANCO	(Pilot plant)	1000
Belgium	Brecht	DRANCO	VFG + paper	15 000
Denmark	Helsingor	BTA	Biowaste	20 000
Finland	Vaasa	DBA-VABIO	Biowaste plus sewage sludge	14 000
France	Amiens	VALORGA	MSW	100 000
France	La Buisse	VALORGA	Biowaste	16 000
Germany	Garching	BTA	Pilot plant	1000
Germany	Kaufbeuren	BTA	Biowaste	2500
Germany	Zobes	Zobes	Biowaste plus agric. waste	10 000
Germany	Bremen	AN-Maschinenfabrik	Pilot plant	
Germany	Landkreis Oldenburg	AN-Maschinenfabrik	Biowaste	3500
Italy	Bellaria	Solidigest	Biowaste plus sewage sludge	20 000
Italy	Verona	Snamprogetti	MSW	50 000
Italy	Avezzano	Snamprogetti	Sewage sludge	10 000
Italy	Bellaria	ITALBA	MSW	30 000
Netherlands	Breda	Prethane	Veg. market waste	20 000
Switzerland	Rümlang	Bühler AG KOMPOGAS	Biowaste	3500

Box 9.1   CLASSIFICATION OF BIOLOGICAL TREATMENT

	BIOLOGICAL TREATMENT	

COMPOSTING		BIOGASIFICATION

Windrows	Enclosed vessel	Wet process	Dry process

Static	Dynamic	1-stage	2-stage

Examples	HERHOF (D)	ALTVATER (D)	ITALBA (I) WMC (GB) ROTTWEILER- MODELL (D)	BTA (D) AN- PROCESS(D)	DRANCO (B) VALORGA (F) KOMPOGAS (CH) T.U.H.H. (D)

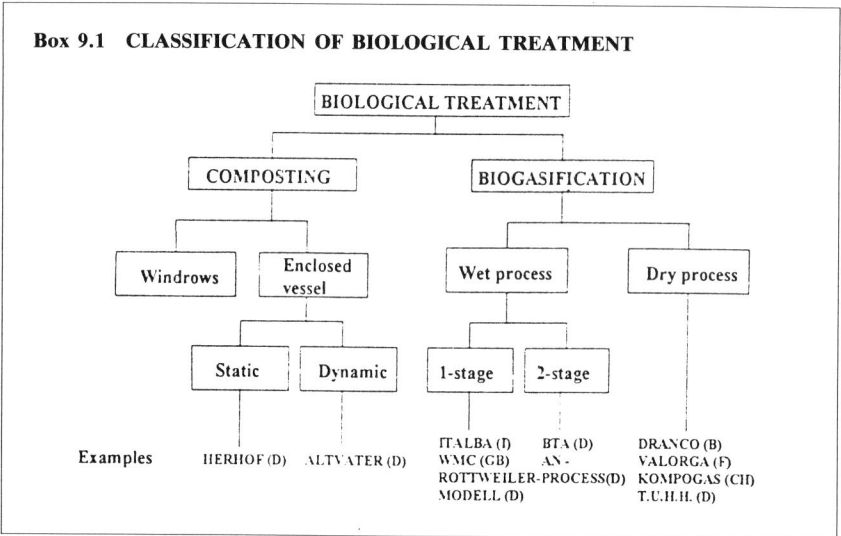

9.4   Biological treatment processes

A classification of the types of biological treatment processes is given in Box 9.1. Each consists of a pre-treatment stage followed by a biological decomposition process.

9.4.1   Pre-treatment

Pre-treatment has two basic functions, the separation of the organic material from other fractions in the feedstock, and the preparation of this organic material for the subsequent biological processing. Clearly, the amount of pre-treatment will depend on the nature of feedstock; the more narrowly defined the incoming material, the less separation will be required, although pre-treatment for size reduction, homogenisation and moisture control will still be needed.

Where the plant input is mixed waste, e.g. MSW, the non-organic material (plastic, glass, metal, etc.) needs to be removed at this stage (unless the overall objective is volume reduction alone). In some plants, part of this material can be recovered for use as secondary materials. In the Duisburg-Huckingen composting plant in Germany, for example, incoming mixed waste is passed under a magnet to remove ferrous metals, and then along conveyor belts where glass bottles, non-ferrous metals and plastic items are hand picked and recovered (Ernst, 1990) (Figure 9.3).

Preparation for the actual composting or biogasification usually involves some form of screening to remove oversize items, size reduction and

```
            ┌─────────┐
           (   MSW    )
            └─────────┘
                │
                ▼
        ┌───────────────┐
        │ Storage Bunker│
        └───────────────┘
                │
                ▼
        ┌─────────────────────┐      ┌──────────────┐
        │ Magnetic Separation │─────►( Ferrous metal)
        └─────────────────────┘      └──────────────┘
                │
                ▼
        ┌─────────────────┐          ┌──────────────┐
        │ Sorting Conveyor│─────────►( Recyclables  )
        └─────────────────┘          └──────────────┘
                │                 └──►( Residue      )
                ▼                     └──────────────┘
        ┌─────────────────┐
        │    Rotating     │
        │   Composting    │
        │      Drum       │
        └─────────────────┘
                │
                ▼
        ┌─────────────────┐          ┌──────────────┐
        │  Coarse Screen  │─────────►(   Residue    )
        └─────────────────┘          └──────────────┘
                │
                ▼
        ┌─────────────────┐
        │   Fine Screen   │
        └─────────────────┘
            │         │
            ▼         ▼
         ( Raw )   ( Raw )
         (compost) (compost)
         ( (fine) ) ((coarse))
            │         │
            ▼         ▼
        ┌─────────────────┐
        │ Sec. Fermentation│
        └─────────────────┘
                │
                ▼
        ┌─────────────────┐
        │ Compost Storage │
        └─────────────────┘
                │
                ▼
```

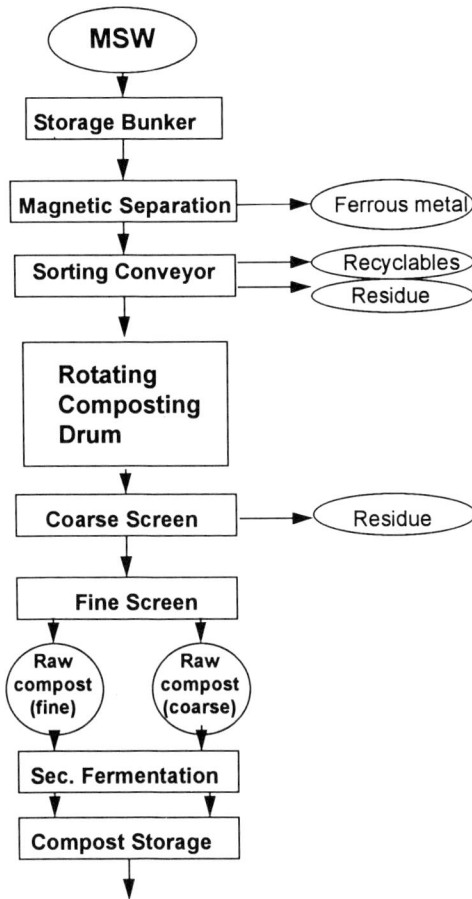

Figure 9.3 Simplified flow chart for composting process for commingled municipal solid waste (Duisburg-Huckingen Plant). *Source*: Ernst (1990).

homogenisation. For biogasification, it is necessary to produce a pumpable feedstock. Size reduction and mixing are achieved either by shredding the feedstock in a mill, or by the use of a large rotating drum. Shredding the feedstock removes the need for a screening stage prior to processing, but means that nuisance materials are also shredded. This makes them much more difficult to separate from the compost in the later refining stage. A drum achieves some degree of size reduction and homogenisation as it rotates, but does not shred nuisance materials. These can then be removed intact by a screen (which can be incorporated into the rotating drum) prior to the biological process, so that the later refining stages can be simplified.

Nuisance materials can therefore be removed either before or after the biological treatment stage. There is an advantage in removing them as early

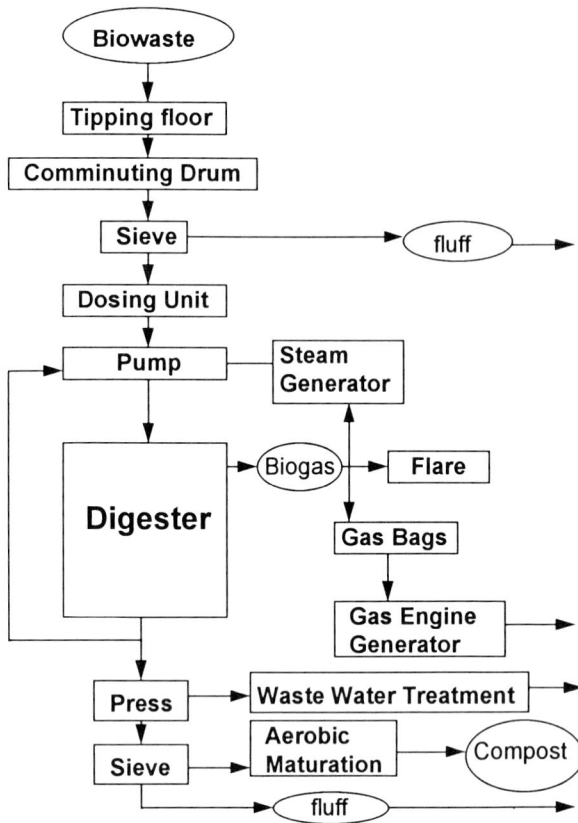

Figure 9.4 Flow chart for dry biogasification process at plant in Brecht, Belgium.
Source: De Baere (1993).

as possible, since the longer they are in contact with the organic material, the greater the likelihood of cross-contamination. Additionally, it reduces the amount of material entering the biological treatment process, by removing material that would not be degraded.

The amount of nuisance material removed in the pre-treatment stage depends on the waste used as feedstock. Even where the feedstock ariscs from source-separated collection of biowaste, separation of nuisance material during pre-treatment is advisable, especially if the feedstock comes from urban areas (see Chapter 6). A biogas plant in Brecht, Belgium, for example, using a feedstock of separately collected VFG plus paper (Table 9.1) discards up to 19% of its input during the pre-treatment stage (De Baere, 1993) (Figure 9.4).

9.4.2 Aerobic processing: composting

Biological treatment can be described as the biological decomposition of organic wastes under controlled conditions; for composting, these conditions need to be aerobic and elevated temperatures are achieved due to the exothermic processes catalysed by microbial enzymes. Three main groups of micro-organism are involved in the composting process: bacteria, actinomycetes and fungi (Warmer, 1991). Initially bacteria and fungi predominate, and their activity causes the temperature to rise to around 70°C in the centre of a pile. At this temperature, only thermophilic (heat-tolerant) bacteria and actinomycetes are active. As the rate of decomposition and hence temperature subsequently falls, fungi and other heat-sensitive bacteria become active again (Lopez-Real, 1990). Temperature, therefore, is one of the key factors in composting plants that needs to be constantly monitored and controlled.

To maintain the high rate of decomposition, oxygen must be constantly available. In the simplest process type, as with a garden compost heap, this is achieved by regular turning of the composting material in long piles or windrows. The alternative method is forced aeration, whereby air is forced through a static pile using small vents in the floor of the composting area. Air can either be forced out of the vents, or drawn down through the composting pile by applying a vacuum to the vents. The former method aids dispersion of the heat from the centre of the pile to the outside, making for a more uniform process. The latter helps in controlling odours as the air passing through the pile is effectively filtered to reduce many odours before release. Aeration also helps remove carbon dioxide and volatile organic compounds, such as fatty acids, and buffers the pH.

Percolation of composting piles by air depends on the structure and water content of the input material. The water content for aerobic composting needs to be over 40%, otherwise the rate of decomposition will start to fall, but if it is too high the material will become waterlogged and limit air movement. If the input material is too wet, water-absorbing and bulking agents such as woody garden waste, wood chips, straw or sawdust can be added to improve the structure, and increase the air circulation.

Windrow composting is the most common technology used, being least capital intensive. However, when it is open to the elements, control over moisture content, temperature and odour emissions is limited. One way round this is to have the entire area enclosed, and the exhaust air filtered. In the Netherlands, open air composting is generally not allowed for plants with a capacity exceeding 2000 tonnes/year, but due to the expense, Fricke and Vogtmann (1992) recommend this only for plants in excess of 12 000 tonnes/year. Further control over both composting conditions and emissions are possible in more advanced technologies, using a variety of enclosed vessels (boxes and drums) for totally enclosed processing (Table

Table 9.5 Types of composting systems in Germany

Windrow composting of which	54
Open, no roof	33
Open, partly or fully covered with a roof	21
Fully enclosed	0
Composting in piles	5
Box composting	6
Container composting	1
Rotating drum systems	6
'Brikollare'-system	1

Source: Fricke and Vogtmann (1992).
Data as at May 1991.

9.5). Open or semi-enclosed windrows are often still used in these systems, for the maturation stage.

The duration of the composting process varies with the technology employed, and the maturity of the compost required. Compost maturity can be assessed by the carbon/nitrogen (C/N) ratio of the material, which falls from around 20 in raw organic waste to around 12 in a mature compost after some 12–14 weeks. Application to soils of immature composts with high residual microbial activity and high C/N ratios can result in uptake of the nitrogen from the soil by the compost, which will reduce, rather than enhance the soil fertility.

Before marketing of the compost, further maturation and refining are needed. Additional maturation may be necessary to break down complex organic materials toxic to some plants, which may still be present in the compost. Refining involves size classification of the compost particles and the removal of nuisance materials by sieving, ballistic separation or air classification, ready for the chosen end use. Nuisance materials may include oversized material, stones, metal fragments, glass, plastic film and hard plastic. Oversize organic fragments can be recycled into the composting process, but the rest of the residue will need to be incinerated or landfilled.

An alternative way to both mature and refine compost has been developed using earthworms. The beneficial effects of earthworms in garden compost heaps has long been recognised (Barret, 1949), but recent research in France has developed 'lumbricomposting' or 'vermicomposting' into an industrial process to deal with household waste. The plant takes in mixed waste and the pre-treatment stage involves removal of glass, plastic and metal for materials recovery. The remaining waste is graded, but not shredded, and then composted aerobically to start off the decomposition process and kill off pathogens. At this stage, the immature compost material is added to vertical cages known as 'lumbricators' containing cultures of earthworms. Various species have been tested for their suitability, but due to its prolific reproduction, *Eisenia andre* (Bouche) is the species usually cultured. The earthworms eat and partially digest the organic material, leaving any contaminants intact as residue. Part of the eaten organic material is metabolised by the worms and converted into worm biomass, the rest is eliminated as worm faeces, which are collected at the base of the lumbricators. The worms are particularly efficient at eating all of the biodegradable material, but will not ingest any of the inert and contaminating material. The resulting compost has a very fine and consistent particle size, and so can be separated from the residual inert material with relative ease, using a 5 mm sieve. This is facilitated by the fact that the feedstock is not shredded in the pre-treatment stage.

9.4.3 Anaerobic processing: biogasification

Conditions for biogasification need to be anaerobic, so a totally enclosed process vessel is required. Although this necessitates a higher level of technology than some forms of composting, containment allows greater control over the process itself and also of emissions such as noxious odours. Greater process control, especially of temperature, allows a reduction in treatment time when compared to composting. Since a biogas plant is usually vertical, it requires less land area than a composting plant.

Biogasification is particularly suitable for wet substrates, such as sludges or food wastes, which present difficulties in composting as their lack of structural material will restrict air circulation. The anaerobic process has been used for some time to digest sewage sludges (Noone, 1990) and organic industrial wastes (Bundesamtes für Energiewirtschaft, 1991), and this has been extended more recently to fractions of household solid waste (Coombs, 1990).

The various biogasification processes can be classified according to the solids content of the material digested, and the temperature at which the process operates. Dry anaerobic digestion may be defined as taking place at a total solids concentration of over 25% (De Baere et al., 1987); below this level of solids, the process is described as wet digestion. With regard to temperature, processes are either described as mesophilic (operating between 30 and 40°C) or thermophilic (operating between 50 and 65°C). It has been well established that different anaerobic micro-organisms have optimum growth rates within these temperature ranges (Archer and Kirsop, 1990). In contrast to aerobic processing (composting), the biogasification process is only mildly exothermic. Thus, heat needs to be supplied to maintain the process temperature, especially for thermophilic processes. The advantage of the higher temperature, on the other hand, is that the reactions will occur at a faster rate, so shorter residence times are needed in the reactor vessel.

'Wet' anaerobic digestion. In its simplest form, this process consists of a single stage in a completely mixed mesophilic digester, operating at a total

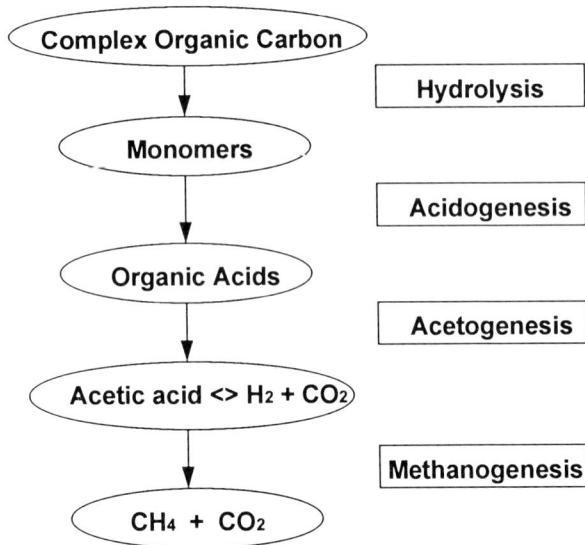

Figure 9.5 Metabolic stages in the biogasification of organic wastes.

solids content of around 3–8% (De Baere *et al.*, 1987). To produce this level of dilution, considerable water has to be added (and heated), and then removed after the digestion process. This method is routinely used to digest sewage sludge and animal wastes, but has also been used to treat household waste in Italy and Germany. During a retention time of 12–30 days, the organic materials are broken down in a series of steps that first hydrolyse them into more soluble material, then break these down into short chained organic acids before converting them to methane and carbon dioxide (Figure 9.5).

The single stage wet process can suffer from several practical problems, however, such as the formation of a hard scum layer in the digester, and difficulty in keeping the contents completely mixed. A basic deficiency is that the different reactions in the process cannot be separately optimised. The acidogenic micro-organisms will act to lower the pH of the reaction mixture, whereas the methanogens, which reproduce more slowly, have a pH optimum around 7.0. This problem has been solved by the development of the two-stage process. Hydrolysis and acidification are stimulated in the first reactor vessel, kept at a pH of around 6.0. Methanogenesis occurs in the second separate vessel, operated at a pH of 7.5–8.2. Variations of the two-stage wet (mesophilic) digestion process have been developed and implemented in Germany. The whole process can be run with a retention time of 5–8 days (De Baere *et al.* 1987).

'Dry' anaerobic digestion. Several processes have been developed that digest semi-solid organic wastes (over 25% total solids) to produce biogas in a single stage. The processes can be either mesophilic, or thermophilic, and can use organic material from mixed wastes such as MSW, or separated biowaste. The dry fermentation process means that little process water has to be added (or heated). No mixing equipment is necessary and crust formation is not possible due to the relatively solid nature of the digester contents. This anaerobic process usually takes from 12–18 days, followed by several days in the post-digestion stage for residue stabilisation and maturation (De Baere *et al.*, 1987).

Maturation and refining. The residues of both wet and dry biogasification processes require further treatment before they can be used as compost. They contain high levels of water; even the dry process residue contains around 65% water, compared to German maximum recommended water levels for compost of 35% and 45% for bagged and loose compost respectively (ORCA, 1992b). Excess water can be removed by filtering or pressing, to produce a cake-like residue; further drying can be achieved using waste heat from the gas engines if the biogas is burnt on site to produce electricity (De Baere *et al.*, 1987). Some of the waste water can be recirculated and used to adjust the water content of the digester input, the rest represents an aqueous effluent requiring treatment prior to discharge.

The digested residue, initially anaerobic, will also still contain many volatile organic acids and reduced organic material. This needs to be matured aerobically to oxidise and stabilise these compounds, in a process similar to the maturation of aerobic composts, prior to sale as compost, or disposal as a residue.

9.5 Compost markets

It is the presence or absence of a viable market that determines whether the composted output from biological treatment represents a valued product or a residue for disposal. Consequently much effort has been put into the definition and development of markets for waste-derived composts both in Europe and the United States by organisations such as the Organic Reclamation and Composting Association (ORCA) and the Solid Waste Composting Council (SWCC), respectively.

Compost can fulfil one or more of four basic functions:

1. *Soil conditioner or improver*: By adding organic matter to the soil, compost will improve the structure of the soil and replace the organic material lost during sustained intensive cultivation.
2. *Soil fertiliser*: The actual value of compost as a fertiliser will depend

on its content of nutrients in general and of nitrate and phosphate in particular. This is normally much lower than for inorganic fertilisers, and because these nutrients are bound to the organic matter, their release is slow and sustained. This is an advantage that is becoming increasingly important in countries such as Denmark, Belgium, Netherlands and Germany where strict limitations on nitrate application are being implemented to reduce ammonia emissions and possible groundwater contamination (ORCA, 1992a).

3. *Mulch*: Compost can be applied to the soil surface to reduce evaporation losses and weed growth.
4. *Peat replacement*: Use of peat is facing growing public opposition, being seen as the exploitation of an irreplaceable natural biotope. In both the UK and Germany, some sectors of the trade have specified that no peat be used in products for home gardening. Whilst the use of waste-derived composts instead of peat may be limited in the potted plant industry due to very strict phytosanitary regulations in Europe, to control the spread of plant diseases, there appears to be a market as a peat replacement in the home gardening and landscaping sectors (ORCA, 1992a).

As well as fulfilling different functions, composts from biological treatments come in different forms. Many processes produce more than one grade of compost (e.g. coarse and fine) at the final refining stage. The essential marketing step is to match up these products to the market requirements. In some cases, new markets may need to be developed; for example, where composting plants produce a novel product, such as the very fine textured and uniformly graded compost produced by the lumbricomposting process. The market potential can be assessed by considering both current and potential future usage of composts.

Surveys of current compost consumption show that, in Switzerland, of the 100 000 tonnes of compost produced each year, 46% is used in agriculture and vineyards, 30% in horticulture and tree nurseries, 13% in hobby gardening and 11% in recultivation (Schleiss, 1990). A more detailed analysis of German usage is given in Table 9.6. Several assessments undertaken of market potential suggest that there is considerable potential for increasing this level of usage. ORCA have conducted a market survey for Germany, and concluded that the potential demand for compost exceeded the likely supply, even if all of the organic waste were subjected to biological treatment (Table 9.7) (ORCA, 1992a). Other surveys have reported essentially the same result, with large untapped markets for compost in many areas, but particularly in agriculture (Fricke and Vogtmann, 1992). Penetration of these new market areas will depend on effective marketing of waste-derived compost as a quality product, i.e. that it is safe and fit for use, and gives clear benefits compared to competing products at

Table 9.6 Utilisation of compost in Germany

	Compost from biowaste (%)	Compost from greenwaste (%)
Hobby gardening	30	25
Commercial horticulture	10	13
Departments of parks and cemeteries	12	29
Roads Departments	1	7
Landscape gardening	29	13
Viniculture	1	1
Agriculture	10	8
Technical use	2	–
Landfill cover	5	4
Total	100	100

Source: Fricke and Vogtmann (1992).

Table 9.7 Compost market potential assessment for Germany

	Amount of compost (million tonnes/year)
Potential compost supply	4.5
Potential continuous demand:	
Agriculture	3.5
Viniculture	0.75
Substrates and soils	1.0
Forestry	0.25
Total continuous demand	5.5
Potential one-time demand (million tonnes)	
Landscaping/restoration (new German states)	3.5

Source: ORCA (1992a) (updated).

an affordable cost (ORCA, 1992a). Whilst these are general pre-conditions, there are additional specific compost quality requirements that will vary between different compost uses (Table 9.8).

The failure to obtain widespread acceptance of waste-derived composts, especially in the agricultural sector has most likely been due to concerns over its safety and quality. Failure of many early plants to completely separate visual contaminants (e.g. plastic film) from the final compost reinforced the idea of waste-based composts as inferior products. The connotation of waste also raises concerns over safety, in particular the possible presence of pathogens, although these should be effectively destroyed in the biological treatment process. More recently, the level of heavy metals in waste-derived composts has become a concern. The high levels of some heavy metals in some fractions of household waste (Table 5.6) can result in contamination of the final compost, if biological treatment of mixed

Table 9.8 Market requirements for compost quality in France[a]

Market Outlet	Impurities (glass, plastic)	Maturity	Organic material	Particle size	Salinity	Humidity
Agriculture	1 xxx	3	2 xxx	4	6	5
Market gardening	1 xxxx	1 xxxx	3 xxx	5	3 xxx	6
Produce farming	1 xxxx	2 xxxx	3 xxx	5	4 xxx	6
Viniculture	1 xxx	2	2	4	6	5
Arboriculture	1 xxx	2	2 xxx	4	6	5
Mushroom farms	2 xxx	1 xxxx	4	5	6	2 xxx

Source: ANRED (1990).
[a] Numbers 1–6 give quality criteria in descending order of importance; xxx/xxxx, customers sensitive/very sensitive to this criterion.

wastes is used. What are needed, if waste-derived composts are to become fully accepted and used more extensively, are (a) widely accepted quality standards that can reassure potential users that the compost is both safe and fit for use, especially with repeated applications and (b) plant growth studies demonstrating a commercial benefit from the application of compost or compost-based products.

9.6 Compost standards

Compost market development would be facilitated by the application of consistent quality standards across Europe, but present standards vary widely between countries both in approach and detail. ORCA have recently published a review of compost standards for 12 European countries (ORCA, 1992b). Of these, 9 countries (Austria, Belgium, Denmark, France, Germany, Greece, Italy, Netherlands and Switzerland) have implemented or proposed standards, Sweden and the UK are in the process of drafting standards while Spain uses standards relating to fertilisers. In countries such as Germany, these criteria take the form of marketing standards, whereas in other countries they actually comprise a legally defined standard (Table 9.9).

The objective of the standards is to protect land from contamination and to ensure that the composts marketed are fit for use. Since there are many uses for compost, however, different compost quality criteria need to be applied for each separate application. Several countries, such as the

Table 9.9 Summary of criteria specified in existing compost standards

	Legal definition (LD), marketing standard (MS)	Number of grades of compost	Minimum processing requirements	Heavy metal limits	Limits on other characteristics	Analytical methods	Quality control procedures
Austria	LD	2	No	Yes	Yes	Yes	Yes
Belgium	LD	2	Yes	Yes	Yes	Yes	No
Denmark	LD	1	Yes	Yes	Yes	Yes	Yes
France	LD and MS	4	Yes	Yes	Yes	Yes	Yes
Germany	3 MS	1 (for each standard)	No	Yes	Yes	Yes	Yes
Greece	LD	1	No	No	Yes	No	No
Italy	LD	2	Yes	Yes	Yes	Yes	Yes
Netherlands	LD	3	No	Yes	No	Yes	No
Spain	LD	1	No	Yes	Yes	Yes	No
Sweden	None	0	No	No	No	No	No
Switzerland	LD	1	No	Yes	No	No	No
United Kingdom	None	0	No	No	No	No	No

Source: ORCA (1992b).

Netherlands and Austria, define several different grades of compost with different maximum levels of contaminants for each (Table 9.10). Belgium also specifies which grades can be used for different applications such as growing food or fodder crops.

Most standards relate to the physical and chemical properties of the compost, although there is normally more emphasis on what should not be in the compost (i.e contaminants) rather than what should be in the compost (e.g. nutrients). Heavy metal levels come in for close scrutiny, but as Table 9.10 demonstrates, limit levels vary widely between countries. To a large extent, this reflects differences in the interpretation of the available scientific data on the heavy metal levels that constitute a significant health or environmental risk. Another cause for variability, however, is the use to which the compost may be put. Many of the most restricted heavy metal limits refer to composts that can be freely applied; some of the more relaxed standards are supplemented by restrictions on their level or time of use, frequency of application, application during wet weather, soil type or proximity of water supply plants (ORCA, 1992b). Measured levels for contaminants such as heavy metals will also depend on the analytical methods used. Whilst most national standards include details of the analytical methods required, some, such as Switzerland, do not. Clearly this lack of uniformity can only hinder the development of free markets for compost across Europe.

Not all standards systems even take the same basic approach. Since the quality of a compost is largely determined by the feedstock used and the processing method, some standards set criteria for these rather than the quality of the resulting compost itself. Criteria for some of the high quality composts specify that unsegregated household waste cannot be used as a feedstock. Criteria for compost processing methods are commonly used to determine microbiological safety. Several standards include both the temperature that must be achieved and its duration for destruction of pathogens during aerobic composting.

The French compost standard takes yet another approach. Rather than set criteria in terms of compost/composting conditions considered safe and fit for use, the criteria reflect the engineering capabilities of existing plants. In this case, some plants should be capable of meeting the requirements, but elsewhere the problem of compost not meeting quality criteria seems widespread. Taking a sample of 27 biowaste composting plants in Germany, for example, ORCA (1991a) calculated that the output of 25% of them would not meet the German Bundesgütesgemeinschaft quality limits, and almost 70% would not meet the stricter proposed Dutch limits.

In the United States, the Environmental Protection Agency (EPA) has taken yet a different approach, based on a risk assessment of soil to which compost has been added. This approach is related to the US (and UK) approach towards application of sewage sludge to fields, and does have a certain scientific logic.

Table 9.10 Limit values for elements in compost, in current standards (mg/kg dry matter)

Country	A	A Class 1	A Class 2	B Agricultural land	B Parkland	DK	F NF Urbain	D RAL	I	NL Compost (to end 1994)	NL Clean compost (to end 1994)	E	CH
Arsenic	-	-		-	-	-[a]	-	-	10	25	15	-	-
Boron	100	-		-	-		-	-		-	-	-	-
Cadmium	4	0.7	1	5	5	1.2	8	2	10	2	1	40	3
Chromium	150	70	70	150	200	-	-	100	-[c]	200	70	750	150
Cobalt	-	-	-	10	20	-	-	-	-	-	-	-	25
Copper	400	70	100	100	500	-	-	100	600	300	90	1750	150
Lead	500	70	150	600	1000	120[b]	800	150	500	200	120	1200	150
Mercury	4	0.7	1	5	5	1.2	8	1.5	10	2	0.7	25	3
Molybdenum										5			
Nickel	100	42	60	50	100	45	200	50	200	50	20	400	50
Selenium	-	-	-	-	-	-	-	-	-	-	-	-	-
Zinc	1000	210	400	1000	1500	-	-	400	2500	900	280	4000	500

Source: ORCA (1992b).

[a] 25 for private gardens.
[b] 80 for private gardens.
[c] 500 for chromium(III) and 100 for chromium(VI).

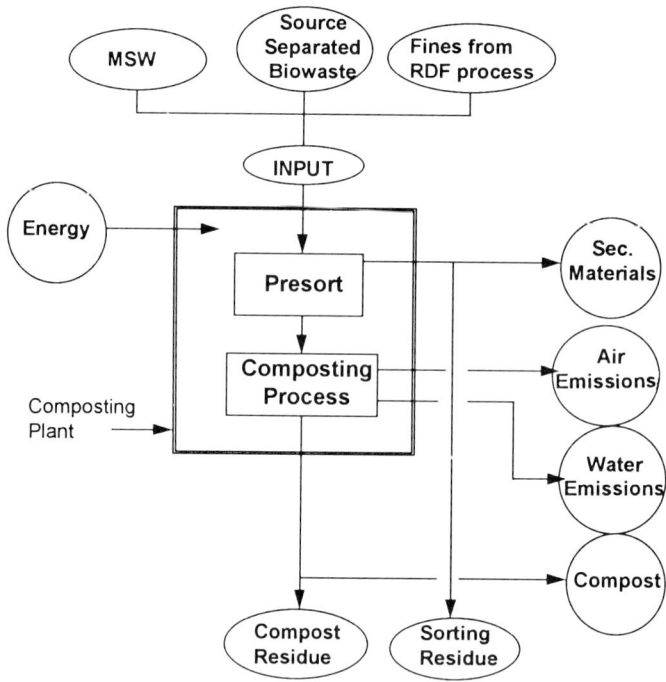

Figure 9.6 Flow diagram for typical composting plant.

In conclusion, criteria are needed to provide reassurance that marketed composts are fit and safe for use, but such criteria should be set uniformly across Europe on the basis of good scientific data, considering all aspects of compost usage. Such standards should define the quality and quantity of compost that can be used for different applications ranging from horticulture and agriculture to the reclamation of derelict land and erosion control.

9.7 Environmental impacts: input–output analysis

9.7.1 Defining the system boundaries

The system boundaries for biological treatment are defined here as the physical boundaries of the plant. Thus, both the pre-sorting treatment and the biological process are included (Figure 9.1). Materials enter the system as waste inputs and leave as compost, recovered secondary materials, residues (from sorting and composting), or as air or water emissions (Figures 9.6 and 9.7). Energy enters the system as either electrical energy from

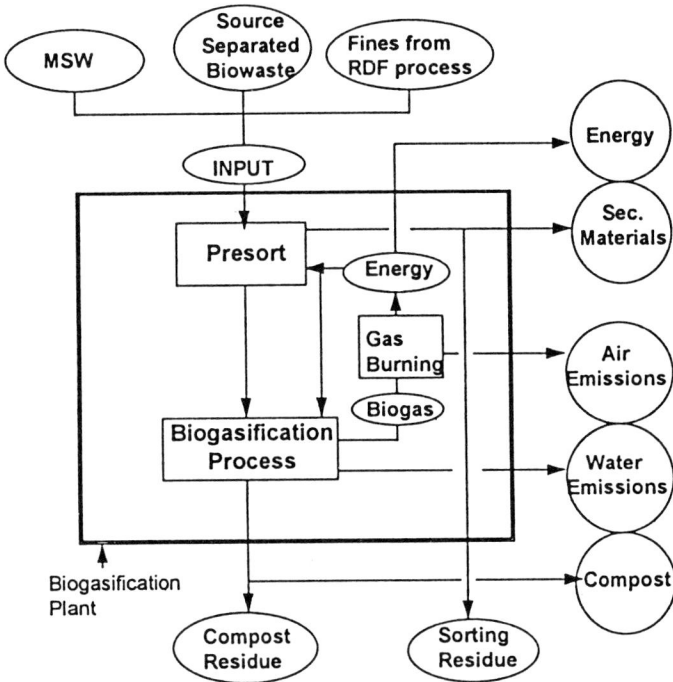

Figure 9.7 Flow diagram for typical biogasification plant.

the national grid, or as fuels (e.g. diesel). In the case of biogasification, some of the energy recovered in the biogas is consumed on site for process heating, and it is assumed that the rest of the biogas is burned on site in a gas engine-powered generator to produce electricity. Again some of this is used on site, but the rest is exported from the site as electrical energy.

9.7.2 Inputs

Waste input. The feedstock for biological treatment can arise from at least three different sources: separately collected organic/paper material, mechanically separated putrescibles from an RDF process or mixed and unsorted MSW (Figure 9.1).

The current trend in Europe is towards a separated collection of organic material from households. The exact composition of this feedstock will vary according to the definition of 'green waste' or 'biowaste' used by the collection scheme (see Chapter 6 which deals with different collection schemes), but will generally consist of kitchen and garden waste, plus in many cases non-recyclable soiled paper and paper products. Even in narrowly defined feedstocks there will always be a level of nuisance materials, requiring a

Table 9.11 Energy consumption of various composting plants

Plant type	German range	Windrow composting	Box composting	Tunnel reactor	Drum composting
Capacity (tonnes/year)	10 000	13 700	6800	18 000	35 000
Feedstock[a]	BW	BW/GW	BW/GW	WW	BW/GW
Energy consumption (kW-h electricity/tonne)	20–50	18	18	50	40
Source[b]	1	2	2	2	6

[a] WW, 'wet waste'; BW, 'biowaste'; GW, 'greenwaste'.
[b] 1, German Government (1993); 2, Bergmann and Lentz (1992).

pre-sorting stage. Such nuisance materials arise from the inclusion of (a) bags (often plastic) used to contain organic material, (b) other materials which form a small part of otherwise organic materials, (c) materials included in the biowaste by mistake. There will also be some organic material not suitable for biological treatment, e.g. woody garden waste. Mixed waste inputs, such as MSW will need extensive pre-sorting to remove all of the non-organic material, which is not suitable for biological treatment. In contrast, finely sorted putrescible feedstock from an RDF type process will have already undergone a sorting stage, so will not require another pre-sort prior to the biological treatment process.

Energy consumption. The energy consumption of the pre-treatment process will depend on the feedstock used. Mixed feedstocks, such as MSW will need more extensive sorting per tonne of input, with associated energy requirements, than more narrowly defined feedstocks or those that have already been mechanically sorted as part of an RDF process, irrespective of the subsequent method of treatment. The energy consumption of the biological treatment process itself will depend on the technology employed.

Composting involves a net consumption of energy, consuming process energy and not producing any energy in a usable form. The German Government (1993) report a typical energy consumption of from 20 to 50 kW-h (electrical energy) per input tonne for plants capable of processing 10 000 tonnes of biowaste per year. Available data on the overall energy consumption for various methods of composting are given in Table 9.11, and suggest a range from 18 to 50 kW-h (electrical) per tonne of input. This variability will reflect both the different feedstocks used, the different sizes of the composting plants, and also the maturity of the compost produced. Kern (1993) looked at several different composting methods and calculated an average energy consumption of 21 kW-h/tonne input; for plants producing less mature compost (rotte grades I–II) the average consumption was 18.3 kW-h/tonne, whilst for plants producing mature compost (rotte

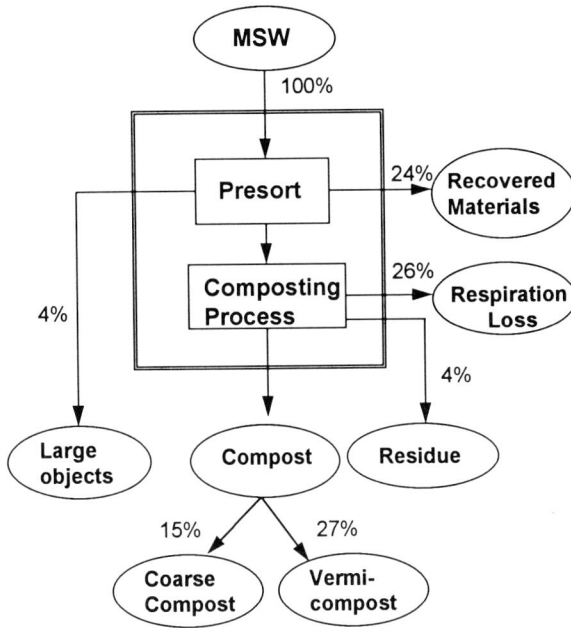

Figure 9.8 Typical mass balance (on the basis of dry weight) for a lumbricomposting plant (La Voulte, France). *Source*: SOVADEC; Schauner (1994).

grades III–IV) the average was 30.7 kW-h/tonne. For the purposes of the LCI model, an energy consumption of 30 kW-h of electrical energy per tonne of input to the composting plant is assumed.

Biogasification involves both consumption of energy during processing, and the production of useful energy as biogas. Since some of the biogas can be burned to produce steam to heat the digester, and more can be burned in a gas engine to produce electricity, the energy requirement for the process can be met from within the biogas produced. The remaining biogas can either be exported as biogas (i.e. as fuel) or burned on site to provide heat or to generate electricity (both for export). For the purposes of the present study, it is assumed that the biogas is burned on-site for power generation, and that surplus energy is exported as electrical energy.

The electrical energy requirement for biogasification has been reported as 50 kW-h and 54 kW-h per input tonne for two different processes (Schneider, 1992; Schön, 1992). This represents around 32–35% of the gross electricity produced by the plant. In another example, a biogas plant operating using the dry process consumes from 30 to 50% of the electricity produced (De Baere, 1993). Thermal energy is also required for the process, but this can be obtained by using waste heat left after electricity generation,

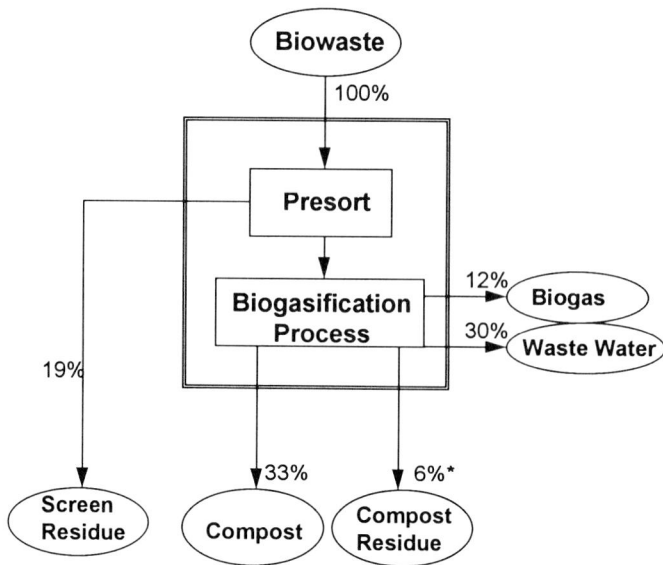

Figure 9.9 Mass balance (on the basis of wet weight) for a dry process biogasification plant. * estimated. *Source*: De Baere (1993).

or by burning some of the biogas. Therefore no additional energy needs to be imported into the site for this.

In the LCI spreadsheet, biogasification is assumed to consume 50 kW-h of the generated electricity, for every input tonne (including nuisance materials and recoverable materials).

9.7.3 Outputs

Mass balances for both aerobic and anaerobic processing plants are given in Figures 9.8–9.10.

Secondary materials from pre-sorting. The amount of secondary materials that will be produced by a composting plant depends on the composition of the input stream, and on the pre-sorting equipment installed. A narrowly defined input, such as biowaste or VFG will contain a certain level of contamination, but this material would not be suitable for recovery. A mixed waste stream input (MSW or household waste) will contain considerable amounts of glass, plastic and metal that could be recovered for use as secondary materials, but levels of contamination are likely to be high, and the quality of the material recovered is likely to be lower than that from source-separated collection of recyclables (Chapter 6). Recovery of recyclables from the input requires suitable sorting equipment or manual

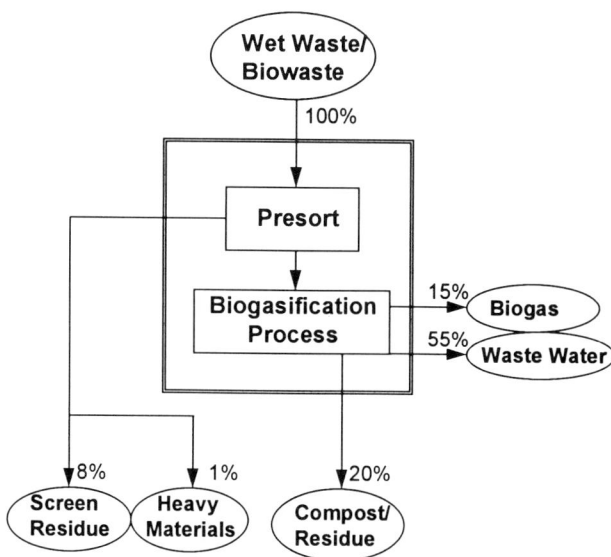

Figure 9.10 Mass balance (on the basis of wet weight) for a two-stage wet biogasification plant (Garching, Germany). *Source*: German Government (1993).

sorting; in most cases this is limited to magnetic separation, which can remove up to 90% of incoming ferrous material.

Biogas/energy. The amount of biogas produced during anaerobic digestion will depend on the nature of the organic material used as feedstock, as well as the process used. Biogasification of a range of industrial organic wastes from vegetables to dairy and brewing wastes in Switzerland produce from 200 to 600 N m³ per tonne of input (dry weight) (Bundesamtes für Energiewirtschaft, 1991). Biogas from household wastes is normally expressed per tonne as received (i.e. wet weight); grass clippings, for example, may be expected to produce 100 N m³ per input tonne (German Government, 1993). Putrescible material mechanically sorted has been shown to produce 130–160 m³ of biogas per tonne (Van der Vlugt and Rulkens, 1984).

Production figures for the different processes reflect the amount of organic decomposition that is achieved. The more complex two-stage process converts more organic material to biogas (around 65–70% by dry weight) than single stage processes (around 45% by dry weight), giving typical production rates of 115 and 75 m³ per tonne of biowaste, respectively (Korz and Frick, 1993). Biogas composition, especially methane content, also varies with process type (Table 9.12), again being generally higher in the two-stage process. The methane content generally varies from 50 to

Table 9.12 Energy requirements and biogas generation for biogasification plants

Process type	Dry 1-stage	Dry 1-stage	Dry 1-stage	Dry 1-stage	Wet 1-stage	Wet 2-stage	Wet 2-stage
Capacity (tonnes/year)	10 000	100 000	Pilot	Pilot	30 000	20 000	1000
Feedstock[a]	BW	HW	BW	VFG	MSW + sewage sludge	BW/WW	BW/MSW
Energy requirement/ tonne input	Total 840 MJ Th	270 MJ El ? Th	76 MJ El 263 MJ Th	144 MJ El ? Th		288 MJ El 130 MJ Th	
Biogas production (N m^3/tonne)	90	100–140	100		50	115	90–150
Methane content (%)	55	60–65	60–65	54		70–80	60–75
Source[b]	1	2	2,3	2	2	2,4	2

[a] BW, 'biowaste'; HW, household waste; WW, 'wet waste'; MSW, municipal solid waste; VFG, vegetable, fruit and garden waste.
[b] 1, De Baere (1993); 2, Bergmann and Lentz (1992); 3, Schön (1992); 4, Korz and Frick (1992). Th = thermal; El = electrical.

Table 9.13 Biogas composition

Source	Biogas	Biogas	Biogas after combustion
	Lentz *et al.* (1992)	BTA (1992)	IFEU (1992)
% vol			
CO_2	26.8%	45%	
CH_4	71.4%	54%	
N_2	1.4%		
O_2	0.3%		
mg/m^3			
NO_x			100
SO_x			25
sum chlorine	0.6	0.9	11
sum fluorine	0.1		0.021
HCl			
HF			
H_2S	700	420	0.33
Total HC		<1.5	0.023
Chlorinated HC		<1.5	7.3 E−3
Dioxins/furans (TEQ)			1.0 E−7
Ammonia			
Arsenic			
Cadmium			9.4 E−6
Chromium			1.1 E−6
Copper			
Lead			8.5 E−6
Mercury			6.9 E−8
Nickel			
Zinc			1.3 E−4

75%, the rest of the biogas comprising carbon dioxide and some trace components (Table 9.13).

If it is assumed that 100 N m^3 of biogas are produced per tonne of input to the digestor (i.e. after nuisance materials have been removed in the pre-treatment stage), with a methane content of 55% (methane has a calorific value of 37.75 MJ/N m^3 (Perry and Green, 1984)), this will give a gross energy potential of 2076 MJ thermal energy per tonne digested. If this is burned in a gas engine to produce electricity with an efficiency of around 30% (Schön, 1992; Schneider, 1992), this will give a gross electricity production of 173 kW-h per tonne digested. Figures reported by Schneider (1992) and Schön (1992) for actual plant performances agree with this; they report gross electricity generation rates of 140 kW-h/tonne and 169 kW-h/tonne, respectively. For the LCI spreadsheet, a gross production of 160 kW-h/tonne of digestor input will be used.

Compost. The quantity and quality of the compost produced by biological treatment are clearly not independent. The more the product is refined to improve the quality, the less the final quantity (and hence the

Table 9.14 Physical characteristics and plant nutrient contents of different types of compost produced by aerobic process

	Biowaste compost	Biowaste with paper compost	Greenwaste compost	Wetwaste compost[a]	Total waste compost
H_2O% wet wt.	37.7	45.0	34.8	44.2	35.6
pH value	7.6	7.5	7.6	7.5	7.3
Salt (g/l wet wt.)	3.9	3.6	2.3	5.8	7.3
OS (% dry wt.)	33.3	42.0	32.5	55.4	39.7
C/N ratio	17.0	21.8	20.0	18.8	17.8
N total (% dry wt.)	1.2	1.1	0.8	1.7	1.1
P_2O_5 (% dry wt.)	0.6	0.6	0.4	0.9	0.9
K_2O (% dry wt.)	1.0	0.9	0.8	1.2	0.6
MgO (% dry wt.)	0.8	0.8	0.6	2.0	0.7
CaO (% dry wt.)	4.0	4.1	3.0	10.0	4.9

Source: Fricke and Vogtmann (1992).
[a] Fraction remaining after separate collection of dry waste (e.g. recyclables like glass, paper, metal, wood, etc.).

Table 9.15 Heavy metal content of different composts produced by aerobic process (mg/kg dry wt.)

Element	Biowaste compost	Biowaste with paper compost	Greenwaste compost	Wetwaste compost	Total waste compost	BGGK limits[a]
Based on material as produced						
Pb	77.6	78.6	60.8	449	513	
Cd	0.8	0.7	0.7	2.6	5.5	
Cr	33.7	31.7	27.0	72	71.4	
Cu	43.2	58.2	32.7	228	274	
Ni	19.1	16.1	17.5	30	44.9	
Zn	232.8	273.8	167.8	850	1.570	
Hg	0.3	0.4	0.3	1.0	2.4	
Based on standardised organic matter content of 30% (dry wt.)						
Pb	83.1	116.2	63.1	705	596	150
Cd	0.8	1.0	0.7	4.1	6.4	1.5
Cr	35.8	39.8	28.4	113.0	82.9	100
Cu	46.8	76.2	34.5	357.8	318	100
Ni	20.5	21.4	18.6	47.1	52.1	50
Zn	249.1	350.3	176.9	1334	1823	400
Hg	0.4	0.5	0.3	1.6	2.8	1.0

[a] BGGK, Bundesgütesgemeinschaft Kompost (German Federal Association for Quality Compost).

greater the residue). In many cases, different grades of compost will be produced, the important factor being the existence of a market for the material. Put simply, if there is no market for the compost, regardless of its quality, it will be a residue rather than a valuable product.

Compost quantity. For composting, the final amount of compost produced (wet weight) is in the region of 50% of the input of organic material (organics plus paper) (ORCA, 1992a). The other 50% is lost due to evaporation and respiration. Where further refining of the compost occurs, the amount of compost actually marketed may be considerably less than this. For biogasification, the amount of final compost-like material will depend on the extent to which the organic material is broken down into biogas. Production can account for 33% by weight of a plant's input (equivalent to 41% of the input to the digester after the pre-sort) (De Baere, 1993). By contrast, the two-stage wet process produces more biogas with a higher methane content than the dry process, leaving around 20% of the plant input (22% of the digester input) as composted residue (Figure 9.10).

For the purposes of the LCI spreadsheet, it is assumed that in composting, the final compost accounts for 50% of the input to the composting process (i.e. after any pre-sorting); for biogasification, an average figure of 30% is used.

Compost quality. Compost quality is the key factor that determines whether the output from biological treatment processes is a valuable product or a residue. A valuable material is one that has a market, hence the need to develop markets for different grades of compost. Producers will then either be able to produce large amounts of lower grade composts, or smaller amounts of higher grade material.

Compost quality is determined by the feedstock type, technology used and level of process control. The physical characteristics, plant nutrient and heavy metal contents of a range of composts derived from different feedstocks are given in Tables 9.14 and 9.15. These can be compared with the standards discussed above. It can be seen that the major variability occurs in the heavy metal content. Not surprisingly, the more mixed the feedstock, the higher the heavy metal content of the compost. The same pattern is shown for the compost-like residues from biogasification (Table 9.16). So, while it is possible to make compost from mixed waste streams, the high level of contamination may mean that no market for this material can be found. It is accepted in Germany, for example, that the composted residue from biogasification is not marketable as compost (German Government, 1993) and that it needs to be disposed of. Once the T.A. Siedlungsabfall ordinance comes into effect, it will no longer be possible to landfill this material directly in Germany, so incineration will need to be used.

Table 9.16 Quality of compost produced by a variety of biological treatment processes

	Composting			Biogasification					
	Windrow	Box composting	Tunnel reactor	Dry 1-stage	Dry 1-stage	Dry 1-stage	Dry 1-stage	Wet 2-stage	Wet 2-stage
Feedstock	BW	BW	WW	VFG	VFG	BW	GW	BW	BW
Nutrients (% by wt)									
N-total		1.1	n.a	1.8–2.1		1.2	0.8–0.9	1.92	1.24
P_2O_5		0.73	n.a	0.15–0.20		0.7	0.6	0.9	0.6
K_2O		1.4	n.a	1.0–1.2		1.1	2.3	0.49	0.5
$CaCO_3$		1.9	10.0	5.2–6.4		2.7	2.7	6.6	3.82
C/N ratio		17–18	16	12–15		n.a	15	20	11–15
pH		7.8	7.5	8.0–8.6		n.a	8	7.6	n.a
Heavy metals (mg/kg TS)									
Zn	324	247	850	138	173	253	122	135	491
Pb	139	84	449	67	75	100	43	85	155
Cu	61	36	228	20	27	54	27	52	27
Cr	32	55	72	n.a	30	36	15	44	34
Ni	17	41	35	25	9	20	7	27	16
Cd	1.8	0.7	2.6	1.8	0.8	1.3	0.4	1.0	1.1
Hg	0.8	0.2	1	n.a	n.a	0.7	0.3	<0.25	0.2

Source: Bergmann and Lentz (1992).

Residues

Sorting residue. This will consist of two types of materials: (a) non-biodegradable materials arriving as nuisance materials in biowaste, or materials in mixed waste that have not been recovered as secondary materials and (b) degradable material (organic or paper) that is either unsuitable for biological processing (e.g. too large) or is removed adhered to nuisance materials. There is little data available that distinguish between these types, however. Where the feedstock is source-separated biowaste, a nuisance level of around 5% is typical (see Chapter 6). Where the feedstock is mixed waste such as MSW then the level of sorting residue is likely to be much higher, although where recovery of other materials occurs (e.g. La Voulte, Figure 9.8), residue rates as low as 5% may be found.

In the LCI spreadsheet, it is assumed that all of the categories other than paper and organics are removed as residue during the pre-sort. In addition, 5% of the organic and paper fractions are added to the residue to account for material that is not readily biodegradable, or that adheres to the nuisance materials as they are removed.

Compost residue. This represents the composted/digested output that is not marketed. The amount will range from zero, if a use can be found for all of the compost, to 100% of the output if no market can be found.

Air emissions. The major air emission by volume from biological processing will be carbon dioxide, which is a contributor to the greenhouse effect. In aerobic processing, the organic material is broken down directly to carbon dioxide and water. In anaerobic processing, biogas containing methane and carbon dioxide is produced, of which the methane also forms carbon dioxide when burned. The amount of emissions per tonne of process input will depend on the moisture content of the incoming material. In the following calculations, an average moisture content of 50% is assumed. The actual level will depend on the ratio of paper to wet organic material present, but ORCA (1992a) suggest that 50% is the optimum moisture content level for composting feedstocks.

For composting, the dry weight loss during composting is around 40%, giving a dry weight loss of 200 kg per wet input tonne. Assuming that most of the organic material decomposed is cellulose, with a carbon content of 44% (from formula), composting will evolve approximately 323 kg of CO_2 (164 N m³) per tonne of wet organic feedstock.

For biogasification, the dry weight loss varies with process type, and reports vary from 45% to 70%. Assuming a mid-range dry weight loss of 55%, means that 275 kg of organic matter are converted into gas. If all this was converted to carbon dioxide the total emitted would be 444 kg (226 N m³) per tonne of digester input. This does not agree with published

Table 9.17 Effect of including paper in biowaste on leachate production during windrow composting

Feedstock	Biowaste (narrow definition)	Biowaste + 10% (by wt.) paper	Biowaste + 20% paper	Biowaste + 30% paper
Leachate production (litres/tonne)	13.5	1.6	0	0
Leachate composition				
COD (mg O_2/l)	33 100	30 200	–	–
BOD$_5$ (mg O_2/l)	19 000	19 000	–	–

Source: Verstraete *et al.* (1993).

biogas production figures, however. Given the composition of biogas in Table 9.13, combustion will convert the methane to carbon dioxide and water in the reaction

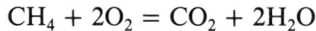

$$CH_4 + 2O_2 = CO_2 + 2H_2O$$

Complete combustion of the biogas will therefore produce 0.982 N m^3 of CO_2 per N m^3 of biogas burned, equivalent to 1.93 kg of CO_2. Given a production of around 100 N m^3 of biogas per tonne of organic material feedstock, this produces a CO_2 emission of 193 kg per input tonne, considerably less than that predicted from the dry weight loss during the process. This discrepancy probably reflects some process losses, and more importantly, the aerobic maturation stage that follows the anaerobic stage. During this stage, the material needs to be aerated and heats up, demonstrating considerable aerobic microbial activity, during which further carbon dioxide is likely to be released.

For the purposes of the LCI spreadsheet, overall carbon dioxide emissions are assumed to be 320 kg and 440 kg per tonne of wet organic material, for composting and biogasification, respectively.

No reliable data were found on other air emissions from composting processes, although the odour problems that can occur around compost plants demonstrate that other air emissions do occur. Air emissions resulting from the combustion of biogas are given in Table 9.13.

Water emissions. The aqueous effluents reported for biological treatment vary widely in both amounts and composition, depending on both the process used and the feedstock. In composting, considerable evaporation will take place during the process. Any run off collected is often sprayed back onto the composting material to maintain sufficiently high moisture contents. If waste paper is included in the feedstock, this will absorb much of the water, and so little or no leachate is actually produced (Table 9.17).

In biogasification, water is produced when the digested material is pressed or filtered. Large amounts will be produced, especially in the wet

Table 9.18 Water emissions from biological treatment processes

Process	Composting				Biogasification			
	Worm composting	Box composting	Drum composting	Tunnel reactor	Dry 1-stage	Dry 1-stage	Dry 1-stage	Wet 2-stage
Amount (litres/tonne)	0	300	n.a.	n.a.	290	490	540	500
Composition (mg/l)								
BOD_5		270–485	50–600	3300–7050	<65	n.a.	740	60
COD		458–808	150–7000	6200–15100	<250	n.a.	1400	200
NH_4		48–117	n.a.	n.a.	<100	n.a.	250	100
N total		0–1	6–36	0–3	<100	n.a.	6	n.a.
pH		7.9	7.1–7.8	7.1–8.1	n.a.	n.a.	8.0	n.a.
Source[a]	1	2	2	2	2	2	2	2

[a] 1, Schauner, unpublished data; 2, Bergmann and Lentz (1992).

Table 9.19 Space requirements for biological treatment plants

	Composting			Biogasification					
	Windrow	Drum	Tunnel reactor	Dry 1-stage	Dry 1-stage	Wet 2-stage	Dry 1-stage	Wet 2-stage	Dry 1-stage
Space m^2/t capacity	1.45	0.6	0.5	0.12	0.4	0.32	0.23	0.57	0.14

Source: Bergmann and Lentz (1992).

(low solids) process type. Some of this water will be recirculated to adjust the water content of the incoming feedstock, the rest needs to be treated prior to discharge. Typical amounts and compositions of the leachates produced by both composting and biogasification are given in Table 9.18.

9.7.4 Other considerations

Land usage. Table 9.19 compares the land usage of composting and biogas processing. Whilst there is likely to be an effect of plant capacity on land usage (larger plants will have proportionately less free space at any time than smaller plants), it can be seen that generally composting is a more space intensive process than biogasification. This is because biogas plants are built vertically, whilst composting plants are built horizontally. Also, in composting, a greater percentage of the input is produced as compost, which requires maturing, and so occupies space for some time.

9.8 Economic costs

Data on the economic costs of biological treatment are not always reported on a consistent basis, so comparisons are difficult to make. In many cases, biological treatment is considered as a final disposal option, and consequently costs given are as an all-inclusive 'gate fee' or 'tip fee'. This cost will include allowance for any revenues collected from the sale of recovered materials, compost, and energy from biogas utilisation, and include disposal costs for any residues requiring incineration or landfilling. The problem with this level of accounting is that the cost of biological treatment will vary with the market prices of energy, compost and recovered materials and the cost of landfill. Alternatively, cost data for biological treatment can refer to the biological processing itself (e.g. as in ORCA, 1991b). This is more useful when modelling the economics of the overall waste management system, since it is independent of the cost of other parts of the system, but this type of data is not widely available.

Table 9.20 Operating costs for biological treatment (in ecu)

Country	Operating costs (per input tonne)		Source
	Composting	Biogasification	
Europe (average)	50–70	50–70	ORCA (1992a)
Austria	37		pers. comm. 1994
France	42–58		ORCA (1991a)
Germany	56–77	56–77	ORCA (1991b)
Netherlands	37		pers. comm. (1994)
Spain	9–18		pers. comm. (1994)
Sweden	33–55		pers. comm. (1994)
Switzerland	50	77	Schleiss (1990); Bundesamtes für Energiewirtschaft (1991)
USA	62 ($70) tip fee	n.a.	P&G survey (1993)

Table 9.21 Revenue from sale of compost (in ecu)

Country	Price/tonne	Reference
France	5	ORCA (1991b)
Germany	Biocompost[a]	Fricke and Vogtmann (1992)
	loose 19	
	bagged 64	
	Garden waste compost	
	loose 23	
	bagged 85	

[a] Loose compost represents 81% of biocompost sales in Germany, bagged compost 19% (Fricke and Vogtmann, 1992).

A guide to the relative cost of biological treatment for a range of countries is given in Table 9.20. An attempt is made to distinguish between gate fees and processing costs. Likely revenues for the sale of compost are given in Table 9.21.

9.9 Operation of the biological treatment module of the LCI spreadsheet

The layout of the biological treatment module of the computer spreadsheet is shown in LCI Box 5.

BIOLOGICAL TREATMENT
Input:

insert '1' in box if biological treatment of unsorted restwaste occurs:

LCI BOX 5

5. BIOLOGICAL TREATMENT

Input:

Insert "1" in box if Biological Treatment of unsorted waste occurs [_____] (If not insert "0")

Amounts:
- From source-separated collection   [_____] 000 tonnes
- From RDF sorting process   [_____] 000 tonnes
- As unsorted restwaste   [_____] 000 tonnes
- TOTAL PLANT INPUT   [_____] 000 tonnes

Paper	Glass	Metal-Fe	Metal non-Fe	Plastic-film	Plastic-rigid	Textiles	Organic	Otner

Presort:

Recovery of materials

	Glass	Metal-Fe	Metal non-Fe	Plastic-film	Plastic-rigid
Amount (as % of input)					

(If no recovery, leave blank)

Biological Treatment Process:

Define process used   Insert "1" in box for COMPOSTING, "2" for BIOGASIFICATION   [_____]

Outputs

Compost produced   [_____] 000 tonnes   Electricity exported: [_____] MWh (net)

Compost - % marketable   [_____]

Residue: Treatment

	Incineration	Landfill	Calorific value of presort residue
Sorting residue			= ___ GJ/tonne
Compost residue			
Transport distance (km)			

Costs

Processing Cost per input tonne   [_____] ecu

Transport cost for residues (per tonne)

	Incineration	Landfill	
			ecu

Revenue from sale of products:

Recovered materials	Glass	Metal-Fe	Metal non-Fe	Plastic-film	Plastic-rigid	
						ecu per tonne

Market price for compost   [_____] ecu per tonne
Sale of electricity   [_____] ecu per MWh

Go to Box 6.

This will add the remaining restwaste to the input of the biological treatment plant.

> *Amounts:*   *From source-separated collection:*
> *From RDF sorting process:*
> *As unsorted restwaste:*
> *TOTAL PLANT INPUT*

Composition:

The model then automatically displays the amounts of input material from each of the three possible sources, and the overall composition of the input.

Pre-sort:

Recovery of materials:

Amount (as % of input):

The user then inserts the level of recovery for glass, metals and plastics in the pre-sort. If no recovery occurs, these cells are left blank. The model assumes that all of the material except for the paper and organics fractions are removed as pre-sort residue, along with 5% of the organic material and paper (to account for material not biodegradable or adhering to the nuisance materials.)

LCI Data Box 5 Table of impact data used in the biological treatment module of the LCI spreadsheet. Data are per tonne of input to biological treatment process, unless otherwise stated.

	Composting	Biogasification
Compost production (kg)	500	300
Energy consumption[a] (electrical)	30 kW-h	50 kW-h
Energy Production[a]	–	160 kW-h
Air Emissions (g)		
Particulates		
CO		
CO_2	320 000	440 000
CH_4		
NO_x		10
N_2O		
SO_x		2.5
HCl		0.011
HF		2.1 E−3
H_2S		3.3 E−2
HC		2.3 E−3
Chlorinated HC		7.3 E−4
Dioxins/furans (TEQ)		1.0 E−8
Ammonia		
Arsenic		
Cadmium		9.4 E−7
Chromium		1.1 E−7
Copper		
Lead		8.5 E−7
Mercury		6.9 E−9
Nickel		
Zinc		1.3 E−5
Water Emissions (g)		
BOD	81	19
COD	137	73
Ammonium	14	29

[a] Energy consumption calculated per tonne of input to pretreatment process, whilst energy production (from biogas) calculated per input tonne to biological treatment process.

Biological treatment process:

Define process used: Insert '1' in box for COMPOSTING, '2' for BIOGASIFICATION:
This defines which set of data in LCI Data Box 5 are used to calculate the energy consumption and emissions for the biological treatment process.

Outputs: Compost produced: Electricity exported:
The model automatically enters the amounts of compost and electricity that will be produced for export from the site.

Compost – % marketable:

The user specifies the percentage of final compost that is sold as opposed to treated as a residue.

Residue treatment:
  Sorting residue:
  Compost residue:
The user specifies how much of each residue is treated by incineration or landfilled directly.

Transport distance:
The user inserts the one way distance in km from the biological treatment plant to the incinerator or landfill. The model calculates the fuel consumption assuming a 20 tonne load and that no return load is carried, and adds the fuel used to the fuel consumption column.

Costs:
  Processing cost per input tonne:
This is the overall cost for pre-sorting and biological treatment, not including revenues from sale of products, or costs of transporting residues. (If only an inclusive cost is available, this can be inserted so long as the remaining revenue and cost cells below are left blank.)

  Transport cost for residue (per tonne):
  Revenue from sale of products:
    Recovered materials (per tonne):
    Compost (per tonne):
    Electricity (per kW-h):
These costs and revenues are inserted by user to calculate the overall cost of biological treatment. Note the revenue from sale of biogas/electricity is only applied to the surplus of biogas, after the amount used to run the process has been subtracted.

References

ANRED (1990) Sorting and composting of Domestic Waste. *ANRED Publication 27.* ANRED, France.
Archer, D.B. and Kirsop, B.H. (1990) The microbiology and control of anaerobic digestion. In: *Anaerobic Digestion: a Waste Treatment Technology*, ed. A. Wheatley. Elsevier Applied Science, London.
Barret, T.J. (1949) *Harnessing the Earthworm.* Faber and Faber, London.
Barton, J. (1986) The application of mechanical sorting technology in waste reclamation: options and constraints. Warren Spring Laboratory. Paper presented at the Institute of Waste Management and INCPEN Symposium on Packaging and Waste Disposal Options, London.
Bergmann, M. and Lentz, R. (1992) Vorstudie zu einer Ökobilanz über biologische Abfallbehandlungsverfahren. Procter & Gamble GmbH, Schwalbach, Germany. Internal report.
Bundesamtes für Energiewirtschaft (1991). Vergärung biogener Abfälle aus Haushalt,

Industrie und Landschaftspflege. Schriftreihe des Bundesamtes für Energiewirtschaft Studie Nr. 47.

Coombs, J. (1990) The present and future of anaerobic digestion. In: *Anaerobic Digestion: A Waste Treatment Technology*. Elsevier Applied Science, London, pp. 1–42.

De Baere, L. (1993) DRANCO, a novel way of composting food and wastepaper: results of a first full-scale plant. Paper given at 2nd CIES Conference on the Environment, Barcelona.

De Baere, L., van Meenen, P., Deboosere, S. and Verstraete, W. (1987). Anaerobic fermentation of refuse. *Resources Conservation* **14**, 295–308.

Ernst, Λ.-A. (1990) A review of solid waste management by composting in Europe. *Resources, Conservation Recycling* **4**, 135–149.

Fricke, K. and Vogtmann, H. (1992) *Biogenic Waste Compost, Experiences of Composting in Germany*. IGW, 33 pp.

German Government (Deutscher Bundestag) (1993). Anaerobe Vergärung als Baustein der Abfallverwertung. *Drucksache* **12**, 4905.

IFEU (1992) Vergleich der Auswirkungen verschiedener Verfahren der Restmüllbehandlung auf die Umwelt und die menschliche Gesundheit. Institut für Energie-und Umweltforschung, Heidelberg, GmbH, Heidelberg.

Kern, M. (1993) Grundlagen verfahrenstechnischer Vergleiche von Kompostierungsanlagen. In: *Biologische Abfallbehandlung; Kompostierung -Anaerobtechnik -Kalte Vorbehandlung; Reihe (book series) Abfall-Wirtschaft-Neues aus Forschung und Praxis*, eds K. Wiemer and M. Kern. University of Kassel, pp. 11–40.

Korz, J. and Frick, B. (1993) Das BTA-Verfahren. In: *Biologische Abfallbehandlung; Kompostierung – Anaerobtechnik – Kalte Vorbehandlung ; Reihe (book series) Abfall – Wirtschaft – Neues aus Forschung und Praxis*, eds. K. Wiemer and M. Kern. University of Kassel, pp. 831–49.

Lentz, R., Carra, R. and Scherer, P. (1992) Composite hygiene papers in anaerobic digestion of biowaste – a case study at the biogas pilot plant of BTA, Munich. In: *Proc. International Symposium of Anaerobic Digestion of Solid Waste* (Italy).

Lopez-Real, J. (1990) Agro-industrial waste composting and its agricultural significance. Paper presented at Fertiliser Society of London. *Proc. Fertiliser Society*, No. 293.

Noone, G.P. (1990) The treatment of domestic waters. In: *Anaerobic Digestion: A Waste Treatment Technology*, ed. A. Wheatley. Elsevier Applied Science, London, pp. 139–170.

OECD (1991) *Environmental Data Compendium*. Organisation for Economic Cooperation and Development, Paris.

ORCA (1991a) The role of composting in the integrated waste management system. In: *Solid Waste Management, An Integrated System Approach. Part 5*. Organic Reclamation and Composting Association, Brussels.

ORCA(1991b) Composting of Biowaste – the important role of the waste paper fraction. In: *Solid Waste Management, An Integrated Approach. Part 8*. The Organic Reclamation and Composting Association, Brussels, Belgium.

ORCA (1992a) *Information on Composting and Anaerobic Digestion. ORCA Technical Publication No. 1*. Organic Reclamation and Composting Association, Brussels.

ORCA (1992b) *A Review of Compost Standards in Europe. ORCA Technical Publication No. 2*. Organic Reclamation and Composting Association, Brussels, 102 pp.

OWS Organic Waste Systems, Company Literature, OWS n.v., Dok Noord 7b, B-9000 Gent, Belgium.

Perry and Green (1984) *The Chemical Engineer's Handbook*, 6th edition. McGraw-Hill, New York.

Schauner, P. (1994) Industrial vermicomposting of home-sorted biowaste – the naturba process. District of Bapaume, France. In press.

Schleiss, K. (1990) Übersicht über die Kompostanlagen in der Schweiz. *Abfall-spektrum* April, 12–14.

Schneider, D. (1992) Das Wabio-Verfahren der Deutschen Babcock Anlagen. In: *Biologische Abfallbehandlung; Kompostierung-Anaerobtechnik – Kalte Vorbehandlung; Reihe (book series) Abfall-Wirtschaft – Neues aus Forschung und Praxis* eds. K. Wiemer and M. Kern. University of Kassel, pp. 863–875.

Schön, M. (1992) Das Kompogas-Verfahren der Firma Bühler. In: *Biologische Abfallbehandlung; Kompostierung-Anaerobtechnik – Kalte Vorbehandlung; Reihe (book series) Abfall-Wirtschaft - Neues aus Forschung und Praxis*, eds. K. Wiemer and M. Kern. University of Kassel, pp. 821–829.

Six, W and De Baere, L. (1988) Dry anaerobic composting of various organic wastes. *Proc. 5th International Symposium on Anaerobic Digestion*, Bologna, Italy, pp. 793–797.

Van der Vlugt, A.J. and Rulkens, W.H. (1984) Biogas production from a domestic waste fraction. In: *Anaerobic Digestion and Carbohydrate Hydrolysis of Waste*, eds. G.L. Ferrero, M.P. Ferranti and H. Naveau. Elsevier Applied Science, London, pp. 245–250.

Verstraete, W., De Baere, L. and Seeboth, R.-G. (1993) *Getrenntsammlung und Kompostierung von Bioabfall in Europa*. Entsorgungs Praxis.

Warmer (1991) Compost. Warmer factsheet, Warmer Campaign, UK.

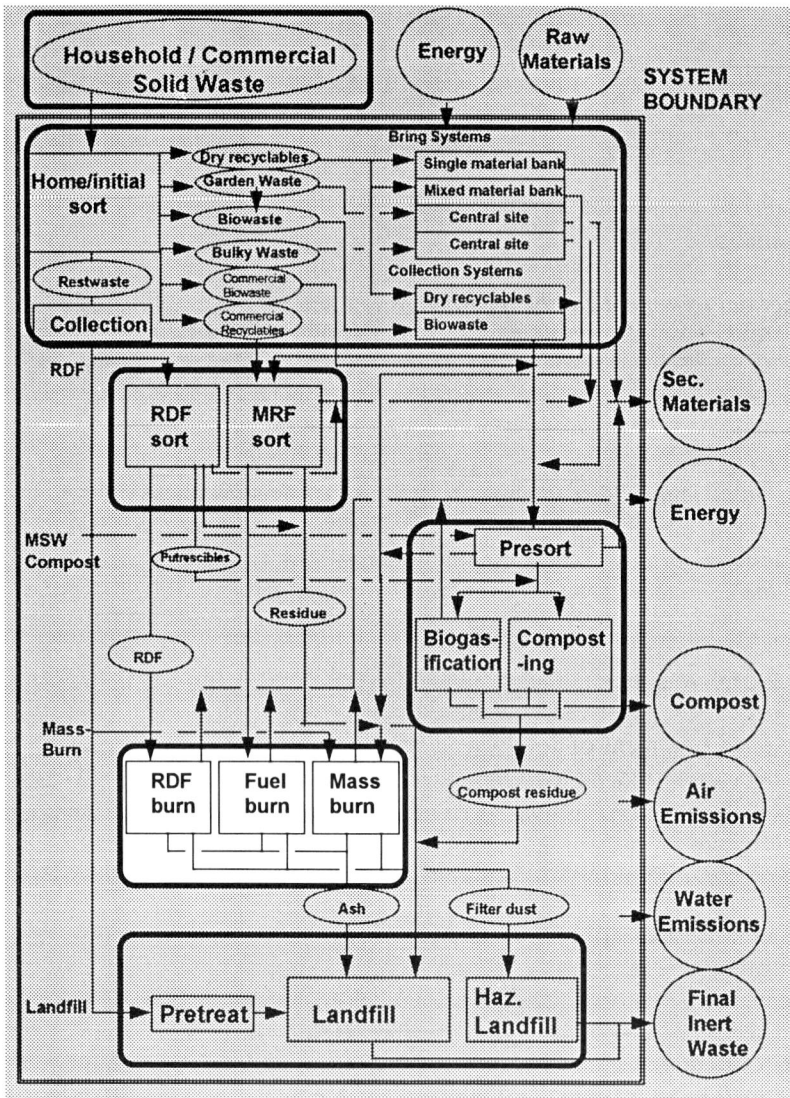

Figure 10.1 The role of thermal treatment in an integrated waste management system.

10 Thermal treatment

Summary

Thermal treatment can be regarded as either a pre-treatment of waste prior to final disposal, or as a means of valorising waste by recovering energy. It includes both the burning of mixed MSW in municipal incinerators and the burning of selected parts of the waste stream as a fuel. These different methods reflect the different objectives that thermal treatment can address. This chapter describes the various thermal treatment processes, and their use across Europe. It then attempts to quantify the environmental factors associated with thermal treatment, in terms of energy consumption and emissions, and the economic costs. Finally the thermal treatment module of the LCI spreadsheet is presented and explained.

10.1 Introduction

Thermal treatment of solid waste within an integrated waste management system can include at least three distinct processes (Figure 10.1). Most well known is the mass burning, or incineration, of mixed municipal solid waste in large incinerator plants, but there are two additional 'select burn' processes whereby combustible fractions from the solid waste are burned as fuels. These fuels can be separated from mixed MSW either mechanically to form refuse-derived fuel (RDF), or can be source-separated materials from household collections such as paper and plastic, that have been recovered but not recycled (see Chapter 7).

The range of methods reflects the different objectives that thermal treatment can address (Box 10.1). The process can be considered as a waste-to-energy (i.e. valorisation) technique or as a pre-treatment to final disposal. Although their objectives may differ, all methods of burning solid waste are dealt with together in this chapter, due to the similarity of the underlying physical processes and issues involved.

10.2 Thermal treatment objectives

Burning of solid waste can fulfil up to four distinct objectives:

1. *Volume reduction.* Depending on its composition, incinerating MSW

Box 10.1 THERMAL TREATMENT OPTIONS.

1. Mass-burn incineration of Municipal Solid Waste (MSW) and residues
   from other waste treatment options.
   - reduces volume of waste for final disposal
   - produces a stable residue that will produce little or no gas on landfilling
   - produces a sanitised residue for landfilling
   - energy recovery possible, but limited by high moisture content of MSW, presence of non-
       combustible material and corrosive contaminants.
   - high level of airborne emissions from combustion, requiring extensive gas-cleaning
       equipment to meet emission standards.

2. Burning Refuse-Derived Fuel (RDF).
   - uses the combustible part of the waste stream specifically for energy production
   - either loose cRDF or pelletised dRDF can be burned
   - cRDF must be burned as produced. dRDF can be stored and transported, but needs more
       energy to produce.
   - RDF has a higher calorific value than MSW, so energy recovery is higher
   - RDF contains less non-combustible material than MSW, so less ash is produced
   - RDF has lower heavy metal content than MSW, so less gas cleaning is needed.
   - Combustion characteristics of RDF are more consistent than MSW, so combustion can be
       better controlled

3. Burning Source-Separated Paper and Plastic as Fuel.
   - Can use the paper and plastic collected separately from households which is in excess of
       recycling capacity
   - Due to low moisture content can be stored and transported
   - High calorific value, low ash content and consistent combustion advantages apply to this fuel
       as they do to RDF
   - Heavy metal content likely to be lower than for RDF as produced by a positive sort (rather
       than the negative sort for RDF), so gas cleaning requirements should be even lower
       than for RDF.

reduces the volume of solid waste to be disposed of by, on average, 90%. The weight of solid waste to be dealt with is reduced by around 70%. This has both environmental and economic advantages since there is less demand for final disposal to landfill, as well as reduced costs and environmental impacts due to transport if a distant landfill is used.

2. *Stabilisation of waste*. Incinerator output (ash) is considerably more inert than incinerator input (MSW), mainly due to the oxidation of the organic component of the waste stream. This leads to reduced

landfill management problems since the organic fraction is responsible for landfill gas and leachate production.

3. *Recovery of energy from waste.* This represents a valorisation method, rather than just a pre-treatment of waste prior to disposal. Energy recovered from burning waste is used to generate steam for use in on-site electricity generation or export to local factories and district heating schemes. Combined heat and power (CHP) plants increase the efficiency of energy recovery by producing electricity as well as utilising the residual heat. Often viewed as a 'renewable resource' (van Santen, 1993), burning solid waste can replace use of fossil fuels for energy generation. As a large part of the energy content of MSW comes from truly renewable resources (biomass), there should be a lower overall net carbon dioxide production than from burning fossil fuels, since carbon dioxide is absorbed in the initial growing phase of the biomass.

4. *Sanitisation of waste.* Whilst this is of primary importance in incineration of clinical waste, incineration of MSW will also ensure destruction of pathogens prior to final disposal in a landfill.

10.3 Current state of thermal treatment in Europe

The prevalence of thermal treatment, and the actual approach that it takes, reflects the relative importance attached to the different objectives listed above, and varies from country to country across Europe (Table 10.1). Countries with an acute shortage of landfill capacity, such as Switzerland, the Netherlands (and Japan), incinerate a high proportion of MSW principally for volume reduction, with some energy recovery. By comparison, countries with plentiful and currently cheap landfills, such as the UK and Spain, have little MSW incineration (8% and 5% of MSW, respectively).

Volume reduction, for both environmental and economic reasons, and sanitisation of waste have historically been important objectives for incineration. These were the prime reasons for the MSW incinerators built in the UK, for example, in the 1960s. It is also likely that the future will see more emphasis on using incineration for stabilising wastes for subsequent landfilling. This will increase the proportion of MSW incinerated, particularly in Germany and Austria. Due to growing concern over the production of landfill gas and leachate from landfills receiving untreated MSW, stabilisation of waste prior to disposal is becoming an important additional objective in some countries. Landfill gas and leachate arise principally from the organic fraction of MSW, which can be effectively converted to gases and mineralised ash by incineration.

Concern over emissions from landfill sites has led Germany to pass a new National Technical Directive for the future management of communal waste (TA Siedlungsabfall), which states that only inert materials can be

Table 10.1 The current state of MSW incineration in Europe

Country	No. of MSW incineration plants	% of MSW incinerated	Energy recovery (% of capacity)	Energy recovery: Type[a]
Austria	2	8.5%	100%	
Belgium	25	54%	30%	HW/ST/E
Czech Republic		4%	77%	
Denmark	38	65%	100%	mostly district heating
Finland		2%	100%	
France	170	42%	72%	ST/HW/E
Germany (unified)	49	34%	43 plants	E/ST/HW
Greece	0	0%	–	
Hungary	0	0%	–	
Ireland	0	0%	–	
Italy	94	18%	33%	HW/E
Luxembourg	2	69%	100%	E
Netherlands	8	35%	6 plants	E/ST/HW
Norway		20%	89%	
Poland	0	0%	–	
Portugal	0	0%	–	
Slovak Republic		6%		
Spain	23	6%	5 plants	ST/E
Sweden	23	56%	100%	mostly district heating
Switzerland	30	80%	72%	
UK	34	8%	37%	HW/E
USA	168	16%	128 plants	
Japan	1873	74%	most plants	district heating/E

[a] HW = hot water, ST = steam generation, E = electricity generation (given in order of level of use)
Sources: European Energy from Waste Coalition (1993); *Shell Petrochemicals* (1992); Warmer Campaign (1990); RCEP (1993); MOPT (1992); OECD (1993).

landfilled. Inert materials are defined as those that have less than 1% (for normal community waste landfills) or 5% (for 'special' community landfills) weight loss on incineration. Note that the directive does not make incineration mandatory for material going to landfill, but at present there is no other technology available that can achieve this level of inertness. This may require the building of 40 or more new incinerators to meet the new legal requirements. Austria is also considering similar legislation, and it seems likely that this trend will spread.

Scandinavian countries have generally exploited incineration of MSW for energy recovery (e.g. Denmark and Sweden use 65% and 56% of their MSW, respectively, in this way) usually via district heating schemes (van Santen, 1993). Energy recovery has been taken further in the development of the refuse-derived-fuel process. RDF technology has been developed extensively in the UK and Sweden, although plants have also been set up in France, Germany, Switzerland and the United States (see Chapter 7 for details of the process and for locations of RDF plants). The other process discussed in this chapter, the burning of source-separated paper and plastic

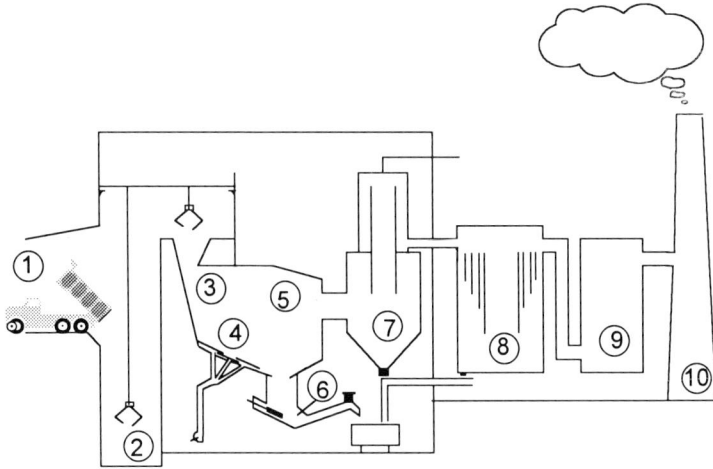

Figure 10.2 How does incineration work? 1, Reception hall; 2, waste pit; 3, incinerator feed hopper; 4, combustion grate; 5, combustion chamber; 6, quench tank for bottom ash; 7, heat recovery boiler; 8, electrostatic precipitator; 9, acid gas scrubbing equipment; 10, incinerator stack.

as an alternative fuel has not been fully developed. However, as the amount of paper and plastic collected by materials recovery schemes across Europe increases and exceeds reprocessing capacity, interest in this area is likely to grow.

10.4 Mass burn incineration of MSW

This form of thermal treatment can be divided into several distinct stages: the incineration process, energy recovery, emission control and treatment of solid residues.

10.4.1 The incineration process

The layout of a modern incinerator is shown in Figure 10.2. Mixed waste for incineration is delivered into a reception hall or tipping bunker, from which it is fed into the furnace feed hopper, usually by mechanical grab. The reception area can be a source of local nuisances such as noise, odour and litter, which need to be controlled.

The majority of MSW incinerators have a furnace with a moving grate design, also known as a 'stoker' type incinerator. The moving grate keeps the waste moving through the furnace as it burns and deposits the unburned residue (bottom ash) into a quench tank (Figure 10.2). Primary air for combustion is pumped through from under the grate, and secondary air is introduced over the fire to ensure good combustion in the gas phase.

Box 10.2   COMPARISON OF FLUIDISED BED AND MASS BURN (GRATE) SYSTEMS FOR MSW INCINERATION

	Fluidised bed incinerator	Grate incinerator
Structure		
Orientation	vertical	horizontal
Moving parts?	no	yes
Max capacity of single stream	*ca.* 350 tonnes/day	*ca.* 1200 tonnes/day
Combustion		
Mixing	turbulent	mild agitation
Rate	rapid	slow
Burn out	complete	often incomplete
Air ratio	1.5–2.0	1.8–2.5
Load	400–600 kg/m^2/h	200–250 kg/m^2/h
Fuel size	50 cm	75 cm
Combustion residue		
Unburnt carbon	0.1% by wt	3–5% by wt
Volume	smaller	larger
State	dry	wet
Iron recovery	easy	difficult
Fly ash		
Volume	larger	smaller
Unburnt carbon	1% by wt	3–7% by wt
Flue gas		
Volume	smaller	larger
NO$_x$ control	by air ratio control	by added chemicals
Operation		
Stop	few mins (no unburnt fuel)	few hours (unburnt fuel remains)
Restart after 8 h stop	5 mins	1 h
Restart after weekend stop	30 mins	2 h

Source: Patel and Edgcumbe (1993).

Fluidised-bed incinerators (FBI) represent an alternative design, which has been developed and implemented, particularly in Japan. These use a combustion chamber containing a fluidised bed, created by air forced up through a bed of inert material such as sand, into which the waste is introduced. The hot fluidised-bed ensures a high level of combustion, and also benefits from having no moving parts, so generally fewer mechanical problems. A comparison of moving-grate and fluidised-bed incinerators is given in Box 10.2.

A third technology is the rotary combustor, in which the waste is rotated slowly in a cylindrical kiln as it burns; the kiln has perforations along its length to allow air to be pumped in to ensure good combustion. This type of incinerator is not widely used for municipal solid waste, however.

Table 10.2 Crude gas values of two different waste incineration plants showing effect of furnace design on combustion efficiency

	Old	Modern
Dust (g/N m^3)	6.5	1.7
C in dust (%)	2.7	1.4
PCDD (ng/N m^3) (dioxins)	270	55
PCDF (ng/N m^3) (furans)	1100	110
TEQ (ng/N m^3)	25	2.5

Source: Vogg (1992).

The key factors for high levels of combustion and destruction of organic pollutants in the incoming waste are temperature (high), residence time (long) and turbulence (high) (Vogg, 1992). The EC directives (1989a, b), for example, require a residence time of at least 2 s at temperatures in excess of 850°C, in the presence of at least 6% oxygen to ensure maximum oxidation of dioxins and other organic pollutants. Furnace design and operating efficiency are the crucial factors which determine the levels of pollutants in the crude gas that enters the flue gas cleaning system (Table 10.2).

The hot gases then enter the energy recovery boiler, where they cool rapidly. If no energy is recovered, the flue gases must be cooled using air or water sprays before they enter the gas cleaning systems.

10.4.2 Energy recovery

Conventional energy recovery involves passing the hot flue gases through the boiler, whose walls are lined with the boiler tubes. Water circulated through these tubes is turned to steam, which can be heated further, using a superheater, to increase its temperature and pressure to make electricity generation more efficient. The thermal efficiency of modern boilers is around 80%. If the steam is used to generate electricity, however, the overall energy recovery efficiency (from calorific content of fuel to electricity generated) is around 20% (RCEP, 1993).

If wet feedstocks are used (e.g. high in food and garden waste), much of the gross calorific content of the waste is used up in evaporating this moisture. This latent heat contained within the flue gas is not normally recovered. Recent advances in energy recovery techniques, in particular the flue gas condensation (FGC) process, however, allow some of this latent heat to be recovered as well, so increasing the efficiency of energy recovery. An additional benefit of FGC systems is the relatively high rate of removal of acid gases from the flue gas during the process, leaving less to be removed by subsequent emission control equipment. FGC systems are not widespread, but have been installed in five large MSW incinerators in Sweden.

10.4.3   Emission control

The flue gases that leave the energy recovery boiler are subjected to a range of treatment processes to remove both particulate and gaseous pollutants before they are released via the incinerator stack. The number of different treatment processes used varies widely from plant to plant, reflecting the emission standards that are required. Only the most common processes are described briefly here.

Dust filters.   In many plants the first (and sometimes only) treatment process involves electrostatic precipitators to remove particulate matter. Cyclones can also be used for this purpose. Where wet acid gas scrubbing occurs, filtering occurs before gas scrubbing. With dry or semi-dry processes, filtration occurs after acid gas removal. In the later stages of the emission control process, filtration often involves the use of a bag filter.

Acid gas removal.   Three basic alternatives exist for removing the acid gases SO_2, HCl and HF from the flue gas:

  (a) *Wet method*. Pollutants are removed by large quantities of cleaning liquid (usually slaked lime or aqueous sodium hydroxide) in a Venturi scrubbing system. This gives a high degree of acid gas removal, and the waste products from the cleaning process (gypsum, sulphuric acid) may be re-usable. The disadvantages are that extensive additional equipment is necessary and waste water is produced, which requires treatment prior to discharge. In some plants, this water is recirculated after treatment, leaving only a sludge for disposal. Also, as the flue gases are cooled during the process they may need to be reheated prior to discharge (although this can be achieved by heat exchange from the incoming hot flue gases) to promote dispersal of the emissions and to reduce the visibility of the stack plume caused by condensing water vapour.
  (b) *Dry method*. Limestone, for neutralisation, is either mixed with the waste before combustion or introduced separately into the furnace. This method has the advantage that it requires little additional equipment, produces no waste water, and the exhaust gases require no reheating. However, the level of acid gas removal is relatively low, there is a high limestone demand and the reaction product is mixed with ash, so cannot be re-used.
  (c) *Semi-dry method*. A higher level of acid gas removal than the dry method is achieved by spraying lime (CaO) mixed with water into the flue gas flow. The water evaporates and the pollutants combine with the absorbent to form a dry solid. Here again no waste water is produced, and the flue gases are only slightly cooled, so need no reheating. Additional equipment and surplus lime are needed for this

method, however, and the reaction products, as with the dry method, are not directly re-usable.

NO$_x$ control. In fluidised bed incinerators, production of oxides of nitrogen (NO$_x$) can be limited by controlling the amount of air supplied to the combustion process. In moving-grate incinerators, production of NO$_x$ cannot be prevented, but it can be removed by the injection of urea or ammonia into the flue gases during the emission control process.

Dioxin control. Organic pollutants in the MSW entering an incinerator include chlorinated benzols and phenols, polychlorinated biphenols (PCBs), polychlorinated dibenzo-*para*-dioxins (PCDDs) and polychlorinated dibenzofurans (PCDFs). Of these, the PCDDs ('dioxins') and PCDFs ('furans') have generated the most concern, due to their generally high levels of toxicity and persistence in the environment. As different isomers of these compounds differ considerably in their toxicity, amounts of these two classes are usually presented in terms of their toxicity equivalents (TEQ).

Provided that the incineration process is run efficiently (residence time after last air injection of at least 2 s at $\geqslant$850°C at $\geqslant$6% oxygen (EC Directives, 1989a, b)), most of these organic pollutants in the incoming waste will be broken down. However, the issue is complicated by the production of more dioxins and furans by *de novo* synthesis in the flue gases in the post-combustion zone at temperatures between 250°C and 450°C. Although the precise mechanisms for dioxin formation are unknown, fly ash (in particular the presence of catalysts such as copper oxide in the fly ash), oxygen content and the level of water vapour in the cooling flue gas are known to be important factors (Vogg, 1992).

Despite such *de novo* synthesis, however, the levels of dioxin emitted from a MSW incinerator are considerably less than the levels in the input (Figure 10.3).

To meet current or future stringent emission controls for dioxins (0.1 ng toxic equivalent (TEQ)/N m^3) (see below) further treatment of the flue gases is likely to be necessary. One method is adsorption onto activated carbon filters. This does not destroy the dioxins, however, so the contaminated carbon presents a further disposal problem. An alternative method recently developed is catalytic destruction (ENDS, 1993). The process uses a mixed metal oxide catalyst and destroys the dioxins by reaction with oxygen. It is claimed that this process can reduce typical dioxin levels from 1 to 10 ng TEQ per m^3 to below 0.1 ng/m^3.

10.4.4   Treatment of solid residues

The bottom ash, once it has been cooled in the quench tank, is usually passed under an overhead magnet to recover any ferrous metal. Around 90% of the

Table 10.3 Distribution of heavy metals and other elements in MSW incinerator residues

Fraction	Stack gas	Bottom ash	Filter dust	Salts and sludges from gas cleaning
Cadmium	0.04%	11%	85%	3.6%
Chlorine	0.12%	9%	15%	76%
Chromium	0.01%	94%	5.8%	0.27%
Copper	0.01%	95%	4.9%	0.53%
Fluorine	1.5%	69%	3.0%	26%
Lead	0.01%	75%	24%	0.9%
Mercury	2.1%	7%	5.1%	86%
Nickel	0.04%	87%	13%	0.61%
Sulphur	0.47%	50%	10%	40%
Zinc	0.05%	49%	51%	0.7%

Source: IFEU, 1992.

ferrous metal can be recovered this way (IFEU, 1992). The remaining bottom ash, and the fly ash and residues from gas cleaning then require disposal.

MSW contains inorganic pollutants, of which heavy metals form an important group (as listed in Table 10.3 and above in Table 5.6.) The chemical nature of these may be modified by heat, but they are not destroyed. They will therefore leave the incinerator either in the air emissions, bottom ash, filter dust or sludge from gas cleaning. The distribution of common inorganic pollutants between these different outputs is shown in Table 10.3. Due to the volume reduction involved in incineration, considerable concentration of these materials will occur in the resulting residues, especially in the fly ash. This has both advantages and disadvantages. Concentration of pollutants means that the residues may need to be handled as hazardous wastes, necessitating disposal in special hazardous waste landfills, or further treatment. The converse of this is that being concentrated, there is less material to render inert and treatment to remove mobile heavy metals

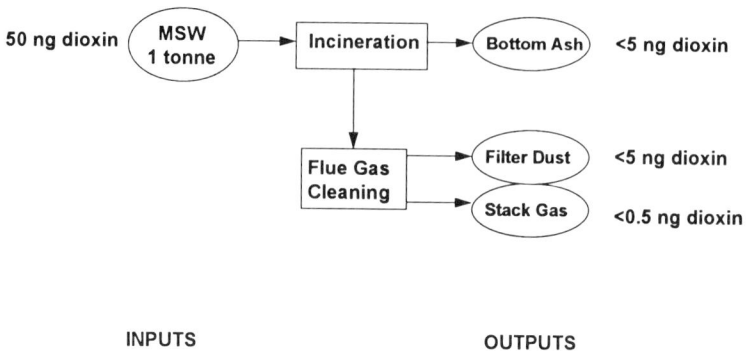

Figure 10.3 Dioxin balance for MSW incineration. *Source*: Vogg (1992).

(e.g. cadmium, mercury, copper, zinc) for re-use is possible (Vogg, 1992). Processes involving solidification, washing or melting techniques to render flue dusts safe for final disposal are under investigation (Vogg, et al., 1989).

The bottom ash generally has a low carbon content (1–2%), making this material suitable for disposal to landfill, either direct or after further processing. In Japan, for example, ash melting in a carbon arc furnace is used to reduce the volume of the bottom ash. This produces a vitrified slag with half the volume of the original ash, but the process is very energy intensive, using up to 1000 kW-h per tonne of ash.

An alternative to disposal is the re-use of either the ash or the vitrified slag as a building material (Toussaint, 1989). By including 30–50% slag in concrete blocks, it is possible to produce a useful product as well as reduce the demand on landfill space.

10.5 Burning of refuse-derived fuel

Refuse-derived fuel (RDF) is produced by a specific process designed to select out the combustible from the non-combustible fraction of mixed MSW (as described in Chapter 7). The RDF consists mainly of the lighter materials in MSW (paper and plastic), which are separated out and shredded to produce 'floc' or non-densified cRDF. This can then be pelletized to improve handling, producing densified RDF (dRDF). Pelletizing the RDF is costly, however, with a high energy consumption, so attention has begun to move to the use of floc RDF, handled by bulk handling techniques (Ogilvie, 1992). A similar waste derived fuel is used in some industries, notably the cement industry, where shredded MSW can be used as a partial replacement for coal (Warmer, 1993b).

The coarse cRDF can be burned in a fluidised bed incinerator (ETSU, 1993), densified dRDF can be burned in either a fluidised bed or on a conventional grate.

The benefits of using RDF, and cRDF in particular have been summarised as follows (RCEP, 1993):

(a) cRDF has a higher calorific value than raw MSW and is more uniform in combustion characteristics;
(b) cRDF has a lower heavy metal content than MSW so reducing the demands on the gas cleaning equipment;
(c) cRDF contains much less non-combustible material than MSW, so there is less ash left for disposal;
(d) overall efficiency is higher since the combustion characteristics can be tailored more precisely to the fuel specification.

10.6   Burning of source-separated paper and plastic

Source-separated fuel is similar in some ways to floc RDF (above), but would arise from source-separated collections for materials recycling, rather than produced from a mechanical screening process (Chapter 7). Current kerbside collection schemes for dry recyclable materials have demonstrated that high recovery rates for most materials can be achieved (e.g. IGD, 1992). If such schemes are to be extended country- or even Europe-wide, large amounts of materials will be collected (see Chapter 6). Whilst there is likely to be recycling capacity for metals and most glass, the amounts of paper and plastic collected are likely to be well in excess of what can be recycled in an environmentally and economically sustainable way. In the case of paper, it will always be necessary to inject virgin pulp into the system to compensate for the degradation of fibre length with each successive use. There will always be excess paper in the waste stream over recycling requirement. In the case of plastic, recycling capacity is low because it is still a costly operation, resulting in a premium being paid for recycled material over virgin. In the case of very lightweight plastic packaging items (e.g. yoghurt pots, carrier bags, sweet wrappers) recycling is unlikely to be environmentally sustainable because of the environmental impacts associated with washing.

A logical use of this source-separated paper and plastic that cannot or should not be recycled is as a fuel, since it has a high calorific value (see LCI Data Box 1 in Chapter 5), and should be relatively free of contaminants, since it is produced by a positive sort at source (compared to a negative sort from mixed MSW as in the case of dRDF). This alternative could either be burned in power stations along with conventional fuel (using coal-burning grate technology) or in small dedicated boilers in industrial plants to provide steam or heating. Trials are underway to assess the feasibility of such fuel usage.

10.7   Emission limits

Strict emission limits for solid municipal waste incineration are currently in force in several countries across Europe, and similar standards will be introduced to all EC countries as a result of the EC Directives for new and existing MSW incineration plants (EC, 1989a, b) (Table 10.4). Plants equipped only with electrostatic precipitators (e.g. in the UK; Clayton et al., 1991) will not be able to meet these standards, and so will be forced to either upgrade their emission control, to include at least gas scrubbing equipment, or to shut down (see ENDS, 1992). Some plants are likely to fail the requirement for a minimum residence time of 2 s at over 850°C, and here retrofitting is not an option. As a result of the EC Directive, there is likely to be

Table 10.4 Emission values for municipal solid waste incineration plants (mg/m^3)

Contents of regulation	EC Directive for new plants > 3 tonnes/h (1989) (7 day average value)	Germany		Netherlands (1990)
		TA air emission values (1986) (24 h average value)	Limit values of the new Ordinance (17.BImSch V) (December 1990)	
HCl	50	50	10	10
HF	20	2	1	1
SO$_x$	300	100	50	40
NO$_x$	–	500	200	–
CO	100[a]	100	50 (100)[a]	50
Organics[b]	20	20	10	10
Dust	30	30	10	5
Heavy metals	Cd + Hg 0.2 Ni + As 1.0	Cd + Hg + Tl 0.2 Class II 1.0 (As + Co + Ni + Se + Te) Class III 5.0 (Sb + Pb + Cr + Mn + V + Sn + Cu)	Cd + Tl 0.05 Hg 0.05 Class III 0.5 (Sb + As + Pb + Cr + Co + Cu + Mn + Ni + V + Sn)	Cd 0.05 Hg 0.05 Class III 1.0
	Pb + Cr + Cu + Mn 5.0			
PCDD + PCDF	–		0.1 ng I-TE	0.1 ng
Values relate to	11% O$_2$ or 9% CO$_2$ dry flue gas	11% O$_2$ dry flue gas	11% O$_2$ dry flue gas	11% O$_2$ dry flue gas

[a] 1 h average value.
[b] Total carbon.

a fall-off in incinerator capacity across Europe in the mid-1990s as old plants which cannot meet the new emission standards close down. There is evidence from both manufacturers' guarantees and from test firings, however, that both retrofitted old plants (e.g. Zirndorf, Bavaria) and new plants can meet or exceed the new stricter standards for atmospheric emissions (Table 10.5).

For the burning of RDF and source-separated materials, the position is less clear. In several countries (e.g. Germany, Belgium, the Netherlands and the UK), these materials are legally considered to be waste, and are therefore subject to the same emission controls as mass burn incinerators. Since they involve burning only part of the waste stream, both the variety and levels of many pollutants might be expected to be lower than for mass burn incineration.

10.8 Public acceptability

During the 1980s, public concern over emissions from incinerators was very high. This period saw the growth both of the NIMBY (Not In My Back Yard) syndrome, and of genuine concern about the environmental effects of airborne emissions from the incinerators then operating. To some extent, the public debate has failed to separate issues relating to incineration of hazardous chemicals from those relating to the incineration of municipal solid waste. Also, typically at that time, dust collecting electrostatic filters were the only emission abatement techniques used in most European countries, giving rise to high levels of HCl, heavy metals and dioxins in stack gases released. More stringent emission regulations, both on a national and EC scale now mean that all new facilities will need sophisticated gas cleaning equipment to achieve the regulated levels. Emission levels from such incinerators may be sufficiently low to reduce concern over their environmental effects.

Public concern has turned instead to the perceived incompatibility of incineration with materials recycling (ENDS, 1992). If incinerators attempt to maintain a certain throughput for profitability, this will obstruct schemes to remove certain materials from the waste stream. Whilst this may be true for existing incinerators, it should be possible to match incinerator capacity to the waste generated from an area, after a given level of materials recovery has occurred. Figure 10.4 shows that recycling metals and glass and composting of organics has a very positive effect on the calorific value of the residue, and another application of waste-to-energy can fully complement materials recycling by utilising collected paper and plastic that is in excess of recycling capacity.

Table 10.5 Effect of improved gas cleaning equipment on performance of MSW incineration

	Limit value TA Luft (clean air act)	17th Emission regulation	Plant emissions before retrofitting (mg/m^3)	Plant emissions after retrofitting (mg/m^3)
Total dust	30	10	50	<0.3
Organic substances (Total carbon)	20	10		<2
Gaseous inorganic chlorine compounds (HCl)	50	10	35–40	<1
Gaseous inorganic fluorine compounds (HF)		1	1	0.9
Sulphur dioxide and trioxide (SO$_2$ and SO$_3$)	100	50	300	<1
Carbon monoxide (CO)	100	50	20–30	11.2
Cadmium, thallium (Cd, Tl)	–	0.5		<0.001
Mercury (Hg)	0.05	0.05	0.2	<0.02
Total of all other heavy metals (Sb, As, Pb, Cr, V, Sn, Co, Cu, Mn, Ni)	100	0.5		0.02[a]
Dioxins/furans (PCDD/PCDF) Total TEQ in ng/N m^3	–	0.1	8	0.0059

Table gives limit values and actual values before and after retrofitting of incineration plant at Zirndorf, Fürth, Germany.
< Signifies emission below level of detection;
[a] Most heavy metals are below level of detection
Source: Warmer (1993a).

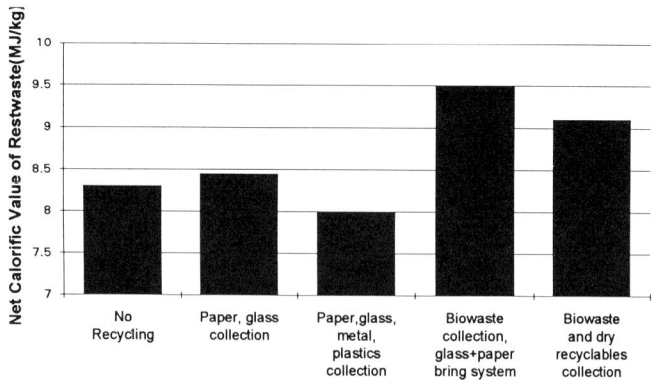

Recycling rates:					
paper	0%	75%	75%	50%	75%
glass	0%	80%	80%	50%	80%
metal	0%	0%	50%	0%	50%
plastic/textiles	0%	0%	40%	0%	40%
biowaste	0%	0%	0%	94%	94%
Overall	0%	20%	25%	50%	60%
Net calorific value of restwaste (MJ/kg)	8.3	8.45	8.0	9.5	9.1

Figure 10.4 Variation in the calorific value of restwaste with different recycling scenarios. Based on MSW composition in West Germany in 1985. *Source*: Habig (1992).

10.9   Environmental impacts: input–output analysis

10.9.1   Defining the system boundaries

Before the inputs and outputs of incineration can be assessed, it is necessary to define the system boundaries for the process. These are shown in Figure 10.5, and include the emission abatement processes within the system. The actual emissions to air will therefore depend on the type of emission abatement equipment installed, and the efficiency of its operation. For this study, emission data from modern, state-of-the-art equipment will be used, since the overall objective is to assess possible effects of future IWM schemes, rather than what is currently being achieved across Europe. As with all aspects of this LCI approach, if local data on existing or planned incinerators are available, these should be used in preference to generic figures.

The amounts of solid and aqueous wastes generated will also depend on the type of gas cleaning process used (see above): a dry scrubbing system

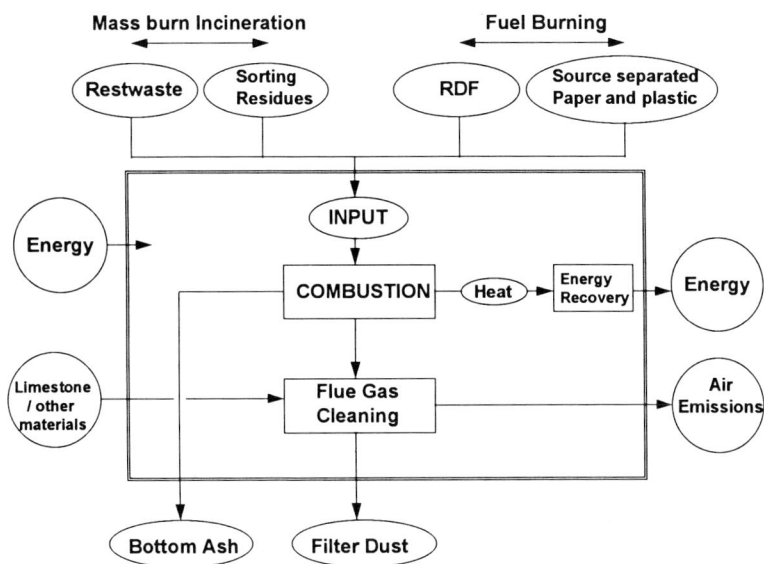

Figure 10.5 Inputs and outputs of thermal treatment processes.

will generate only solid waste whereas a wet gas scrubbing system will produce both solid and aqueous wastes.

For the purpose of the LCI, it is assumed that recovered energy is used to generate electricity only, and that this is exported from the system.

10.9.2 Data availability

The outputs of the incineration process clearly also depend on the inputs, i.e. what is burned. Most incinerators are mass burn facilities, with an input of mixed MSW, and data are generally available on the associated outputs. However, as fractions of the waste stream are removed for recycling or composting, or when RDF or source separated materials are burned as fuel, the input to the incineration process may be markedly different (Figure 10.4). To predict the outputs from such different combustion scenarios, it is necessary to be able to attribute energy production and emissions to individual materials or fractions within the incinerator's input. Whilst this is possible for energy production, emissions cannot be reliably attributed for two reasons. Firstly, there is a serious lack of data in this area, since most experience has been with the mass burn approach. Secondly, there are interactions between different materials in the combustion process, such as in the *de novo* production of dioxins, and these mechanisms are not fully understood.

Available data are presented below for incineration of mixed MSW, dRDF and for combustion of individual materials. It needs to be borne in mind, however, that not all emission data sets are complete, and that incineration of mixtures may not give rise to the sum of their parts, since interactions may occur.

10.9.3 Inputs

Waste inputs.   The input to the thermal treatment system will be either MSW, RDF or source-separated paper and plastic depending on the process involved. Burning RDF or paper and plastic as fuels involves a relatively consistent and well-defined feedstock. The input to MSW incinerators, however, comprising of restwaste and residues from other waste treatment processes, is much more variable. This feedstock will vary with the waste composition, which has been shown to vary both geographically and seasonally (Chapter 5).

Further variability in the input to MSW incinerators will also occur when materials are recovered for recycling or composting, so affecting the composition of the remaining restwaste. This will affect the overall calorific value of what is burned, and has led to concern over the compatibility of recycling with waste-to-energy schemes (e.g. ENDS, 1992). Whilst any materials recovery and recycling scheme will reduce the overall throughput of a mass burn incinerator, the average calorific content of the throughput may rise or fall, depending on which fractions are removed. Several studies have looked at this effect (Figure 10.4). In general, recovery of glass and steel for recycling will increase the average calorific content, as will separate collection of putrescible material for composting.

The calorific value of the feedstock, and the variability thereof, needs to be considered when planning an MSW incinerator. Problems with the calorific value of the waste input rising above the heat capacity of the incinerator plant have been encountered in both Switzerland and Japan. In both these countries, plastics now need to be excluded from the feedstock to prevent damage to some incinerators. In Greece, by contrast, the high level of putrescible waste has resulted in a calorific value too low to give effective burning.

Energy consumption.   The thermal treatment process, as well as liberating energy from the feedstock, also consumes energy. This is required for operating the cranes, moving grates or fluidised beds, fans for air injectors, and emission control equipment, as well as for general heating and lighting. The on-site energy consumption will particularly reflect the level of flue-gas treatment. Table 10.6 gives the energy consumption of various parts of the gas-cleaning equipment in a hazardous waste incinerator in Vienna. In this case, almost all of the electrical energy recovered from the waste is

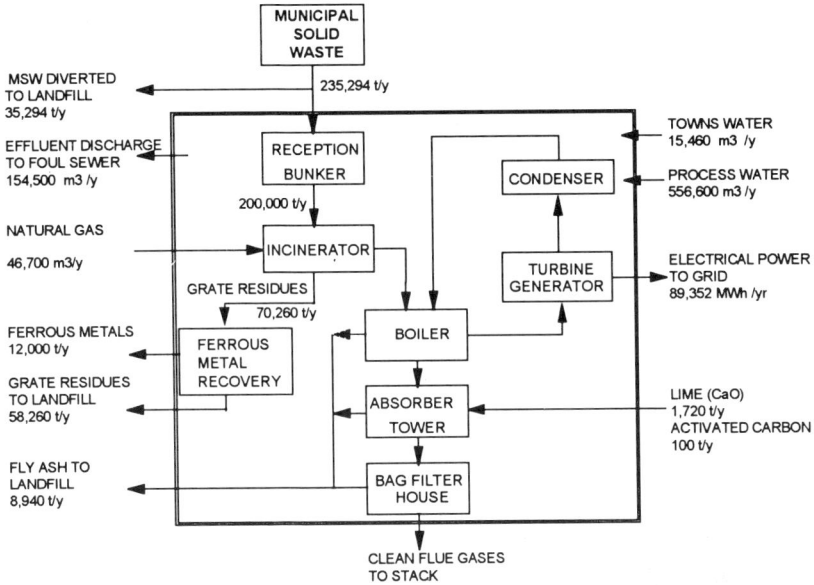

Figure 10.6 Mass balance for a MSW incinerator with a capacity of 200 000 tonnes/year. *Source*: ETSU (1993b).

Table 10.6 Energy consumption of gas cleaning systems for hazardous waste incinerator located in Vienna

Equipment	Power consumption kW electricity	% of total power consumption	% for environmental protection
Rotary kiln	638	14.2%	–
Precipitator	86	1.9%	1.9%
Gas scrubbing systems	1933	43.1%	43.1%
Water purification	117	2.6%	2.6%
Carbon filter	1200	26.8%	26.8%
Others	511	11.4%	–
Total	4485	100.0%	74.4%

Source: Hackl (1993).

consumed on site. For MSW incineration, ETSU (1993) suggest that around 14% of the electrical power generated is consumed on-site, with a specific consumption of around 70 kW-h per tonne incinerated. Energy is also required in the form of natural gas, to heat up the incinerator during start-up. For this ETSU (1993) give an average figure of 0.23 m^3/tonne.

For RDF and alternate fuel (paper/plastic) burning, there will be some energy consumed by the flue-gas cleaning equipment, although no data are available for this. It is likely to be lower than for MSW incineration, since there is less plant to operate and the feedstock is less contaminated, so the flue gas should need less cleaning. For the purposes of the LCI spreadsheet, it will be assumed that there is an energy consumption of 20 kW-h/tonne of RDF or paper/plastic fuel burned.

10.9.4   Outputs

Energy production.   Energy released during incineration may be used for several purposes (heat and steam production, electricity generation) each with its own conversion rate for the amount of useful energy produced. For the purposes of this LCI, however, it will be assumed that the recovered heat is used to generate electricity only.

Mass burn. Energy production will depend on the composition of MSW, and will be higher in countries with high levels of paper and especially plastic (generally Northern Europe), and lower where there is a high level of wet organic/putrescible waste (Southern Europe) (see Chapter 5). For Germany values in the range 8.3–9.2 MJ/kg are given (Bechtel and Lentz, unpublished), Sweden 7–11 MJ/kg (Svedberg, 1992), and for the UK a gross calorific value of 8.4–10 MJ/kg has been reported (Barton, 1986; Porteous, 1991). It is necessary to take account of the water content of MSW as received at the incinerator, since this will reduce the overall calorific value per kilogram, due to both its weight (dilution effect) and latent heat of evaporation. In the case of the UK, this reduces a typical gross calorific value of MSW from 8.4 MJ/kg to 7.06 MJ/kg (see LCI Data Box 1 in Chapter 5). For comparison, the calorific value for fuel oil is 40–42 MJ/kg, for hard coal 25–30 MJ/kg, for peat 9–13 MJ/kg and for wood chips 7–13 MJ/kg (Svedberg, 1992).

In the LCI spreadsheet, to allow for variability in the feedstock, the calorific value of the input to MSW incinerators will be calculated from the composition, using the material-specific calorific values given in LCI Data Box 6 below.

RDF. Ranges given for the calorific value of RDF vary from 11 to 18 MJ/kg (Swedish data; Svedberg, 1992) to 18–20 MJ/kg (UK data; Ogilvie, 1992). The exact value will vary with the composition of the original waste, and the process used to produce the RDF. For the purpose of the LCI spreadsheet, a calorific value of 18 GJ/tonne will be assumed for dRDF.

Source separated fuel. The calorific value of this alternative fuel will depend on the ratio of paper to plastic burned. The net calorific values of each fraction given earlier in LCI Data Box 1 are used to calculate the calorific values of any specified mixture.

Energy recovery. Clearly not all of the primary heat energy released in an incinerator can be recovered in a useful form, and the level of recovery depends on the use to which the energy is put. Since boilers attached to municipal waste incinerators must operate at lower steam temperatures to reduce corrosion, incinerators producing electricity only have a conversion efficiency of around 20% (RCEP, 1993). ETSU (1993) predict a gross power production of 520 kW-h per tonne of waste with a net calorific value of 8.01 GJ/tonne, giving a conversion efficiency of 23% for a plant of 200 000 tonnes per year capacity. Once the on-site power consumption of 70 kW-h/tonne has been subtracted, this gives a net export of 450 kW-h/tonne from the site.

Energy recovery for district heating schemes, such as used extensively in Japan, recover around 70% of the energy released, whereas combined heat and power (CHP) schemes, that utilise the residual heat after generation of electricity achieve an overall conversion efficiency of around 70–80%, This means that a plant such as in Frankfurt, burning 420 000 tonnes of waste per year can provide 30 000 people with both electricity and district heating, replacing the need for some 64 000 tonnes of fuel oil per year (Bechtel and Lentz, unpublished).

For the purposes of the LCI model, an electricity production efficiency of 20% will be assumed for MSW incinerators. For plants burning RDF or a paper/plastic fuel, a higher efficiency of 30% will be assumed due to the more homogenous and controlled nature of the feedstock.

Air emissions. It is important to remember that due to the conservation of mass, whatever materials enter the incineration process will leave it in one state or another. Although organic pollutants may be broken down into harmless molecules, heavy metals within the waste stream will either be left in the ash (clinker), removed as filter dust or will escape to the air. The total amount will remain the same, so as flue gas cleaning becomes more effective and less inorganic pollutants are emitted to the atmosphere, the amounts collected from the flue gas cleaning process will correspondingly increase.

Mass burn. The most significant pollutants emitted from MSW incineration are the acid gases (hydrogen chloride, sulphur dioxide, nitrogen oxides), carbon dioxide, heavy metals (mercury, cadmium, lead), particulate matter and dioxins/dibenzofurans (PCDD/PCDFs) (Clayton *et al.*, 1991). As legislative limits on emissions are set in terms of concentrations of pollutants in the stack gas emitted from incinerators, most data are collected in this form. Data on emissions from MSW incinerators have already been given in Table 10.5, but since the level of air emissions will vary according to the flue gas cleaning equipment used, it is impossible to give typical figures. An alternative approach for the purpose of the LCI spreadsheet is to assume that the emissions from incinerators are within the limits of the EC Directive, since all incinerators in the European Union countries will have to achieve these standards by 1996. Table 10.7 gives the predicted emissions for MSW incineration, based on this approach (IFEU, 1992). To convert these concentrations into amounts emitted per tonne, it is necessary to multiply by the volume of flue gas produced; IFEU (1992) use a figure of $5000\,N\,m^3$ of flue gas per tonne of MSW incinerated (taken from Fichtner, 1991).

The amount of carbon dioxide emitted cannot be calculated this way, since there are no legal limits for this gas. The amount emitted can be calculated, however, from the carbon content of the fractions of MSW given earlier in LCI Data Box 1, assuming that all of the carbon is given off as carbon dioxide.

To account for variations in the composition of MSW, the emissions arising from each fraction of the waste stream need to be identified. Such data are not available, and in any case some emissions, such as dioxins formed *de novo* in the incineration process cannot be attributed to any waste fraction in particular. Because of this, for the LCI calculations, the emissions for MSW in Table 10.7 are attributed equally to all the fractions entering the mass burn incineration process. The one exception is carbon dioxide, which is attributed according to the carbon content of each fraction (see LCI Data Box 6).

RDF and source-separated fuels. Data for burning RDF and paper and plastic are also included in Table 10.7. Note that the data for RDF burning are from UK trials with only dust precipitation equipment. Use of gas scrubbing equipment would reduce airborne emissions, but would increase solid waste generation from the filters.

Water emissions.   This arises only from use of wet gas scrubbing equipment. Figures given for the amount of sewage produced range from 200 to 770 l/tonne of waste input (SPMP, 1991; Bechtel and Lentz, unpublished; ETSU, 1993). In the LCI spreadsheet, it is assumed that all water emissions are treated on-site, with only resultant sludges leaving the site.

Solid waste.   Solid residues from incineration arise from two main sources: combustion residues (bottom ash and fly ash) and solid residues from the gas cleaning system. The latter will include both filter dusts and sludge residues resulting from water treatment (where wet gas scrubbing systems are installed).

Mass burn of typical MSW results in 250–300 kg of bottom ash per tonne of waste (SPMP, 1991; IFEU, 1992: ETSU, 1993). To allow for variability

Table 10.7 Air emissions from combustion of MSW, refuse derived fuel (RDF) and source separated paper and plastic

Source	Municipal Solid Waste (g/tonne burned) IFEU (1992)[1]	Municipal Solid Waste (g/tonne burned) EC Directive 89/369/EEC (1989)[2]	Refuse Derived Fuel (RDF) (g/tonne burned) Hickey and Rampling (1989)[3]	Paper/Board (g/tonne burned) Habersatter (1991)[4]	Plastic (no PVC) (g/tonne burned) Habersatter (1991)[4]
Particulates		150	1710	20	50
CO		500	347	400	1250
CO_2			1 421 000		
CH_4					
NO_x	500		2019	1600	5000
N_2O					
SO_x	125	1500	5197	300	360
HCl	25	250	5570	100	0
HF	2.5	100			
H_2S					
Total HC	0.0384	100	46		
Chlorinated HC	0.0836				
Dioxins/furans (TEQ)	0.0000005		0.0000678		
Ammonia					
Arsenic	0.008	2.5*			
Cadmium	0.0135	0.5*	2.8		
Chromium	0.013	6.25*			
Copper	0.055	6.25*			
Lead	0.245	6.25*	63.2		
Mercury	0.0085	0.5*			
Nickel	0.002	2.5*			
Zinc	0.85				

[1] Based on plant meeting or exceeding German 17th Emission Regulation, and assuming a production of 5000 m³ of stack gas per tonne of MSW incinerated.
[2] Calculated for plant meeting EC Directive for new MSW incinerator plants, assuming 5000 m³ of stack gas per tonne incinerated.
* Limits are for sum of heavy metals, so limit has been divided evenly between metals combined.
[3] Data are averages of 3 trials of burning dRDF, on as received basis with average moisture content of 11%. Note gas clean-up was limited to cyclone for particulates only.
[4] Data assume flue gas cleaning capable of removing 90% of SO_x and 95% of HCl.

in the mass burn incinerator feedstock in an IWM system, however, the amount of bottom ash in the LCI spreadsheet is calculated from the ash content of each fraction of the waste (LCI Data Box 6).

Dry gas cleaning systems produce approximately 45–52 kg of dust and residues (calcium chloride and surplus lime) per tonne of waste . The semi-dry or semi-wet processes are similar, producing 40 kg of loose ash mixed with calcium chloride and surplus lime per tonne of waste. Wet gas scrubbing results in 20–30 kg of dust and 2.5–12 kg of sludge residue per tonne of waste (SPMP, 1991: Bechtel and Lentz, unpublished; IFEU, 1992). In the LCI spreadsheet, it is assumed that for each tonne of MSW incinerated, 20 kg of filter dust and 12 kg of sludge residues from the gas scrubbing system are produced (IFEU, 1991).

RDF combustion will typically leave a residue of approximately 86 kg ash/carbon per tonne input, with a further 1.8 kg of ash filtered from the flue gas (Ogilvie, 1992). Data are not available for residues from further emission control, but are likely to be similar to those above for mass burn. It is therefore assumed that a further 12 kg of sludge residues result from gas scrubbing, giving a total of 13.8 kg from the gas cleaning process.

Source-separated fuel. Ash contents of different fractions of the waste stream are given in LCI Data Box 1 (Barton, 1986). To these must be added the solid waste generated from flue gas cleaning, to give total solid waste produced. Figures given by Habersatter (1991) for total solid waste from combustion in Switzerland are typically 35 kg/tonne input for plastics, and 87 kg/tonne for paper/card, but these are more conservative than those given for the ash content alone by Barton (1986) for UK waste. In the LCI spreadsheet, therefore, the bottom ash production is calculated from the ash content levels in LCI Box 1: the gas cleaning residues are assumed to be the same as for RDF combustion above.

10.10   Economic costs of thermal treatment

Mass burn.   The economic costs of incineration are generally considered to be high, because of the capital investment required to set up a plant. Estimates of this cost have been made at 1.18 million ecu to 2.37 million ecu/tonne per h capacity for the UK (Warmer, 1991; DoE, 1993) and 1.34 million ecu to 2.13 million ecu/tonne per h for France (SPMP, 1991). The Saint Ouen II plant in Paris, with a processing capacity of 630 000 tonnes/year, cost around 150 million ecu to build in 1989.

When calculated per tonne of waste input, economic costs of mass burn incineration vary widely across Europe (Table 10.8). Variability within countries results from four key factors: incinerator capacity, level of gas

Table 10.8 Economic costs for mass-burn incineration

Country	Cost per tonne	Comments	Source
Austria	176 ecu		pers. comm. (1994)
Denmark	21 ecu		pers. comm. (1994)
France	25 ecu	Current, no WTE	pers. comm. (1994)
	26 ecu	Current, WTE	
	46 ecu	Future, no WTE	
	61 ecu	Future, WTE	
Germany	63–360 ecu		van Mark and Nellessen (1993)
Netherlands	70–125 ecu		pers. comm. (1994)
Spain	18–31 ecu		pers. comm. (1994)
Sweden	22–33 ecu		pers. comm. (1994)
UK	20–26 ecu	Current, no WTE	DoE (1993)
	26–33 ecu	Future, no WTE	
	26–40 ecu[a]	Current, WTE	
	26–33 ecu[a]	Future, WTE	
USA	$70–120 (62–106 ecu)	With WTE	GBB (1991)

[a] Includes non-fossil fuel subsidy. WTE = Waste to Energy.

cleaning equipment installed, whether energy is recovered or not, and whether economic instruments exist to encourage the generation of power from waste. With the implementation of the EC Directives on new and existing MSW incinerators, all facilities will need to have extensive gas scrubbing equipment, so making incineration more expensive, and less variable in cost. It is likely that all new incinerators will have energy recovery facilities, but as Table 10.8 shows, this may not always be economically advantageous. The economics of incineration with energy recovery are improved, however, when there are fiscal instruments designed to encourage the use of waste as a 'renewable' energy resource. In the UK, the Non-Fossil Fuel Obligation (NFFO) of the 1989 Electricity Act requires electricity supply companies to purchase electricity generated from non-fossil fuel (including MSW). This guarantees a market at a price premium for the energy from waste projects accepted under NFFO, which is worth 6–13 ecu/tonne of waste input (DoE, 1993). A similar scheme operates in Germany, whereby energy generated from waste can be injected into the grid at any time for a guaranteed premium price.

RDF and source-separated materials. The original aim of RDF was to produce a readily transportable fuel that could be sold as a commodity alongside other fuels such as coal. Hence the need for pelletization. This requires markets, however, which have proved difficult to develop for RDF. One reason is that in several countries, when burned, RDF is still legally considered to be waste, and therefore subject to the same emission controls as mass burn incineration, so differs from the combustion of coal and other fuels. As a result, surviving RDF plants burn the pellets on site, and export electricity into the national grid, and/or district heating. The costs of RDF production have been dealt with in Chapter 7. The additional cost involved here is the cost of operating the dedicated boiler, less the revenue from the sale of electricity. Since many RDF plants integrate RDF production and combustion on one site, these costs may not be separated out.

The concept of burning source-separated materials collected for recycling has not been fully explored, and here too markets for the paper and plastic fuel must be developed. If this can be done, this energy producing outlet would fix a lower economic value for this material, i.e. its calorific value. If the recycling market for these materials was weak due to oversupply/lack of demand, as at the time of writing (1994), the material could be sold for its calorific content (given that the calorific content of a 50:50 mix of paper and plastic would have about the same calorific content per kg as industrial coal).

The cost of sorting this material has been covered in Chapter 7. The additional costs that need to be considered here are those for operating the boiler and associated emission control equipment, net of any revenues from the sale of exported electricity. Since this process has yet to be fully developed, no data on such costs are available.

10.11 Operation of the thermal treatment module of the LCI spreadsheet

The layout of the thermal treatment module of the LCI spreadsheet is presented in LCI Box 6. The data used in the thermal treatment module are given in LCI Data Box 6. The input box is split into three parts dealing separately with mass burn incineration of MSW (6a), burning of RDF (6b) and burning of source-separated paper and plastic as an alternate fuel (6c). The user completes the boxes for only those thermal treatment processes that are used in the IWM system. The user only inserts data in shaded boxes; unshaded boxes are calculated by the spreadsheet.

6a. MASS BURN INCINERATION OF MSW/RESTWASTE
Input to Incinerator
 Insert 1 in box if mass burn incineration of restwaste occurs:
If a '1' is inserted, the spreadsheet adds all of the remaining restwaste and the residues of earlier treatment processes to the input of the incinerator.

 Input amount:
 Composition %:
  Calorific content of incinerator input:
The spreadsheet inserts automatically the amount and composition of the incinerator input, and displays the average calorific content.

```
6. THERMAL TREATMENT
         This may consist of any or all of the following options:     (a) Mass-burning of MSW,
                                                                      (b) Burning of RDF as fuel
                                                                      (c) Burning of source-separated paper/plastic as fuel
         Fill in the boxes for those options that are used.
Go to Box 6a
```

```
6a. Mass-Burn Incineration of MSW/Restwaste and Sorting Residues

Input to Incinerator:
         Insert 1 in box if Mass burn incineration of restwaste occurs. If not insert 0     [        0 ]

         Input Amount                    [   0.00 ] 000 tonnes
                        Paper      Glass      Metal      Plastic      Textiles    Organic    Other    Compost
         Composition %  [   0% ]  [   0% ]  [   0% ]   [   0% ]     [   0% ]    [  0% ]   [  0% ]  [   0% ]

                                       [ Calorific value of incinerator input =        0.00  GJ/tonne ]

Incineration Process:

         Energy recovery (insert 1 in box if energy is recovered)        [      1 ] Electricity gen. efficiency %   [   20% ]

Residues.
         Ferrous recovery from ash (insert 1 in box if recovered)         [      1 ] Recovery efficiency %          [   90% ]
         Is any ash/clinker re-used? Insert % re-used                     [   0% ]
         Transport distance to landfill for non-haz. waste (one way)      [       ] km
         Transport distance to haz. waste landfill (one way)              [       ] km

Costs:
         Cost per input tonne for incineration (net of energy sales)      [       ] ecu
         Transport cost to non-haz. waste landfill                        [       ] ecu per tonne
         Transport cost to haz. waste landfill                            [       ] ecu per tonne

Go to Box 6b.
```

```
6b. Burning of Refuse-Derived-Fuel (RDF)
                                                              cRDF        dRDF
Input:          Amount of RDF fuel produced that will be burned:    [  0.00 ]  [  0.00 ] 000 tonnes/yr

Combustion Process:      Efficiency of Electricity generation %:   [   30% ]

Residues:    Transport distance to non-hazardous waste landfill (one way):   [       ] km
             Transport distance to hazardous waste landfill (one way):       [       ] km
Costs:       Cost per tonne for combustion process (net of energy sales)     [       ] ecu (If profit, insert as negative cost)
             Transport cost to non-hazardous waste landfill for ash residues [       ] ecu per tonne
             Transport cost to hazardous waste landfill for filter dust residues [       ] ecu per tonne
Go to Box 6c.
```

```
6c. Burning of Source-Separated Paper and Plastic as fuel.
                                                              Paper       Plastic
Input:          Amount of paper/plastic to be burned:          [  0.00 ]  [  0.00 ] 000 tonnes

Combustion Process:      Efficiency of Electricity generation %:   [   30% ]

Residues:    Transport distance to non-hazardous waste landfill (one way):   [       ] km
             Transport distance to hazardous waste landfill (one way):       [       ] km
Costs:       Cost per tonne input for combustion process (net of energy sales) [       ] ecu (If profit, insert as negative cost)
             Transport cost to non-hazardous waste landfill for ash residues   [       ] ecu per tonne
             Transport cost to hazardous waste landfill for filter dust residues [       ] ecu per tonne
Go to Box 7.
```

Incineration process

Energy recovery (insert 1 in box if energy is recovered):
The spreadsheet calculates the electrical energy produced, using the efficiency shown in the box, assuming that power generation is the only form of energy recovery.

Residues

Ferrous recovery (insert 1 in box if recovered):   Recovery efficiency %:
Spreadsheet calculates the amount of ferrous metal recovered and adds this to the recovered materials total.

Is any ash/clinker re-used? Insert % re-used.
The amount re-used is subtracted from the residue needing disposal.

LCI Data Box 6 Energy production/consumption and emissions due to thermal treatment used in the LCI model.

Data per tonne burned (as received)	Mass burning of MSW[1]									dRDF[2]	Alternate fuel[3]	
	Paper	Glass	Metal-Fe	Metal non Fe	Plastic-rigid	Plastic-film	Textile	Organic	Other		Paper	Plastic
Energy inputs												
Electricity (kW-h)	70	70	70	70	70	70	70	70	70	20	20	20
Natural gas (m³)	0.23	0.23	0.23	0.23	0.23	0.23	0.23	0.23	0.23			
Energy outputs												
Net cal. value (GJ/t)	10.5	−0.5	−0.5	−0.5	28.0	25.0	13.5	3.7	4.0	18	10.5	28.5
Air emissions (g)												
Particulates	150	150	150	150	150	150	150	150	150	150	20	50
CO	500	0	0	0	500	500	500	500	500	347	400	1250
CO₂	1 128 500	0	0	0	2 492 500	2 336 700	1 209 200	563 900	1 025 900	1 421 000	1 129 300	2 415 400
CH₄												
NOₓ	500	500	500	500	500	500	500	500	500	2019	1600	5000
N₂O												
SOₓ	1500	1500	1500	1500	1500	1500	1500	1500	1500	1500	300	360
HCl	250	250	250	250	250	250	250	250	250	250	100	
HF	100	100	100	100	100	100	100	100	100			0

H₂S	100	100	100	100	100	100	100	100	100	46	100	100
HC	100	100	100	100	100	100	100	100	100	46	100	100
Chlorinated HC												
Dioxins/furans (TEQ)	5 E−7	5 E−7	5 E−7	5 E−7	5 E−7	5 E−7	5 E−7	5 E−7	5 E−7	6.8 E−5	5 E−7	5 E−7
Ammonia												
Arsenic	2.5	2.5	2.5	2.5	2.5	2.5	2.5	2.5	2.5	2.5	2.5	2.5
Cadmium	0.5	0.5	0.5	0.5	0.5	0.5	0.5	0.5	0.5	0.5	0.5	0.5
Chromium	6.3	6.3	6.3	6.3	6.3	6.3	6.3	6.3	6.3	6.3	6.3	6.3
Copper	6.3	6.3	6.3	6.3	6.3	6.3	6.3	6.3	6.3	6.3	6.3	6.3
Lead	6.3	6.3	6.3	6.3	6.3	6.3	6.3	6.3	6.3	6.3	6.3	6.3
Mercury	0.5	0.5	0.5	0.5	0.5	0.5	0.5	0.5	0.5	0.5	0.5	0.5
Nickel	2.5	2.5	2.5	2.5	2.5	2.5	2.5	2.5	2.5	2.5	2.5	2.5
Zinc	6.3	6.3	6.3	6.3	6.3	6.3	6.3	6.3	6.3	6.3	6.3	6.3
Solid residues (kg)												
Bottom ash	84	900	850	900	60	90	75	77	420	86	84	75
Filter dusts/sludge	32	32	32	32	32	32	32	32	32	13.8	13.8	13.8

[1] Incineration up to or exceeding EC Directive requirements.
[2] Assuming Gas clean up system for particulates, SO₂ and HCl, and EC Directive levels for heavy metals.
[3] Using EC directive levels for MSW incineration where no other data available.

Transport distance to non-hazardous waste landfill (one way):
Transport distance to hazardous waste landfill (one way):
Spreadsheet calculates the fuel consumption for residue disposal assuming a 20 tonne load and that no return load is carried.

Costs

Cost per input tonne for incineration (net of energy sales):
The cost inserted should include any revenue from sales of electricity and ferrous metal, but not the cost of residue transport or disposal.

Transport cost to non-hazardous waste landfill (per tonne):
Transport cost to hazardous waste landfill (per tonne):
The spreadsheet calculates the transport costs assuming that fly ash, filter dust and gas cleaning residues are sent to a hazardous waste landfill, while bottom ash is sent to a non-hazardous waste landfill.

6b. BURNING OF REFUSE-DERIVED FUEL (RDF)
Input
Amount of RDF produced that will be burned:
The amount of cRDF and dRDF produced earlier in the spreadsheet (Box 3b) is displayed here automatically.

Combustion process

Efficiency of electricity generation %:
The spreadsheet uses this figure to calculate the amount of electricity that will be produced from the burning of RDF.

Residues
Transport distance to non-hazardous waste landfill:
Transport distance to hazardous waste landfill:
The spreadsheet calculates the fuel consumption in the same way as for residues in Box 6a.

Costs
Cost per tonne for combustion process (net of energy sales):
If the revenue from energy sales exceeds the cost of operating the RDF fired boiler and power generation plant, this profit should be inserted as a negative cost. Costs for residue disposal should not be included as they are inserted below.

Transport cost to non-hazardous waste landfill for ash residues:
Transport cost to hazardous waste landfill for ash residues:
The spreadsheet calculates the cost of transporting the residues for landfilling.

6c. BURNING OF SOURCE-SEPARATED PAPER AND PLASTIC AS FUEL

Input

Amount of paper/plastic to be burned;

The amounts of paper and plastic that were recovered in a MRF but not sent for recycling (Box 3a) are inserted here automatically.

The rest of Box 6c is completed in the same way as Box 6b.

References

Barton, J. (1986) The application of mechanical sorting technology in waste reclamation: options and constraints. Warren Spring Laboratory. Paper presented at the Institute of Waste Management and INCPEN Symposium on Packaging and Waste Disposal Options, London.

Bechtel, P. and Lentz, R. Landfilling and Incineration: some environmental considerations and data concerning the disposal of municipal solid waste. P&G Internal report, unpublished.

Clayton *et al.* (1991) Review of municipal solid waste incineration in the UK. Warren Spring Laboratory Report LR 776. (PA). Department of the Environment Research Programme.

DoE (1993) *Landfill Costs and Prices: Correcting Possible Market Distortions.* A study by Coopers & Lybrand for the Department of the Environment. HMSO, London.

EEWC (1993) Waste to energy – an audit of current activity, January 1993. Report prepared for the European Energy from Waste Coalition. Private communication, 1993.

EC (1989a) 89/369/EEC. Council Directive on the Prevention of Air Pollution from New Municipal Waste Incineration Plants. *Off. J. Eur. Commun.* 8/6/89.

EC (1989b) 89/429/EEC. Council Directive on the Prevention of Air Pollution from Existing Municipal Waste Incineration Plants. *Off. J. Eur. Commun.* 21/6/89.

ENDS (Aug 1992) Subsidising the dash to burn trash. *ENDS Report* **211**, 12–14. Environmental Data Services, London.

ENDS (Jan 1993) Catalyst cuts dioxin emissions from incinerators. *ENDS Report* **216**, 9–10. Environmental Data Services, London.

ETSU (1992) Production and combustion of c-RDF for on-site power generation. Energy Technology Support Unit report no. B 1374, by Aspinwall and Company Ltd. Published by the Department of Trade and Industry, 32 pp.

ETSU (1993) An assessment of mass burn incineration costs. Energy Technology Support Unit report no. B R1/00341/REP, by W.S. Atkins, Consultants Ltd. Published by the Department of Trade and Industry, 42 pp.

Fichtner Ingenieurgesellschaft (1991) Gegenüberstellung der Schadstofffrachten bei der thermischen Restmüllbehandlung und einer Restmülldeponie, im Auftrag Fichtner, Landesentwicklungsgesellschaft. Ministerium für Umwelt, Baden-Württemberg.

Habersatter, K. (1991) Ökobilanz von Packstoffen Stand 1990, Bundesamt für Umwelt, Wald und Landschaft (BUWAL) Report No. 132, Bern, Switzerland.

Habig, G. (1992) *Arbeitskreis der Fachgemeinschaft Thermo Prozess- und Abfaltechnik im VDMA (Verband Deutscher Maschinen- und Anlagenbau)*, Frankfurt. March 1992.

Hackl, A.E. (1993) Energy aspects in environment protection. In *Proc. CEFIC Conference: The Challenge of Waste*, Vienna, pp. 127–129.

Hickey, T.J. and Rampling, T.W.A. (1989) *The Hedon Boiler Trials*. Warren Spring Laboratory Report LR 747 (MR).

IFEU (1991) Bewertung verschiedener Verfahren der Restmüllbehandlung in Wilhelmshaven. Institut fur Energie-und Umweltforschung, Heidelberg, GmbH, Heidelberg, 1991.

IFEU (1992) Vergleich der Auswirkungen verschiedener Verfahren der Restmüllbehandlung auf die Umwelt und die menschliche Gesundheit. Institut fur Energie-und Umweltforschung, Heidelberg, GmbH, Heidelberg.

IGD (1992) Sustainable waste management: the Adur project. Report by the Institute for Grocery Distribution, Letchmore Heath, Watford, UK, 85 pp.

MOPT (1992) Residuos sólidos urbanos. Report by Luis Ramon Otero Del Peral for Ministerio de Obras Publicas y Transportes, Madrid. ISBN 84-7433-820-4.

OECD (1993) Environmental Compendium. Organisation for Economic Cooperation and Development, Paris.

Ogilvie, S.M. (1992) A review of the environmental impact of recycling. Warren Spring Laboratory report LR 911 (MR).

Patel, N.M. and Edgcumbe, D. (1993) Observations on MSW management in Japan. *Waste Mgmt.* (*J. Inst. Wastes Mgmt.*) April, 27–31.

Porteous, A. (1991) Municipal waste incineration in the UK – time for a reappraisal? *Proc. 1991 Harwell Waste Management Symposium.*

RCEP (1993) *Incineration of Waste.* Royal Commission on Environmental Pollution 17th Report. HMSO, London, 169 pp.

SPMP (1991) Thermal recycling, cornerstone of rational waste management. Paper prepared by the Syndicat de Producteurs de Matières Plastiques.

Shell Petrochemicals (1992) Cited in EEWC (1993).

Svedberg, G. (1992) Waste incineration for energy recovery. Final report.

Toussaint, A. (1989) *Proc. ENVITEC* 89, Kongressband, S. 61.

van Mark, M. and Nellessen, K. (1993). Neuere Entwicklungen bei den Preisen von Abfalldeponierung und -verbrennung. *Müll and Abfall* January, 20–24.

van Santen, A. (1993) Incineration: Its role in a waste management strategy for the UK. Paper presented at the Institute of Wastes Management annual meeting, 1993. Torquay, UK.

Vogg, H., Merz, A., Stieglitz, L. and Vehlow, J. (1989) VGB-Kraftwerkstechnik 69 S. 795.

Vogg, H. (1992) Arguments in favor of waste incineration. *Annual European Toxicology Forum*, Copenhagen.

Warmer (1990) Waste incineration. *Warmer Bulletin Factsheet.* January.

Warmer (1991) Fuel from waste. *Warmer Bulletin Factsheet.* January.

Warmer (1993a) Retrofitting waste incineration plant – below detectability limits. *Warmer Bull.* **36**, 19.

Warmer (1993b) Refuse-Derived Fuel. Warmer Information Sheet, Warmer Bulletin 39, Nov 1993.

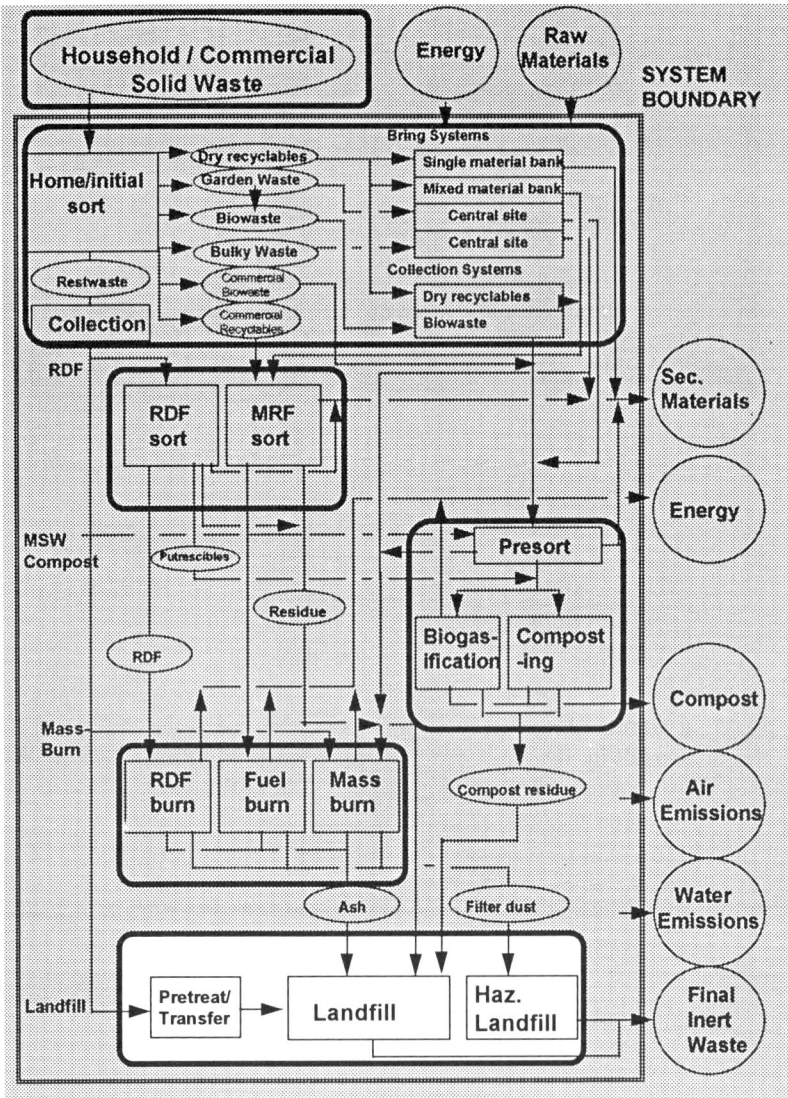

Figure 11.1 The role of landfilling in an integrated waste management system.

11 Landfilling

Summary

Landfilling is considered as a waste treatment process, with its own inputs and outputs, rather than as a final disposal method for solid waste. Landfilling essentially involves long-term storage for inert materials along with relatively uncontrolled decomposition of biodegradable waste. The use of landfilling across Europe is described and landfilling methods are discussed, including techniques for landfill gas and leachate control, collection and treatment. Using available data, an attempt is made to quantify the inputs and outputs of the landfilling process in both environmental and economic terms. Finally, the landfilling module, which completes the computer LCI spreadsheet for solid waste systems, is presented.

11.1 Introduction

Landfilling stands alone as the only waste disposal method that can deal with all materials in the solid waste stream (Figure 11.1). Other options such as biological or thermal treatment themselves produce waste residues that subsequently need to be landfilled. Consequently, there will always be a need for landfilling in any solid waste management system. Landfilling is also considered the simplest, and in many areas the cheapest, of disposal methods, so has historically been relied on for the majority of solid waste disposal. In several European countries (UK, Ireland, Spain) landfilling continues to be the principal waste disposal method, although as land prices and environmental pressure increase, it is becoming more difficult to find suitable landfill sites, and so this position shows signs of changing.

Not all cases of 'landfill' actually involve filling of land. Although the filling of exhausted quarries and clay pits occurs in many countries, and in the UK in particular, above ground structures are also common. In countries such as Japan, 'landfilling' can also take the form of 'sea-filling', where material is used to construct man-made islands in Tokyo Bay and Osaka Bay.

The concept of landfilling as a final disposal method for solid waste can also be challenged. A landfill is not a 'black hole' into which material is deposited and from which it never leaves. Like all the other waste options discussed in this book, landfilling is a waste treatment process, rather than a method of final disposal (Finnveden, 1993). Solid wastes of various compositions form the majority of the inputs, along with some energy to run the process. The process itself involves the decomposition of part of the landfilled waste. The outputs from the process are the final stabilised solid waste, plus the gaseous and aqueous products of decomposition, which emerge as landfill gas and leachate. As in all processes, process effectiveness and the amounts and quality of the products depend on the process inputs and the way that the process is run and controlled. The same applies to landfilling: what comes out of a landfill depends on the quantity and composition of the waste deposited, and the way that the landfill is operated.

This chapter considers both the objectives of the landfilling process and the methods used to achieve them. It then goes on to describe the inputs and outputs of the process, in both environmental and economic terms, so that the lifecycle inventory of solid waste systems can be completed.

11.2   Landfilling objectives

The principal objective of landfilling (Box 11.1) is the safe long-term disposal of solid waste, both from a health and environmental viewpoint; hence the term sanitary landfill which is often applied. As there are emissions from the process (landfill gas and leachate), these also need to be controlled and treated as far as possible.

To a limited extent, landfilling can also be considered as a valorisation process. Once collected, the energy content of landfill gas can be exploited, so landfilling could be argued to be a 'waste-to-energy' technology (as shown in Figure 2.3). There are already numerous landfill sites where energy recovery from landfill gas occurs. In Germany, for example, 65% of municipal waste landfill sites recover energy from landfill gas (UBA, 1993). In the UK, there are over 40 such schemes, which recover the thermal energy equivalent of over 350 000 tonnes of coal per year (Maunder, 1992). On a global scale, it has been reported that there were a total of 453 facilities for landfill gas usage worldwide in 1990 (Gendebien et al., 1991), and the number is likely to have risen since then. The same authors further estimated that there are potentially 730 billion (10^9) m^3 of landfill gas produced annually from domestic solid waste, and that this would be equivalent in energy terms to 345 million tonnes of oil. If allowed to diffuse freely from landfill sites, this landfill gas can present a serious risk both to the environment (methane is a potent greenhouse gas) and the health and safety of local residents. A review of the literature has identified over 160 reported cases of damage due to landfill gas, with 55 explosions reported (Gendebien et al., 1991). Collection and control of landfill gas is therefore needed for safety and environmental reasons. Once collected it makes sense

---

BOX 11.1

LANDFILLING: KEY CONSIDERATIONS

- Can deal with all waste materials

- Essentially a waste treatment process with the following outputs:
    - Landfill gas
    - Leachate
    - Inert solid waste

- The waste treatment process parameters can be optimised e.g.
    - Dry containment
    - Leachate circulation
    - Lining technology
    - Landfill gas and leachate collection

- Can be used to reclaim land (or sea)

- Should avoid groundwater catchment and extraction areas.

---

to utilise the energy content of the gas where it is produced in commercially exploitable quantities. It is not designed as a waste-to-energy technology, however, since conditions in the landfill are relatively uncontrolled, and a large part of the gas escapes uncollected.

Landfill can also be considered as a valorisation method when it reclaims land, either from dereliction (e.g. exhausted quarries) or from the sea (e.g. as practised in Japan), and returns it to general use. This is not always the case, however, since in many countries such opportunities are not available, and landfills need to be sited on otherwise useful land. Thus, while landfilling can be a method for land generation, in most instances it consumes land.

So, although landfilling constitutes a means of valorising waste in two limited ways, its prime objective is the safe disposal of solid waste residues, whether direct from households or from other waste treatment processes.

11.3 Current landfilling activity in Europe

Reliance on landfill for solid waste disposal varies geographically across Europe (Figure 11.2). Countries such as the UK have traditionally used landfilling as the predominant disposal route, partly because of its geology

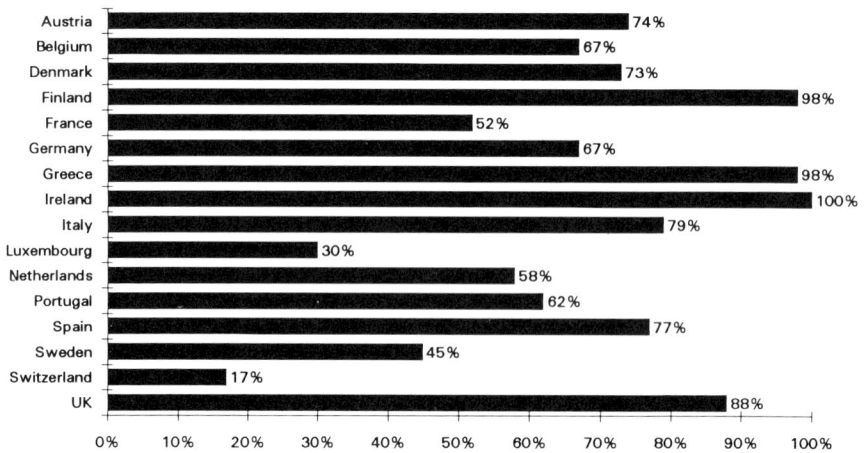

Figure 11.2 Percentage of MSW landfilled in European countries. *Source*: OECD (1991).

and mineral extraction industry which has left many empty quarries that can be filled with waste. Such sites, however, may not always be in suitable locations for minimising their environmental impacts (see discussion below). Conversely, countries such as the Netherlands, where the lack of physical relief and high water table have meant that large void spaces are not available, have had to develop alternative disposal routes.

11.4 Landfilling methods

Basic philosophy.   Historically, landfilling has consisted of dumping waste in deep earthen pits. Over time, and with the percolation of rain water, the degradable fractions of the waste decompose and the resulting products are diluted and dispersed into the underlying soil. On a small scale, this 'dilute and disperse' method of operation can be effective, since soils have a natural capacity to further decompose organic material and to adsorb many inorganic residues. Such sites will still produce landfill gas, however, which will diffuse into the atmosphere, and can cause safety concerns if it is allowed to accumulate. With increasing urbanisation, increased waste generation and increased difficulty in locating suitable and publicly acceptable sites, landfills have increased in size over time. As a result, dilution and dispersion is no longer an effective way of dealing with the landfill site emissions. Leachate produced by large unlined sites can pose a serious risk to groundwater supplies. The Environmental Protection Agency, for example, has estimated that in the United States, around 40 000 landfill sites may be contaminating groundwater (Uehling, 1993).

To address the problems of landfill gas emissions and groundwater contamination, most modern landfills are operated on a containment, as opposed to a dilute and disperse, basis. Sites are lined with an impermeable layer or layers, and include systems for collecting and treating both the resulting landfill gas and the leachate. In Germany, for example, 94% of all municipal waste landfills are now lined and collect and treat leachate prior to release into public sewage systems (UBA, 1993). The quantity of leachate produced in a landfill depends on, amongst other things, the amount of water that percolates into the site from rainfall and groundwater, so there has been a further tendency to seal the landfill capping and make the whole structure water-tight. From October 1993, for example, all new landfills in the United States will have to be kept sealed and dry, with plastic membranes isolating them from percolating rain and groundwater (Uehling, 1993). Dry containment will also reduce the initial production of landfill gas, since a high moisture level is necessary for biodegradation. The methanogenic micro-organisms, for example, require a moisture content of over 50% to be active (Nyns, 1989).

The dry containment method of operating a landfill has been described as long-term storage of waste rather than waste treatment or waste disposal (Campbell, 1991), and does have some significant drawbacks. There will always be pockets of moisture within waste, and it is generally accepted that all lining and capping systems will eventually leak, so rain and/or groundwater will eventually enter the site. Thus, the decomposition of the organic fraction of the waste will eventually occur, with resulting emissions of landfill gas and leachate. Since pipes and pumps buried within the waste eventually clog up and fail, there will be less chance of collecting and treating these emissions if they occur in the distant future. In place of such dry containment, therefore, some experts are advocating almost the opposite: the acceleration of the decomposition process by keeping the waste wet. This can be achieved by recirculating the leachate collected within the landfill to keep conditions suitable for microbial activity. In this way most of the gas and leachate production will occur in the early years of the landfill's life, while gas and leachate collection systems are operating effectively. By the time that the lining system eventually fails, the leachate should be dilute and so reduce the risk of groundwater pollution. Operating the landfill in this way as a large 'bioreactor' also means that the gas is given off at higher rates, so making energy recovery economically more attractive. More experience with the wet bioreactor method is required, however, before it replaces dry containment as the preferred landfilling technique.

Landfill site design and operation.   The structure of one form of lined landfill site for containment of leachate is shown in Figure 11.3. The bottom liner of the site can either be a geomembrane (often butyl rubber or HDPE) or a layer of low permeability material such as clay. Whilst

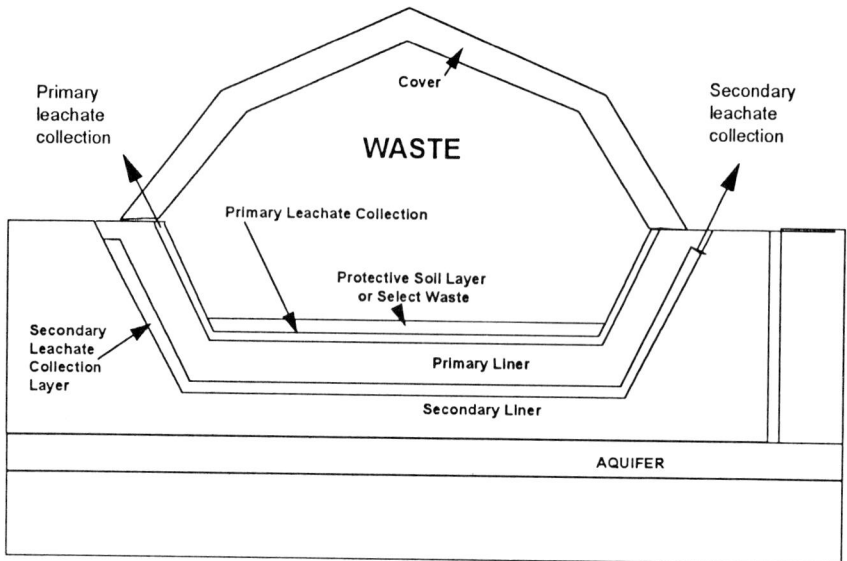

Figure 11.3 Schematic showing a primary liner underlaid by a leak detection secondary leachate collection system. Adapted from Rowe, 1991.

the permeability of the synthetic membranes is lower, they are liable to mechanical puncture and so can then act as a point source for leaking leachate. By comparison, clay barriers are not subject to such localised failure, although they act as a diffuse source of leachate over the whole area of the landfill site (Campbell, 1991).

As well as a choice of material types, there is also a range of options in the way liners are laid down. As liner systems increase in complexity, their costs will increase, but the risk of failure decreases, so there is less likelihood of expensive remediation work following leakages. The simplest form of barrier consists of a single liner, normally with a leachate collection system above the liner. Rather than rely on one type of liner material, a single composite liner system has two or more liners of different materials in direct contact with each other. In this design, it is common to have a leachate collection system above a geomembrane, on top of a low permeability clay layer. The liner system shown in Figure 11.3 is a double liner system: two liners with a leachate collection system above the upper (primary) liner, and a leachate detection system between the two layers. The leachate detection system has a high permeability to allow any leachate that has leaked through the primary liner to be drawn off. Again, each layer in a double liner system may either be a single liner or a composite of two or more materials (Deardorff, 1991). The leachate collection system normally consists of a network of perforated pipes, from which the leachate can be either gravity drained or pumped to a leachate treatment plant. The most significant influence on leachate quantity is the amount of rainfall, which will vary seasonally. A storage sump or pool is often used so that surges in production can be dampened before entering the treatment process.

Landfill management practices greatly affect leachate quality. Acceleration of the early phases of decomposition is needed to produce low concentrations of organic matter and heavy metals in the leachate (Carra and Cossu, 1990). This can be facilitated by having a low waste input rate, moisture control (by leachate recirculation) or by having a composted bottom layer of waste.

Once the liner system has been installed, a cover of clay, soil or other inert material is normally applied to protect it from mechanical damage. Waste is then deposited and compacted, and layers of inert material (soil, coarse composted material) are normally added to sandwich the waste. At many sites, the waste is covered by landfill cover material at the end of every day to reduce the nuisance from wind blown material, and to keep off rodents, birds and other potential pathogen-carrying vermin.

Landfill gas is collected using a system of either vertical or horizontal perforated pipes. Since the gas will migrate horizontally along the layers of waste, vertical collection pipes are likely to collect gas more effectively. The density of pipes will vary across the landfill, with the greatest density needed at the periphery to prevent the migration of the gas laterally from the site. Pumped extraction of gas is needed for efficient collection, and thus less odour and emission problems. Once collected, the gas can either be flared off, to destroy the methane and organic contaminants, or used as a fuel. As produced, landfill gas is saturated with water vapour, and contains many trace impurities. This leads to a highly corrosive condensate, so if the gas is going to be used in a gas engine for energy recovery, some form of gas cleaning is normally required. Similarly, if the gas is to be piped elsewhere for use as a fuel, in many cases it is purified to remove the contaminants and the carbon dioxide, the latter to increase its calorific value.

The rate of gas production also depends on how the landfill is managed. From a commercial point of view, gas utilisation appears to be profitable from about one year after the waste is landfilled, and can be expected to continue to be so for no more than 15–20 years (Carra and Cossu, 1990).

Waste inputs. The above measures to ensure safe disposal of landfilled material are based on containment, collection and treatment of emissions. As with all processes, however, the emissions will depend on the inputs, i.e. on what waste is placed in the landfill in the first instance. Restricting the type of waste entering the landfill can ensure that fewer emissions are produced. This strategy has gained ground in several countries. The

German T.A. Siedlungsabfall (1993) ordinance, for example, defines the characteristics of waste that can be deposited in each of two classes of landfill. Class 1 landfills will only accept waste with a total organic carbon (TOC) content of ⩽1%. This means that the material will have to be thermally treated before landfilling. Class 2 landfills will accept up to 3% TOC, but there are stricter requirements on the construction of such sites. Landfilling of mixed municipal solid waste will therefore no longer be possible in Germany.

In marked contrast to this restriction of landfill inputs, the UK has consistently argued in favour of the co-disposal of certain forms of industrial and potentially hazardous waste with municipal solid waste. The rationale is that the difficult-to-treat materials are diluted and can be decomposed along with the normal solid waste. In a landfill, however, there is little control over this process, and the potential for serious groundwater pollution should such landfills leak is significant, so it is hard to view this as an environmentally sustainable method for the future.

Landfill siting. No discussion of landfilling can neglect the problem of finding suitable, and publicly acceptable, sites. Along with local residents' concerns over traffic, noise, odour, wind blown litter (and resultant effects on local property values), groundwater pollution has also recently become an important issue in selecting suitable sites. Since all landfill sites are likely to leak eventually, new landfills should not be placed within the catchment areas of groundwater abstraction points, where the contamination of drinking water could occur. Ideally, they should not be sited at all on major aquifers, where the potential for groundwater percolation would be greatest. Any new landfills should be located over minor aquifers or non-aquifers (Harris, 1992). On these grounds, the UK National Rivers Authority has produced a national Groundwater Vulnerability Map, which will be used to assess the groundwater contamination potential of any new landfill developments. What has already become clear, however, is that most suitable void space for landfilling in the UK occurs above major aquifers. This is because most quarries, which form the bulk of suitable void space, have been developed to extract chalk, sandstone and limestone, which form the majority of the major UK aquifers. Thus, although the UK may have an abundance of void space for landfilling, in terms of groundwater protection it is not necessarily in the right place. As in other countries, it may be necessary in future to consider greater use of above-ground sites located on non-aquifers (Harris, 1992).

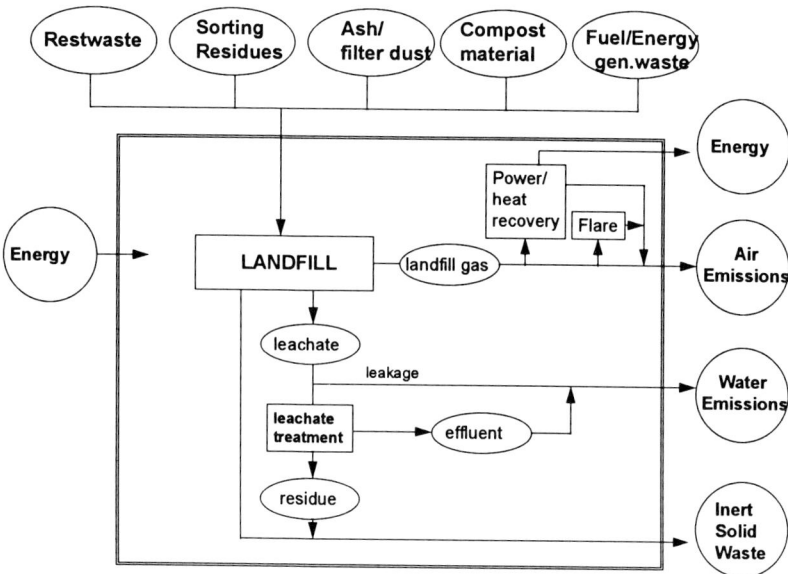

Figure 11.4 Inputs and outputs of the landfilling process.

11.5   Environmental impacts: input–output analysis

Considering landfilling as a waste treatment process, rather than simply a sink for the final disposal of solid waste, the inputs and outputs for the process are shown in Figure 11.4.

The environmental impacts of landfilling waste will depend both on the landfill design and method of operation, and the nature of the waste deposited (Box 11.2). While there has been a lot of attention paid to the details of landfill site location and the design of various systems for gas and leachate containment and/or treatment, there has been less emphasis on the effect of future changes in the composition of the waste materials that will be destined for landfills. As more materials are recovered from waste streams for recycling, both the amount and the composition of the residual waste, from both household and commercial sources, may alter significantly. Although some data are available for the inputs and outputs of the landfilling process, these generally refer to the landfilling of mixed waste streams, such as MSW. Data relating to the individual material fractions within such waste streams are not normally available. Thus, it is difficult to extrapolate from the effects of present waste streams to the environmental effects of future waste streams with different compositions. This will be attempted, however, in this section; by collating data from landfills containing mixed waste streams and the limited amount of data

---

Box 11.2 LANDFILL INPUTS AND OUTPUTS.

INPUTS

      Rest Waste (can be total MSW)
      Sorting Residues
      Biologically Treated Material
      Thermal Treatment Residues (Ash, Filter Dust and other residues from gas
                                        cleaning)

OUTPUTS

      Landfill gas from:
            - biodegradable fraction in MSW $150 \, Nm^3$ tonne
            - biologically treated material $100 \, Nm^3$ tonne
            - ash, filter dust $0 \, Nm^3$ tonne

      Leachate   150 litres/tonne

      Inert solid waste from landfilled waste
      Solid waste residue from leachate treatment

---

on the behaviour of individual materials within landfills, it should be possible to estimate the outputs from landfills for a range of different waste inputs.

11.5.1 Landfill inputs

Waste inputs. Figure 11.1 shows diagrammatically the four main waste streams from municipal solid waste management systems that are land-filled. (Landfills are also likely to receive some industrial wastes, sewage sludges, etc., but these are not included within the boundaries of the system under study here, except those resulting from energy or raw material consumption by the waste management system itself.)

Restwaste, or residual waste, collected from both household and com-mercial sources and landfilled directly. This will vary in composition with geography and time of year and according to what fractions of the waste stream have been recovered or removed for biological or thermal treatment. Where the landfill is located near to the collection area the restwaste will be delivered directly by the collection vehicles; in the case of a distant landfill, the restwaste may be bulked-up at a transfer station (where bulky wastes may also be crushed) ready for transport to the landfill in larger capacity trucks.

Sorting residues from waste sorting processes at materials recovery facilities (MRFs) or RDF plants (Chapter 7), or from pre-sorting processes at biological treatment plants (Chapter 9).

Biologically treated material from composting or biogasification processes. This is the stabilised organic material produced that does not become valorised as compost, due to contamination or lack of a suitable market.

Ash from thermal processing, whether it has been burned as mixed MSW, as RDF or as source-separated material (Chapter 10). This will include the bottom ash (or clinker/slag), which is disposed of in normal landfills and the fly ash and residues from gas cleaning systems, which are disposed of with the bottom ash in some countries but in others are deposited separately in hazardous waste landfills.

There is another source of solid waste resulting from the waste management system itself, which is not shown in Figure 11.1, but is included in Figure 11.4. This is the solid waste generated during the production of energy, fuel and other raw materials (such as the plastic for refuse sacks), which are consumed within the system. In a full 'dustbin-to-grave' analysis, these also need to be taken into consideration. The data given for the amounts of these wastes (see Chapter 6) often do not specify the composition of these wastes, although it is likely that, in most countries, a large part will comprise of ash from energy generation plants. While some of these materials may have possible further uses, or alternative methods for treatment, it is assumed for the purposes of this study that they are all landfilled.

Energy consumption. The landfilling process will consume energy both in the form of vehicle fuel and electricity. For household waste that is landfilled directly, where the distance from the collection area to the landfill site is large, a transfer station may be used to bulk up the waste for more efficient transport by larger truck or rail. No data are available on the energy consumption of transfer stations, although generic fuel consumption data for road transport (Chapter 6) can be used to estimate fuel consumption for onward transport to the landfill site. For all waste types landfilled, fuel and electricity will also be consumed in the operation of the site itself. Preliminary data suggest that the fuel consumption for the landfilling process is around 0.6 l of diesel per m^3 of void space filled (Biffa, 1994).

11.5.2   Landfill outputs

The inputs to a landfill system occur over a limited time period, essentially the working life of the site. The outputs from the system occur over a much longer time span, which may involve at least tens and maybe even hundreds of years. The outputs which are calculated in the following sections are therefore integrated over time, since the gas and leachate produced by each tonne of waste landfilled will eventually be released.

Landfill gas/energy

Gas production. This must be considered separately for the types of waste that are landfilled. As well as posing a local safety concern, landfill gas can have a more global environmental effect. Consisting mainly of methane and carbon dioxide, both 'greenhouse gases' (and especially the former), landfill gas has become significant in the debate over global warming and climate change. Methane has been reported to be responsible for about 20% of recent increases in global warming (Lashof and Ahuja, 1990) and landfills are thought to be a major source of methane. In the UK, landfills are the single largest source of methane, contributing an estimated 23% of total production (ENDS, 1992a), with over 1000 sites reported to be actively producing gas (Brown, 1991). Globally, it has been estimated that methane from decomposition of municipal solid waste, whether in crude dumps or organised landfills, could account for 7–20% of all anthropogenic methane emissions (Thorneloe, 1991, and references therein).

Municipal solid waste/restwaste/sorting residues. In the literature, the amount of landfill gas generated per tonne of municipal solid waste deposited has been estimated by three different methods: (a) by theoretical calculations using the amount of organic carbon present in the waste; (b) from laboratory-scale lysimeter studies; and (c) from gas production rates at existing landfills. Not surprisingly, therefore, there is considerable variability in the estimates of landfill gas production (Table 11.1). Theoretical yields tend to be high (e.g. Gendebien *et al.*, 1991), since they often assume that all of the degradable material does break down, but there may well be pockets within a landfill where little decomposition occurs due to insufficient moisture content. Lab-scale lysimeter studies use actual refuse, but are not likely to reflect fully the conditions existing in a real landfill. Data from existing landfills should be the most appropriate, but these too are difficult to interpret. Gas production rates vary over the active life of a landfill (Figure 11.5), so it is necessary to extrapolate from measured current gas production rates to the total gas production integrated over this active period. Gas yields from landfill sites also do not fully reflect the amount of gas generated within the landfill, since the yield will also depend on the gas collection efficiency. Estimates of collection efficiencies

Table 11.1 Production of landfill gas from MSW and selected waste fractions

Waste fraction	Landfill gas production (N m³/tonne wet material)	Data Type	Source
MSW	372	Theoretical calculation	Gendebien et al. (1991)
MSW	229	Theoretical calculation	Ehrig (1991)
MSW	270	Calculated from Italian data	Ruggeri et al. (1991)
MSW	120–160	Laboratory scale experiments	Ehrig (1991)
MSW	190–240	Measured at landfills	Ham et al. (1979)
MSW	60–180	Measured at landfills	Tabasaren (1976)
MSW	222	Mean UK landfill yield	Richards and Aitchison (1991)
MSW	135	Estimated average	IFEU (1992)
MSW	200	Estimated average	de Baere et al. (1987)
MSW	100–200	Estimated average	Carra and Cossu (1990)
Food waste	191–344	Laboratory scale experiment	Ehrig (1991)
Grass	176	Laboratory scale experiment	Ehrig (1991)
Newspaper	120	Laboratory scale experiment	Ehrig (1991)
Magazines	100–225	Laboratory scale experiment	Ehrig (1991)
Cardboard	317	Laboratory scale experiment	Ehrig (1991)
Composted MSW	133	Laboratory scale experiment	Ehrig (1991)
Composted organic fraction	176	Laboratory scale experiment	Ehrig (1991)

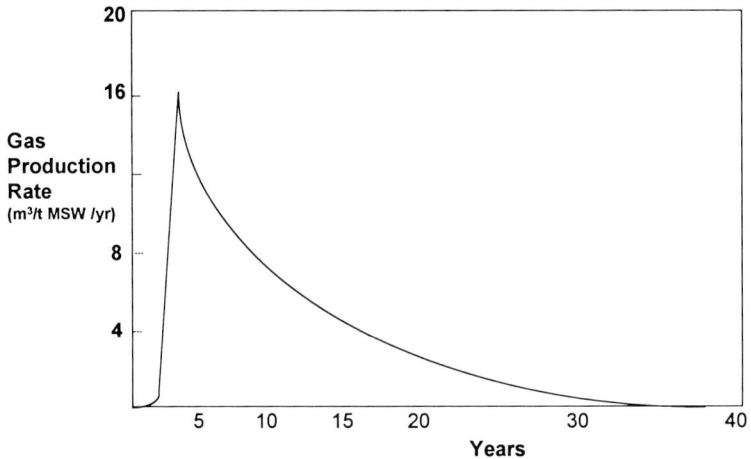

Figure 11.5 Gas production curve for a full scale landfill estimated from laboratory scale experiments. *Source*: Ehrig, 1991.

vary (20–25% (De Baere *et al.*, 1987); 40% (RCEP, 1993); 40–70% (Carra and Cossu, 1990); 40–90% (Augenstein and Pacey, 1991)), and will depend on size, shape and engineering design of the landfill site. For the purposes of the LCI model in this book, a figure of 40% will be assumed.

Given the different methods used to reach the estimates of gas production, Table 11.1 shows some consensus on the amounts of gas produced by landfilling MSW, at around 150 N m^3/tonne of waste as received (i.e. including moisture content). Some laboratory studies produced only 120 N m^3/tonne (Ehrig, 1991), but equally, the average gas yield from landfill sites in the UK has been estimated to be 222 N m^3/tonne (Richards and Aitchison, 1991), so the amount of gas actually produced may well be in excess of 150 N m^3/tonne of municipal solid waste.

Landfill gas is only produced from the biodegradable fractions of MSW, however, which are essentially the putrescible organic fraction, the paper and board fraction and any non-synthetic textiles. Together these typically constitute around 60% by weight of MSW in Europe (Chapter 5). Therefore, gas production for these fractions might be expected to average around 250 N m^3/tonne. (Other fractions such as glass, plastics and metals will probably affect the rate of decomposition, since their presence is likely to facilitate water percolation of the waste and diffusion of gases, but they will not markedly affect the total level of decomposition over time.) Lysimeter studies of these fractions have shown similar levels of gas production from food waste (191–344 N m^3/tonne) and cardboard (317 N m^3/tonne), but lower values for magazines (100–225 N m^3/tonne) and newspaper (120 N m^3/tonne) (Ehrig, 1991; see

Table 11.1). Comparisons with the biogasification process (Chapter 9) are also useful. In a much shorter time span ($\leqslant 20$ days) up to $150 \, N \, m^3$ of biogas can be produced per tonne of biomass in the accelerated process under controlled conditions. As degradation in a landfill occurs over a much longer time, more complete decomposition is likely to occur than in the biogasification process, and there are also likely to be some process losses in biogasification. A figure of $250 \, N \, m^3$ landfill gas per tonne of biodegradable waste (organic, paper and textile fractions) is therefore considered realistic.

Biologically treated material. Where biological treatment has been used to reduce the volume of waste for disposal, or where markets cannot be found for composted or anaerobically digested residues, these materials are also landfilled, and biochemical degradation will continue within the landfill. There are few estimates of the amount of landfill gas produced by such material. In laboratory studies, partially composted MSW ('a few days in a technical compost reactor') produced $133 \, N \, m^3$ of landfill gas (compared to $160 \, N \, m^3$ for untreated MSW in the same test), while a partially composted organic fraction produced $176 \, N \, m^3$ (Ehrig, 1991). For residues from biogasification and composting, IFEU (1992) used an estimate for landfill gas production of $20 \, N \, m^3$/tonne of restwaste entering the biogas/composting process. If, as in some biogasification processes (Chapter 9), the residue represents 20% of the input, this would give a gas production of $100 \, N \, m^3$/tonne of residue landfilled. This figure will be used for landfilled residues from both composting and biogasification processes.

Ash. Given complete combustion, any ash entering a landfill should contain no organic carbon. Therefore no landfill gas should be generated. Although not all combustion processes will completely remove the carbon, it will be assumed for the purposes of this model that no landfill gas is produced from ash.

Landfill gas composition. Landfill gas is produced by the anaerobic decomposition of biodegradable organic material. Much of this material consists of cellulose, which will degrade by the general reaction

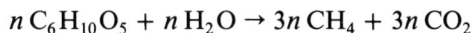

$$n \, C_6 H_{10} O_5 + n \, H_2 O \rightarrow 3n \, CH_4 + 3n \, CO_2$$

Other organic materials will be decomposed by a variety of other reaction pathways. As these different reactions will proceed at varying rates, the composition of the gas released will vary through the different phases of the active life of a landfill site (Figure 11.6). It will also vary with the type of waste contained, but a typical landfill gas composition is given in Table 11.2. The major component is methane, which usually comprises 50–55%, followed by carbon dioxide which makes up most of the remaining volume. In addition, more than 100 different volatile organic compounds

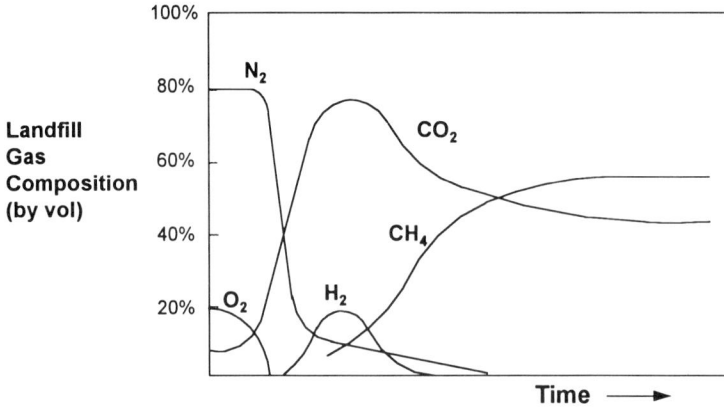

Figure 11.6 Typical production pattern for landfill gas.

Table 11.2 Typical composition of landfill gas

Component	% by volume
Methane	63.8[a]
Carbon dioxide	33.6
Oxygen	0.16
Nitrogen	2.4
Hydrogen	0.05
Carbon monoxide	0.001
Ethane	0.005
Ethene	0.018
Acetaldehyde	0.005
Propane	0.002
Butanes	0.003
Helium	0.00005
Higher alkanes	<0.05
Unsaturated hydrocarbons	0.009
Halogenated compounds	0.00002
Hydrogen sulphide	0.00002
Organosulphur compounds	0.00001
Alcohols	0.00001
Others	0.00005

Source: DoE (1989).
[a] The figure for methane reported in this data set is considered high.
A figure of 55% methane is considered more typical.

have been identified as trace components, many of which are known to be toxic or carcinogenic. The actual trace components vary according to the landfill and the landfilled waste, but the major ones found are hydrogen sulphide, vinyl chloride, benzene, toluene, trichloroethane and mercaptans (Willumsen, 1991).

If no collection system exists for gas control or energy recovery, all of the landfill gas will eventually leak out of the site and enter the atmosphere. Where gas collection occurs, around 40% will typically be recovered (see above), whilst the remaining 60% still enters the atmosphere.

Gas control and energy recovery. In the simplest form of gas control, the collected gas is flared on the site to destroy both the combustible fractions of the gas and most of the organic trace components. Any methane, carbon monoxide and hydrogen in the landfill gas should be converted to carbon dioxide and water if combustion is complete. Comparing the performance of three different flaring systems, Baldwin and Scott (1991) found a large difference in combustion efficiency, however, with one system releasing 16% of the flare exhaust as unburned methane. In an efficient flare, however, all of the methane was shown to be burned. An efficient flare was also shown to significantly reduce, but not remove completely, levels of trace components from 4427 mg/m^3 to 32.8 mg/m^3 (a 99.3% reduction) (Table 11.3). Analysis of the flared gas from several systems also showed low levels (< 10 mg/m^3) of components not present in the unburned gas, such as methyl cyanide, nitromethane, acrolein, ethylene oxide and some alkynes, which must be formed *de novo* in the flaring process.

Where the collected landfill gas is used for energy recovery, this can involve either heating applications (steam-raising via boilers, kiln firing or space heating), or power generation systems. Power generation systems can involve spark ignition or dual fuel diesel engines, or gas turbines (in increasing order of generation capacity). Before the landfill gas is burned in these engines, it is normally compressed and dewatered, which may also remove some of the trace contaminants. Comparison of the exhaust emissions from the different types of gas engines suggests that levels of contaminants can differ widely (Table 11.4) (Young and Blakey, 1991).

The composition of landfill gas (with a typical methane content of around 55%) is similar to that of biogas (Chapter 9), so has a similar heat content. Given a calorific value of 37.75 MJ/N m^3 for methane (Perry and Green, 1984), landfill gas has an energy content of 20.8 MJ/N m^3. The UK Department of Environment (1989) give a figure of 15–21 MJ/N m^3 of landfill gas, depending on its methane content. A heat content of 18 MJ/N m^3 will be assumed for this analysis. This amount of thermal energy is released on combustion; the amount of useful energy resulting will depend on whether the gas is used for heating or power generation purposes. For the purposes of this study, it will be assumed that where energy recovery occurs, it involves the burning of landfill gas in a gas engine to generate electricity, which is then exported into the grid. A conversion efficiency of 30% will be assumed, in line with the value used for the biogas engine in Chapter 9 (Schneider, 1992; Schön, 1992), giving an electrical energy recovery of 1.5 kW-h/N m^3 of landfill gas collected.

Table 11.3 Landfill gas control by the use of flares: trace components in landfill gas (LFG) before and after flare combustion at three different sites

Components	Total concentration (mg/m^3)					
	Flare A		Flare B		Flare C	
	LFG before flaring	Flared gas (adjusted for dilution)	LFG before flaring	Flared gas (adjusted for dilution)	LFG before flaring	Flared gas (adjusted for dilution)
Alkanes	920	0	370	39	510	1.2
Alkenes	400	0	170	1.1	230	18
Alcohols	180	2.1	1.3	55	7.3	6.1
Amines	0	0	0	0	0	0
Aromatic hydrocarbons	1600	6.1	380	58	350	2.6
Alkynes	0	0	0	0	0	3.9
Cycloalkanes	43	0	0	0	5.1	0
Carboxylic acids	0.8	0	0.96	0	0	0
Cycloalkenes	530	0	120	5.6	79	0.96
Dienes	1.7	0	0	0.14	10	0
Esters	290	0	0.2	0	0.5	0
Ethers	1.8	0	0.4	0	0.2	0.32
Halogenated organics	320	1.9	32	18	39	3
Ketones	120	1.6	2.9	2.2	1	0.43
Organosulphur compounds	19	21	2.6	6.7	3.9	4.6
Others	1.4	0	0.5	2.5	0.73	7.29
Totals	4427	32.8	1100	190	1200	48

Source: Baldwin and Scott (1991).

Table 11.4 Emissions from power generation plants using landfill gas[a]

Site[b]	A	B	C
Power generation plant type	Dual-fuel diesel engine	Spark-ignition gas engine	Gas Turbine
Gaseous component	mg/N m^3	mg/N m^3	mg/N m^3
Particulates	4.3	125	9
CO	800	c. 10 000	14
Total unburnt HC	22	>200	15
NO$_x$	795	c. 1170	61
HCl	12	15	38
SO$_2$	51	22	6
	ng/N m^3	ng/N m^3	ng/N m^3
Dioxins	0.4	0.6	0.6
Furans	0.4	2.7	1.2

Source: Young and Blakey (1991).
[a] Differences in emissions reflect both differences between gas engine types and differences in the quality of the incoming landfill gas. At sites A and B the gas was dried, filtered, compressed and cooled before use. At site C the gas was passed through a wet scrubber to remove acid gases, then compressed, cooled, filtered and heated to 70°C before combustion.
[b] All three sites were in UK.

Leachate. As with landfill gas, it is difficult to provide 'typical' figures for the generation of leachate from landfilled wastes, since both the amount and composition of leachate will depend on many factors, including the nature of the waste landfilled, the landfilling method and level of compaction, the engineering design of the landfill, and the annual rainfall of the region.

Leachate production. The amount of leachate produced within a landfill will depend mainly on the rainfall of the area, how well the landfill is sealed (especially the cap), and the original water content of the waste deposited. Data on the amount of leachate produced by actual landfill sites are not commonly reported, but IFEU (1992) estimate that around 13% of the rainfall on a landfill site emerges as leachate. For sites in Germany with an average annual rainfall of 750 mm, this would produce around 100 l of leachate per square metre of landfill site, per year. Using an estimated 20 m depth of landfilled waste, with an approximate density of 1 tonne/m^3, this gives a leachate production of 5 l/tonne of landfilled waste per year. If the active period for leachate production is around 30 years, the total amount of leachate produced would be 150 l/tonne of waste.

Leachate composition. More data are available on the composition of landfill leachate than on the volume produced, but since the leachate composition depends mainly on the nature of the waste landfilled, reported leachate data differ widely. As with landfill gas, leachate composition also varies with the stage of decomposition of the waste; the initial acidification stage is characterised by low pH, along with high levels of organic matter (high BOD and COD values), calcium, magnesium, iron and sulphate, which all decline as the methanogenic stage is reached (Spinosa *et al.*, 1991).

A range of composition data from municipal solid waste leachate are given in Table 11.5, from which the complexity of leachate mixtures can be seen. A range of volatile organic materials plus up to 46 non-volatile organic substances have been analysed from a single landfill site (Öman and Hynning, 1991). Since in many cases, the biochemical pathways involved in the creation of leachate substances are not known, it is not possible to allocate individual pollutants to the different fractions of municipal solid waste. One exception is the organic content of the leachate (BOD/COD), which is derived from the biodegradable fractions, i.e. from the putrescible organic, paper and textile fractions. For the purposes of modelling, therefore, it is assumed that all of the BOD produced originates from these three fractions. All other leachate components are assumed to arise equally from all of the MSW material fractions that are landfilled, since it is not possible to identify their source with any degree of certainty.

Leachate composition data for the other two major types of material from the municipal solid waste stream entering landfills, biologically treated material (composting and biogasification residues) and thermally treated material (ash), are also given in Table 11.5. Note that the residues from thermal treatment will include both bottom ash (clinker), which is relatively inert, and fly ash which often contains high levels of inorganic pollutants such as heavy metals. In some countries, the fly ash is classified as a hazardous waste and subjected to stricter controls during its disposal. No data were available for the leachate composition from fly ash-containing landfills, however. For the calculation of leachate resulting from ash in landfills, the fly ash and bottom ash amounts are combined, and the leachate composition from bottom ash is used.

Leachate collection and treatment. Landfills operated on a 'dilute and disperse' basis will release all of the leachate generated into the surrounding soil and rock strata, where the constituent materials may be further broken down by soil micro-organisms, adsorbed onto soil particles or may enter the groundwater system. Most large modern landfills are lined by a geomembrane or layer of compacted clay, however, and operate on a 'containment' basis. The leachate produced within the sealed landfill can either be recirculated to accelerate the process of decomposition, or drained/pumped out for leachate treatment. Both recirculation and leachate treatment will consume energy, although these are not included in the present model due to lack of suitable data. Leachate treatment can

Table 11.5 Composition of landfill leachates from MSW, ash and biologically treated material (mg/litre, except for dioxins/furans)

Component	MSW restwaste	Compost/biogas residues	MSW bottom ash
Aluminium	2.4	2.4	0.024
Ammonium	210	10	0.06
Antimony	0.066	0.051	0.051
Arsenic	0.014	0.007	0.001
Beryllium	0.0048	0.0048	0.0005
Cadmium	0.014	0.001	0.0002
Chlorine	590	95	75
Chromium	0.06	0.05	0.011
Copper	0.054	0.044	0.06
Fluorine	0.39	0.14	0.44
Iron	95	1.0	0.1
Lead	0.063	0.012	0.001
Mercury	0.0006	0.000 02	0.001
Nickel	0.17	0.12	0.0075
Zinc	0.68	0.3	0.03
AOX	2.0	0.86	0.011
COD	1900	1900	24
1,1,1 trichloroethane	0.086	0.0086	0.000 86
1,2-dichloroethane	0.01	0.001	0.0001
2,4-dichloroethane	0.13	0.065	0.0013
Benzo (a) pyrene	0.000 25	0.000 13	0.000 0025
Benzene	0.037	0.0037	0.000 37
Chlorobenzene	0.007	0.0035	0.000 07
Chloroform	0.029	0.0029	0.000 29
Chlorophenol	0.000 51	0.000 25	0.000 0051
Dichloromethane	0.44	0.044	0.0044
Dioxins/furans (TEQ)	0.32 ng	0.16 ng	0.0032 ng
Endrin	0.000 25	0.000 13	0.000 0025
Ethylbenzene	0.058	0.029	0.000 58
Hexachlorobenzene	0.0018	0.000 88	0.000 018
Isophorone	0.076	0.038	0.000 76
PCB	0.000 73	0.000 36	0.000 0073
Pentachlorophenol	0.045	0.023	0.000 45
Phenol	0.38	0.1	0.005
Tetrachloromethane	0.2	0.02	0.002
Toluene	0.41	0.041	0.0041
Toxaphene	0.001	0.0005	0.000 01
Trichloroethene	0.043	0.0043	0.000 43
Vinyl chloride	0.04	0.004	0.0004

Source: Data used in IFEU (1992).

involve a range of physical (denitrification, evaporation, drying, etc.) and biological (anaerobic digestion, bio-oxidation) processes to produce an effluent that can be discharged to municipal sewage systems or surface waters. Depending on the process used, the treatment of leachate from the methanogenic phase of landfill activity can produce from 9 to 22 kg of solid residue for every cubic metre treated (Weber and Holz, 1991). These residues can themselves be treated by incineration or landfilling, in which case they will produce further residues and emissions. For simplicity in the LCI model, any leachate treatment residues will be added to the total amount of final solid waste.

As discussed in Section 11.4, it is generally accepted that most landfill liners will eventually leak, so that part of the leachate will be discharged directly into the underlying strata, from where it can contaminate the groundwater. For a lined site, therefore, it is difficult to estimate the proportion of the leachate generated that will be collected and treated, as opposed to leaking into the substrata. The level of leakage will depend on many factors, including the type of liner used (single versus multi-layered, mineral versus synthetic membrane, etc.), the geology of the site (permeability of underlying strata), and the efficiency of any leachate collection and treatment system. Although there is evidence that many lined landfills have leaked, there is virtually no data on leakage amounts. In the absence of reliable data, therefore, where lined landfill sites with leachate collection and treatment are used, it will be assumed that over the active life of the landfill (assumed to be 30 years), 70% of the leachate will be collected and treated and 30% will leak out. This is clearly an area where reliable data are urgently needed.

Final inert solid waste.   Although it does not physically leave the site, one of the primary outputs of the solid waste management systems described in this book is the final solid waste left in a landfill at the end of all decomposition processes. This will not be the same weight of waste as was originally landfilled, since some of the waste has been degraded and will be released from the landfill as landfill gas or leachate. The weight of each type of solid waste (mixed MSW, waste sorting residues, biological treatment residues, ash) that enters landfill is known, but the amount remaining after decomposition will depend on how extensive the degradation process has been. Rather than attempt to predict the weight loss of the waste while in the landfill, the input tonnages will be used as amounts of final solid waste. In any case, the important attribute of the final solid waste is its volume, rather than its weight, since landfill sites fill up rather than get too heavy! Using the specific densities of the different waste materials (Table 11.6) and the known input tonnages, it is possible to calculate the volume of material that is consigned to landfill. Whilst some further compaction and settling of the landfilled material may occur as decomposition occurs, this volume will approximate to the final volume of solid waste resulting from landfilling, and will be the figure used for the output of the LCI model.

The environmental consequence of the final solid waste production is land consumption (or land generation if landfilling is used as a means of land reclamation). If we assume an average depth of waste in a landfill (IFEU, 1992, assume a depth of 20 m), it is possible to calculate from

Table 11.6 Specific densities of MSW fractions and waste materials in landfills

Material/fraction	Density (tonnes/m^3)	Specific volume in landfill (m^3/tonne)	Source
MSW	0.9	1.11	Bothmann (1992)
Paper/board	0.95	1.05	Habersatter (1991)
Glass	1.96	0.51	Habersatter (1991)
Metal-aluminium	1.08	0.93	Habersatter (1991)
Metal-ferrous	3.13	0.32	Habersatter (1991)
Plastic	0.96	1.04	Habersatter (1991) (average of resin types)
Textiles	0.7	1.43	(estimated)
Organic	0.9	1.11	(estimated)
Other (MSW)	0.9	1.11	(average for MSW used)
Bottom ash (MSW)	1.5	0.67	Bothmann (1992)
Filter ash/dust	0.6	1.67	Habersatter (1991)
Compost residue	1.3	0.77	Bothmann (1992)
Industrial waste	1.5	0.67	(assumed to be mainly ash)

final solid waste volume to space consumption by landfilling. Since landfills vary widely in geometry, however, depending on whether they are used to reclaim former quarries, clay pits, or as above ground structures, this conversion will not be attempted, and the environmental impact of producing solid waste will be quoted as a volume requirement.

11.6 Economic costs

As with the other waste treatment options discussed earlier, the economic costs of landfilling vary widely across Europe. The variability reflects geographical differences mainly in land costs, landfill design and engineering requirements and labour costs.

The range of costs that have been quoted for various European countries are given in Table 11.7.

The economic costs of landfilling should include the cost of the land, capital costs of site construction, operating costs, closure costs and long-term post-closure monitoring and aftercare costs. It is unlikely that most costs currently quoted fully account for all of these costs, in particular those for post-closure monitoring and aftercare. The duration of post-closure monitoring and aftercare may be extensive. In the United States, a 30-year post-closure monitoring period is mandated, although there can be significant leachate emissions for considerably longer periods. As discussed above, leakages are also quite likely to occur; a recent survey found that 18% of 1000 UK landfill sites studied had suffered 'significant'

Table 11.7 Summary of landfilling costs for Europe

Country	Waste type	Typical cost (ecu/tonne)	Source
Europe average	MSW	47	ORCA (1991a)
Austria	MSW	176	pers. comm. (1994)
Belgium	MSW	30–45	ORCA, pers. comm. (1994)
Denmark	MSW	26	pers. comm. (1994)
Finland	MSW	18–38	Maddox (1994)
France	MSW	26–47	Maddox (1994)
Germany			
old states	Household	69	UBA (1993)
	Bulky household	72	
	Construction	15	
	Hazardous	up to 500	
new states	Household	13	
Greece	MSW	10	pers. comm. (1994)
Italy	MSW	1	pers. comm. (1994)
Netherlands	MSW	34–92	Maddox (1994)
Spain	MSW	3–13	Maddox (1994)
Sweden	MSW	13–57	Maddox (1994)
UK	Household	6–40	DOE (1993)
	Commercial	6–40	

pollution incidents or failures (ENDS, 1992b). Remedial costs following leakages can also be very expensive. Escapes of landfill gas accounted for most (48%) of the pollution incidents in the above survey, typically costing 65 000–130 000 ecu to remediate. Surface water contamination accounted for 27% of incidents, with a typical cost of 6500–20 000 ecu, whilst ground-water pollution, accounting for 15% of incidents, had costs typically ranging from 65 000 to 1.3 million ecu. Several instances of groundwater pollution cost over 1.3 million ecu to rectify.

As a result of the above potential costs for remediation work, it is likely that the real cost of landfilling is higher than currently quoted figures. A more recent report by Pearce and Turner (1993) came to the same conclusion and suggested that additional external costs of between 1 ecu and 5 ecu per tonne should be added to the current figures. The same is likely to hold, to varying degrees, in other European countries.

11.7   Operation of the landfilling module of the LCI spreadsheet

LCI Box 7 shows the layout of the landfilling module for the LCI computer spreadsheet.

RESTWASTE INPUT TO LANDFILL
The spreadsheet automatically enters both the amount and composition of the waste that has not been separated or treated so far in the model. It is assumed that all of this residual waste will be landfilled directly.

LCI BOX 7

7. LANDFILLING	All material that remains untreated is sent to landfill

RestWaste Input to Landfill

Amount [　　　] 000 tonnes/yr

	Paper	Glass	Metal	Plastic	Textiles	Organic	Other
Composition %							

RestWaste Pretreatment/Transfer station (If no transfer station/pretreatment used, leave blank)
Energy/fuel consumption of transfer station:

Electrical [▒▒▒▒▒] kW hr per tonne input
Diesel [▒▒▒▒▒] litres per tonne input

Distance to Landfill site from transfer station [▒▒▒▒▒] km

Waste Treatment Residue input to Landfill

Amount (non-hazardous) [　　　] 000 tonnes   Amount hazardous (fly ash) [　　　] 000 tonnes

Composition (non-haz)%	Paper	Glass	Metal	Plastic	Textiles	Organic	Other	Compost	Ash

Industrial Solid Wastes resulting from Energy Generation, and Production of Fuel and other Raw Materials

Amount (assumed non-haz.) [　　　] 000 tonnes

TOTAL INPUT TO LANDFILL

Total Amount (non-haz.) [　　　] 000 tonnes   Total Amount (haz) [　　　] 000 tonnes

Composition (non-haz.)%									
Paper	Glass	Metal	Plastic	Textiles	Organic	Other	Compost	Ash	Industrial

Landfill management:

Energy consumption of landfill site   Electrical [▒▒▒▒▒] kW hr per tonne input
Diesel [▒▒▒▒▒] litres per tonne input

Landfill Gas
Landfill gas collected?   Insert 1 in box if collected   [▒▒▒▒▒] Collection efficiency %   [40%]
Energy recovered from gas? Insert 1 in box if recovered

Leachate
Landfill site lined with leachate collection and treatment? (1=YES)   [▒▒▒▒▒] Collection efficiency %   [70%]

Costs: Transfer/transport cost /tonne of Restwaste   [▒▒▒▒▒] ecu
Landfill cost per input tonne: non-hazardous waste   [▒▒▒▒▒] ecu
hazardous waste   [▒▒▒▒▒] ecu

Restwaste pre-treatment/transfer station
  Energy/fuel consumption of transfer station:
    Electrical:
    Diesel:
  Distance to landfill:
The collection and pre-sorting module covered transport of the residual waste to either a local landfill site or a transfer station if a distant landfill site is used. If a transfer station is used, the spreadsheet operator needs to insert the amount of energy/fuel used at the station per tonne of waste handled. The spreadsheet calculates the total amounts of energy/fuel consumed to handle all of the waste, and adds this to the fuel/energy consumption totals within the model. For onward transport to the landfill, the user enters the one way distance to the landfill site. The spreadsheet calculates the fuel consumed, assuming a 40 tonne truck capacity and that no return load is carried.

Waste treatment residues input to landfill
The spreadsheet calculates and displays both the total amount and composition of residues from collection, central sorting, biological treatment

LCI Data Box 7a Data for landfill gas and leachate generation, leachate composition and final solid waste used in the LCI spreadsheet.

	Landfill input materials											
	MSW and sorting residues							Compost	Waste treatment residues		Industrial power generated waste
	Paper	Glass	Metal	Plastic	Textiles	Organic	Other		Bottom ash	Filter ash/dust	
Landfill gas generation (N m³/tonne)	250	0	0	0	250	250	0	100	0	0	–
Leachate generation (m³/tonne)	0.15	0.15	0.15	0.15	0.15	0.15	0.15	0.15	0.15	0.15	–
Leachate composition (g/m³)											
BOD	3167	0	0	0	3167	3167	0	1900	24	24	–
COD	6000	0	0	0	6000	6000	0	3800	48	48	–
Suspended solids	100	100	100	100	100	100	100	100	100	100	–
Total organic compounds	2.0	2.0	2.0	2.0	2.0	2.0	2.0	0.39	0.021	0.021	–
AOX	2.0	2.0	2.0	2.0	2.0	2.0	2.0	0.86	0.011	0.011	–
Chlorinated HCs	1.03	1.03	1.03	1.03	1.03	1.03	1.03	0.18	0.01	0.01	–
Dioxins/furans (TEQ)	$3.2\,E-7$	$3.2\,E-7$	$3.2\,E-7$	$3.2\,E-7$	$3.2\,E-7$	$3.2\,E-7$	$3.2\,E-7$	$1.6\,E-7$	$3.2\,E-9$	$3.2\,E-7$	–
Phenol	0.38	0.38	0.38	0.38	0.38	0.38	0.38	0.1	0.005	0.005	–

Ammonium	210	210	210	210	210	210	210	10	0.06	0.06	–
Total metals	96.1	96.1	96.1	96.1	96.1	96.1	96.1	1.37	0.21	0.21	–
Arsenic	0.014	0.014	0.014	0.014	0.014	0.014	0.014	0.007	0.001	0.001	–
Cadmium	0.014	0.014	0.014	0.014	0.014	0.014	0.014	0.001	0.0002	0.0002	–
Chromium	0.06	0.06	0.06	0.06	0.06	0.06	0.05	0.05	0.011	0.011	–
Copper	0.054	0.054	0.054	0.054	0.054	0.054	0.054	0.044	0.06	0.06	–
Iron	95	95	95	95	95	95	95	1.0	0.1	0.1	–
Lead	0.063	0.063	0.063	0.063	0.063	0.063	0.063	0.12	0.001	0.001	–
Mercury	0.0006	0.000	0.000	0.0006	0.0006	0.0006	0.000	2.0 E−5	0.001	0.001	–
Nickel	0.17	0.17	0.17	0.17	0.17	0.17	0.17	0.12	0.001	0.001	–
Zinc	0.68	0.68	0.68	0.68	0.68	0.68	0.68	0.3	0.0075	0.0075	–
Chloride	590	590	590	590	590	590	590	95	0.03	0.03	–
Fluoride	0.39	0.39	0.39	0.39	0.39	0.39	0.39	0.14	75	75	–
Final solid waste volume (m^3/tonne)	1.05	0.51 (Fe) 0.93 (Al)	0.32	1.04	1.43	1.11	1.11	0.77	0.44	0.44	0.67
Solid waste from leachate treatment (kg/m^3 treated)	15	15	15	15	15	15	15	15	0.67	1.67	–

LCI Data Box 7b   Air emissions from landfilling used in the LCI spreadsheet.

	Gas composition	
	Landfill gas (mg/N m^3)	Flare/engine exhaust (mg/N m^3 of input gas)
Air emissions		
Particulates	–	4.3
CO	12.5	800
CO$_2$	883 930	1 964 290
CH$_4$	392 860	0
NO$_x$	–	100
N$_2$O		
SO$_x$	–	25
HCl	65	12
HF	13	0.021
H$_2$S	200	0.33
HC	2000	60
Chlorinated HC	35	10
Dioxins/furans	–	8.0 E$-$7
Ammonia		
Arsenic		
Cadmium	5.6 E$-$3	9.4 E$-$6
Chromium	6.6 E$-$4	1.1 E$-$6
Copper		
Lead	5.1 E$-$3	8.5 E$-$6
Mercury	4.1 E$-$5	6.9 E$-$8
Nickel		
Zinc	7.5 E$-$2	1.3 E$-$4
Energy recovery (kW-h electricity/N m^3)	–	1.5

Sources: Tables 11.2, 11.3, 11.4 and IFEU (1992).

and thermal treatment processes. Hazardous waste for landfilling (fly ash and other gas cleaning residues from thermal treatment) is accounted for separately.

Industrial solid wastes resulting from energy generation and production of fuel and other raw materials
The total amount of solid waste arising as a result of energy and fuel consumption, and the production of materials such as plastic for refuse sacks is inserted here automatically. It is not differentiated into different materials, although a large part will be ash due to power generation in many countries. As the detailed composition is not known, no emissions are ascribed to these inputs to landfill. They are only included as contributing to the total final solid waste amount.

TOTAL INPUT TO LANDFILL
Both total amount (restwaste plus process residues) and overall composition are calculated and displayed. These are the totals used to calculate

the air and water emissions and the final solid waste volume, using the information in Data Boxes 7a, b.

Landfill management.

Energy consumption of landfill site:
Electrical: kW-h/tonne input:
Diesel: l/tonne input:

Using this information, the spreadsheet calculates the total energy/fuel consumption of the landfill operation and adds these amounts to the energy and fuel consumption totals for the overall system.

Landfill gas

Landfill gas collected? (1 = YES):   *Collection efficiency %:*
Energy recovered from gas? (1 = YES):

If landfill gas is not collected, the appropriate amount of gas (derived from Data Box 7b) is added to the overall total of air emissions. If gas is collected, the portion that diffuses out of the site (100% minus collection efficiency %) is added to the air emissions total. The collected portion is assumed to be burned, whether in a flare, furnace or gas engine, with the resultant air emissions given in Data Box 7b. If energy is recovered, the appropriate amount of electrical energy is added to the system energy production total.

Leachate

Landfill site lined with leachate collection and treatment? (1 = YES):
Collection Efficiency %:

If the site is unlined, or lined but with no leachate collection and treatment system, it is assumed that all of the leachate produced leaks from the site and enters the substrata. The amounts of the leachate materials given in Data Box 7a are therefore added to the totals for water emissions. If the site is lined and leachate collected and treated, the amount collected is calculated using the collection efficiency estimate, and the resulting effluent and residues added to the respective totals for water emissions and solid waste. The amount not collected is again assumed to leak from the site, and is added to the water emissions totals.

Costs

Transfer/transport cost/tonne of restwaste:
Landfill cost per input tonne:

The user inserts appropriate local unit costs for these operations, per tonne of waste material handled. The cost of landfilling should be inclusive of site purchase, construction, operation, gas sales (if appropriate), gas and leachate treatment, site closure and subsequent monitoring and aftercare. The spreadsheet multiplies the unit costs by the amounts landfilled, and adds the total to the overall system cost.

References

Augentein, D. and Pacey, J. (1991) Modelling landfill methane generation. In *Biogas Disposal and Utilisation, Choice of Material and Quality Control, Landfill Completion and Aftercare, Environmental Monitoring. Third International Landfill Symposium*, Sardinia, pp. 115-148.

Baldwin, G. and Scott, P.E. (1991) Investigations into the performance of landfill gas flaring systems in the UK. In *Biogas Disposal and Utilisation, Choice of Material and Quality Control, Landfill Completion and Aftercare, Environmental Monitoring. Third International Landfill Symposium*, Sardinia, pp. 301-312.

Biffa (1994) Biffa Wastes Services, personal communication.

Bothmann (1992) Cited in IFEU, 1992.

Brown, P. (1991) Gas risk to homes from 1000 rubbish tips. *The Guardian*, UK 4th April.

Campbell, D.J.V. (1991) An universal approach to landfill management acknowledging local criteria for site design. In *Biogas Disposal and Utilisation, Choice of Material and Quality Control, Landfill Completion and Aftercare, Environmental Monitoring. Third International Landfill Symposium*, Sardinia, pp. 15-32.

Carra, J.S. and Cossu R. (1990) Introduction. In *International Perspectives on Municipal Solid Wastes and Sanitary Landfilling*, eds. J.S. Carra and R. Cossu. Academic Press, New York, pp. 1-14.

De Baere, L., van Meenen, P., Deboosere, S. and Verstraete, W. (1987) Anaerobic fermentation of refuse. *Resources and Conservation* 14, 295-308.

Deardorff, G.B. (1991) Construction inspection of municipal landfill lining systems: a USA perspective. In *Biogas Disposal and Utilisation, Choice of Material and Quality Control, Landfill Completion and Aftercare, Environmental Monitoring. Third International Landfill Symposium*, Sardinia, pp. 741-752.

DoE (1989) Landfill gas. Waste management paper no. 27. Department of the Environment, UK.

DoE (1993) *Landfill Costs and Prices: Correcting Possible Market Distortions.* A Study by Coopers and Lybrand for the Department of the Environment. HMSO, London.

Ehrig, H.J. (1991) Prediction of gas production from laboratory scale tests. In *Biogas Disposal and Utilisation, Choice of Material and Quality Control, Landfill Completion and Aftercare, Environmental Monitoring. Third International Landfill Symposium*, Sardinia, pp. 87-114.

ENDS (1992a) Landfill, oil and gas industries top methane emissions league. *Environ. Data Services Rep.* **206**, 3-4.

ENDS (1992b) Survey puts landfill clean-up costs in perspective. *Environ. Data Services Rep.* **214**, 11.

Finnveden, G. (1993) Landfilling – a forgotten part of life cycle assessment. In *Product Life Cycle Assessment – Principles and Methodology*. Nordic Council of Ministers, pp. 263-288.

Gendebien, A., Pauwels, M., Constant, M., Willumsen, H.C., Buston, J., Fabry, R., Ferrero, G.L. and Nyns, E.J. (1991) Landfill gas: from environment to energy. State of the art in the European Community context. In *Biogas Disposal and Utilisation, Choice of Material and Quality Control, Landfill Completion and Aftercare, Environmental Monitoring. Third International Landfill Symposium*, Sardinia, pp. 69-76.

Habersatter, K. (1991) Ökobilanz von Packstoffen Stand 1990, Bundesant fur Umwelt, Wald und Landschaft (BUWAL) Report No. 132, Bern, Switzerland.

Ham, R.K., Hekimian, K.K., Katten, S.L., Lockman, W.J., Lofy, R.J., McFaddin, D.E. and Daly, E.J. (1979) Recovery, processing and utilisation of gas from sanitary landfills. US Environmental Protection Agency Report EPA-600/2-79-001.

Harris, R. (1992) The groundwater protection policy of the NRA. *Proc. Landfilling Waste – Asset or Liability*. IBC Conference, London.

IFEU (1992) Vergleich der Auswirkungen verschiedener Verfahren der Restmüllbehandlung auf die Umwelt und die menschliche Gesundheit. Institut für Energie- und Umweltforschung, Heidelberg, GmbH, Heidelberg.

Lashof, D.A. and Ahuja, D.R. (1990) Relative contributions of greenhouse gas emissions to global warming. *Nature* **344**, 529-531.

Maddox, B. (1994) Politics ahead of science. *Financial Times*, 2nd March, 20.

Maunder, D. (1992) Exploiting landfill gas: turning the liability into an asset. *Proc. Landfilling Waste - Asset or Liability?* IBC Conference, London.

Nyns, E.-J. (1989) Methane fermentation. In *Biomass Handbook*, eds. O. Kitani and C.W. Hall. Gordon and Breach, New York, pp. 287-301.

OECD (1991) *Environmental Data Compendium*. Organisation for Economic Cooperation and Development, Paris.

Öman, C. and Hynning, P. (1991) Identified organic compounds in landfill leachate. In *Biogas Disposal and Utilisation, Choice of Material and Quality Control, Landfill Completion and Aftercare, Environmental Monitoring. Third International Landfill Symposium*, Sardinia, pp. 857-864.

ORCA (1991) The role of composting in the integrated waste management system. In *Solid Waste Management, An Integrated System Approach. Part 5*. Organic Reclamation and Composting Association, Brussels.

Pearce, D.W. and Turner, R.K. (1993) *Externalities from Landfill and Incineration*. HMSO, London.

Perry and Green (1984) *The Chemical Engineer's Handbook*, 6th edition. McGraw-Hill, New York.

RCEP (1993) *Incineration of Waste*. Royal Commission on Environmental Pollution 17th Report. HMSO, London, 169 pp.

Richards, K.M. and Aitchison, E.M. (1991) Landfilling in a greenhouse world. In *Biogas Disposal and Utilisation, Choice of Material and Quality Control, Landfill Completion and Aftercare, Environmental Monitoring. Third International Landfill Symposium*, Sardinia, pp. 33-44.

Ruggeri, R., Chiampo, F. and Conti, R. (1991) Economic analysis of biogas production from MSW landfill. In *Biogas Disposal and Utilisation, Choice of Material and Quality Control, Landfill Completion and Aftercare, Environmental Monitoring. Third International Landfill Symposium*, Sardinia, pp. 263-276.

Schneider, D. (1992) Das Wabio-Verfahren der Deutschen Babcock Anlagen. In *Biologische Abfallbehandlung*; *Kompostierung - Anaerobtechnik - Kalte Vorbehandlung*; *Reihe (book series) Abfall-Wirtschaft - Neues aus Forschung und Praxis*, eds. K. Wiemer and M. Kern. University of Kassel, pp. 863-875.

Schön, M. (1992) Das Kompogas-Verfahren der Firma Bühler. In *Biologische Abfallbehandlung*; *Kompostierung - Anaerobtechnik - Kalte Vorbehandlung*; *Reihe (book series) Abfall-Wirtschaft - Neues aus Forschung und Praxis*, eds. K. Weimer and M. Kern. University of Kassel, pp. 821-829.

Spinosa, L., Brunetti, A., Lorè, F. and Antonacci, R. (1991) Combined treatment of leachate and sludge: lab experiments. In *Biogas Disposal and Utilisation, Choice of Material and Quality Control, Landfill Completion and Aftercare, Environmental Monitoring. Third International Landfill Symposium*, Sardinia, pp. 961-968.

Tabasaran, O. (1976) Überlegungen zum ploblem deponiegas. *Müll und Abfall* 7, 204.

Thorneloe, S.A. (1991) US EPA's global climate change program - landfill emissions and mitigation research. In *Biogas Disposal and Utilisation, Choice of Material and Quality Control, Landfill Completion and Aftercare, Environmental Monitoring. Third International Landfill Symposium*, Sardinia, pp. 51-68.

UBA (1993) Umweltbundesamt. (Federal German Environmental Agency). Annual statistical report.

Uehling, M. (1993) Keeping rubbish rotten to the core. *New Scientist* **1888**, 12-13.

Weber, B. and Holz, F. (1991) Disposal of leachate treatment residues. In *Biogas Disposal and Utilisation, Choice of Material and Quality Control, Landfill Completion and Aftercare, Environmental Monitoring. Third International Landfill Symposium*, Sardinia, pp. 951-960.

Willumsen, H.C. (1991) The problematics of landfill gas technology. In *Biogas Disposal and Utilisation, Choice of Material and Quality Control, Landfill Completion and Aftercare, Environmental Monitoring. Third International Landfill Symposium*, Sardinia, pp. 77-86.

Young, C.P. and Blakey, N.C. (1991) Emissions from power generation plants fuelled by landfill gas. In *Biogas Disposal and Utilisation, Choice of Material and Quality Control, Landfill Completion and Aftercare, Environmental Monitoring. Third International Landfill Symposium*, Sardinia, pp. 359-368.

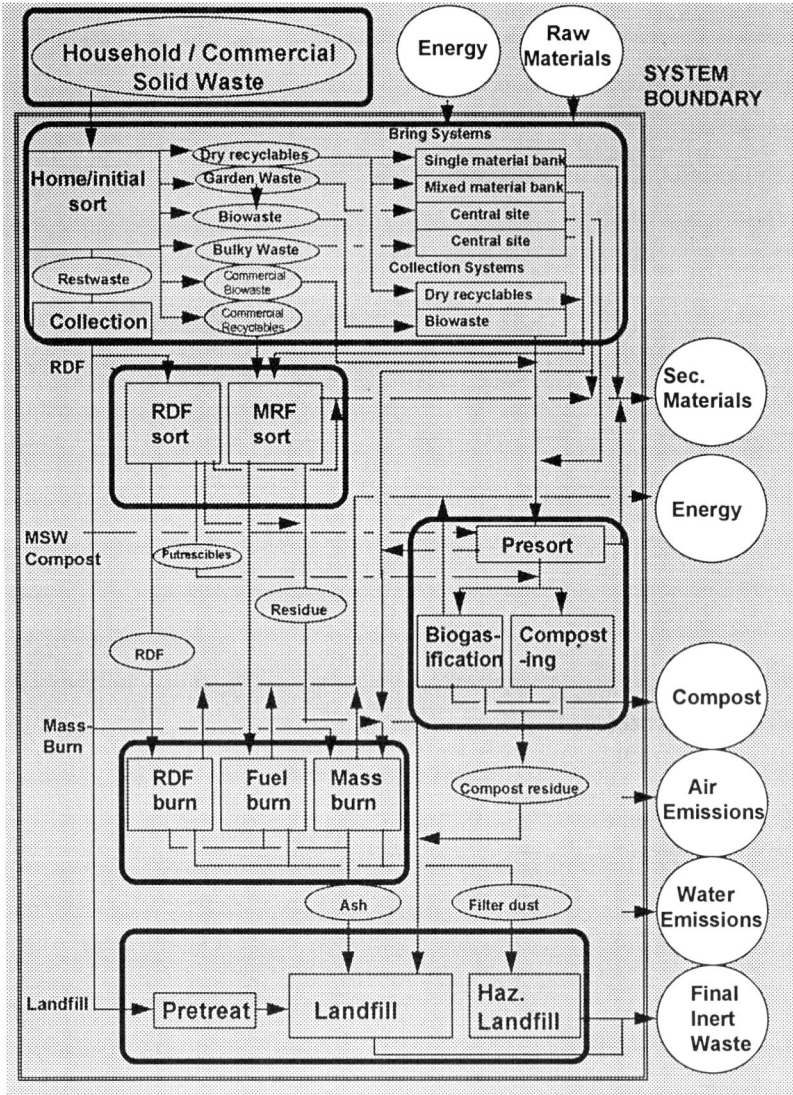

Figure 12.1 Lifecycle inventory results for an integrated waste management system.

12 The overall picture

12.1 Introduction

This book began with a discussion of sustainable development; the need to produce more value from goods and services, with less environmental impact and depletion of resources. When applied to waste management, environmental sustainability requires the production of more value from recovered materials and energy, with the consumption of less energy and the production of less emissions to air, water and land (Box 12.1). The lifecycle inventory (LCI) technique gives us a way to quantify the 'more' and the 'less'; to predict the amounts of materials that will be recovered, the amount of energy consumed and the likely emissions that will be released. This book has constructed a lifecycle inventory (LCI) for municipal solid waste. Starting with a definition of the objectives of the LCI and the system boundaries in Chapter 4, and then definition of the quantity and composition of the waste that is being managed in Chapter 5, each stage in the lifecycle of waste has been described and discussed. For each of the processes from dustbin to grave, i.e. from waste pre-sorting in the home, through collection and waste treatment, to final disposal, the environmental inputs and outputs have been quantified, and generic values suggested. When all of these individual modules are assembled together, it is possible to calculate the overall picture, that is the environmental burden of the whole waste management system.

Environmental sustainability is not the only issue, however. Waste management systems need to be affordable if they are going to be economically sustainable too. Consequently, the LCI has also attempted to calculate the overall economic costs of waste management systems.

This final chapter looks at the results that can be obtained from such a lifecycle inventory of solid waste, and considers ways that such results can be interpreted and used. By way of illustration, a hypothetical case study is used to demonstrate the value of the LCI approach.

12.2 From lifecycle inventory results to sustainability

The results produced by the LCI of municipal solid waste are shown in Box 12.2. This represents a considerable, and perhaps daunting, amount of information, but then the waste management systems that these results

Box 12.1   FROM SUSTAINABLE DEVELOPMENT TO LCI RESULTS.

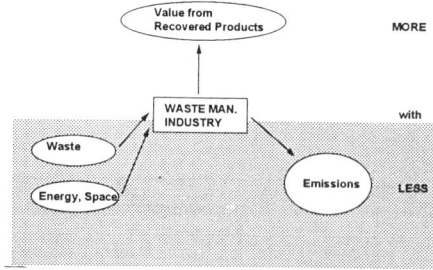

1. The basis for Sustainable Waste Management, *more* value from Recovered Products with *less* consumption of resources and production of emissions (Chapter 1).

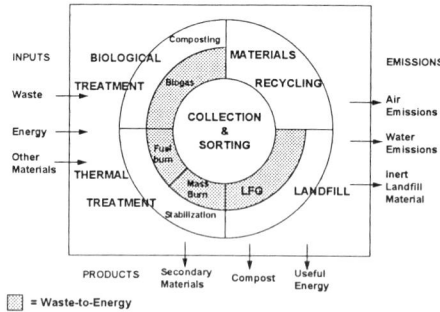

2. Identifying the inputs and outputs of an Integrated Waste Management System (Chapter 4).

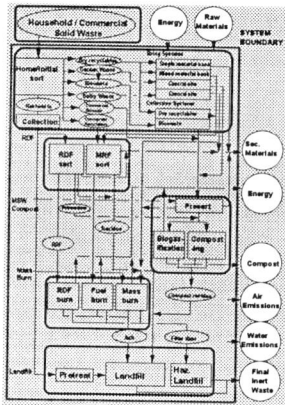

3. Quantifying the '*More*' and the '*Less*'. Using a Lifecycle Inventory to predict the quantity of useful materials recovered and the amounts of energy consumed and emissions released to air, water and land.

Box 12.2   OUTPUTS OF THE LIFECYCLE INVENTORY ANALYSIS.

Environmental factors

> *Energy*
>> *Net thermal energy consumption.*
>
> *Air Emissions*
>> *Amount of each individual material listed*
>
> *Water Emissions*
>> *Amount of each individual material listed*
>
> *Final Inert Waste*
>> *Weight and volume of non-hazardous material landfilled*
>> *Weight and volume of hazardous material landfilled.*

System statistics:

Secondary materials:

Amounts of recovered recyclable materials and marketable compost produced by the system

$$\text{Material recovery rate} = 100\% \times \frac{\text{total amount of recovered recyclable materials leaving IWM system}}{\text{total amount of waste entering the IWM system}}$$

$$\text{Overall recovery rate} = 100\% \times \frac{\text{total amount of recovered recyclable materials plus compost}}{\text{total amount of waste entering the IWM system}}$$

$$\text{Landfill diversion rate} = 100\% \times \left(1 - \frac{\text{Amount of waste entering landfill}}{\text{Total amount of waste entering system}}\right)$$

Economic costs:

> *Overall system cost of solid waste management*
> *Cost per household for solid waste management*

---

describe are themselves complex. The results quantify the input and outputs of an IWM system (shown as white circles in Figure 12.1), namely the amounts of waste and energy (and money) going in, and the amounts of useful products (recovered materials, compost and energy), emissions to air and water, and final inert waste produced. Environmental sustainability involves producing more useful products, but less emissions and final inert waste, with less energy consumed in the process. Economic sustainability is achieved by keeping the amount of money needed to operate the system to an acceptable level.

12.3   Making comparisons

Lifecycle inventories for solid waste, or for any other product or service, are mainly used in a comparative way. The absolute environmental

---

Box 12.3   MAKING COMPARISONS USING LCI RESULTS.

How to use the LCI tool?

Approach 1.
Assume waste stream is constant, and compare the performance of different Integrated Waste Management systems in dealing with this waste.

    – valid comparison
    – basis for the case study in Box 12.4

Approach 2.
Assume the waste management system is constant, and look at the effect of altering the amount and composition of waste that the system handles.

    – valid comparison if comparing geographical differences in waste
    – caution needed if looking at changing the materials entering the waste
       stream in a given area, since these materials performed a function before they
       became waste. Alternative materials for this function may have envi-
       ronmental impacts elsewhere than in the waste management system. A
       product-specific LCI is needed to find the best material for any function,
       throughout the lifecycle, not just in the waste management system.

How to choose between options?

Single Criterion   – Where there is a single over-riding concern (e.g. lack of landfill
                      space)

Multiple Criteria – Where more than one issue is important (e.g. energy
                      consumption and landfill space)

    – 'Less is better' – where one option is lower in all categories
    – Impact analysis – can combine some categories that contribute to the same
       environmental effect such as global warming.

---

performance (in terms of energy consumption and emissions) or economic cost is less informative than a comparison between different options to see which is preferred. This comparison may be between completely different systems or products, or between an existing one and a potential improvement. This section looks at how LCI results for solid waste can be used effectively in a comparative way. But first it is necessary to determine which comparisons are valid (Box 12.3).

The objective of the LCI of solid waste was defined earlier as attempting to predict the environmental performance (in terms of emissions and energy consumption) and economic costs of an integrated waste system, which can manage the waste of a given area. A valid comparison, therefore, is between alternative IWM systems that deal (in different ways) with the specified waste stream of the area in question (Approach 1, Box 12.3). Comparisons of this nature are shown in the case study below.

Rather than keeping the waste stream entering the system constant and looking at the effect of altering the waste management system, an alternative approach is to keep the waste management system constant, and to investigate the effect of altering the amount and composition of the waste stream entering the system (Approach 2, Box 12.3). This could take the form of asking 'what would be the effect of removing all plastic (or paper, glass, metal, etc.) from the waste stream?'. These comparisons, however, need to be treated with considerable caution. The LCI model can predict how present waste management systems would perform with changed waste inputs, and if particular materials were eliminated from the waste stream, but that is not the end of the story. If materials are eliminated from the waste stream by not using them in products or packaging, then there will also be environmental and economic effects earlier in the lifecycle of the product or package themselves, i.e. before they became waste. Thus, while the impacts of waste management may be reduced, there may be greater lifecycle impacts elsewhere, such as in the sourcing, manufacture or transport of any replacement material. Choosing which material to use for a given product or package requires a product-specific LCI, rather than the LCI for waste which is developed in this book. Again this shows the important linkage between product-specific LCIs and the LCI for solid waste; both are important as they answer different questions.

Given that we are making a valid comparison, how can we choose between different waste management system options on the basis of LCI results? The full inventory for each option will consist of all the different categories given in Box 12.2; how can useful comparisons be made between such extensive lists?

Single criteria. If there is a single over-riding environmental or cost consideration, then the basis of choice between alternative waste management systems is straightforward. If the aim is to minimise landfill requirement (as in countries such as Japan), then the system which meets this objective can be chosen. Similarly, if cost is the only important factor, the cheapest option can be selected. There are few instances, however, where a single factor alone is important. Indeed, if the objective of integrated waste management is environmental and economic sustainability, then both environmental impacts and economic costs must be considered. Similarly, within the realm of environmental impacts, there are many individual factors to consider, rather than just one.

Multiple criteria. Where many categories, both environmental and economic are considered important, choosing between alternative waste management options will be straightforward if one option has lower impacts, in all respects, than the other. Given the number of categories calculated in this LCI for solid waste, it is generally unlikely that the energy consumption, individual emissions to air and water, final inert waste and cost of one system will all be lower than those of an alternative system. If all impacts are lower, selecting the preferred system is simple: if, as is more likely, some impacts are higher whilst others are lower, a method to trade-off these differences is needed.

Choosing between different waste management alternatives on the basis of environmental and economic performance is facilitated if the number of categories to be considered can be reduced by aggregation. As discussed in Chapter 3, various methods have been used to aggregate inventory data into a smaller number of impact categories. The method that has been most widely accepted is the aggregation of inventory categories that contribute to the same environmental effect. Methane and carbon dioxide emissions, for example, both contribute to the greenhouse effect and so will lead to global warming. Using weighting factors according to their relative effect on global warming, it is possible to aggregate emissions of these gases together into a global warming impact category. The release of 1 kg of methane will cause an equivalent contribution to global warming as 35 kg of carbon dioxide, taken over a 20-year time span (IPCC, 1992), giving a global warming potential (GWP) for methane of 35 (CO_2 is taken as the reference global warming gas). Similar impact analysis methodologies have been suggested for some, but not all, of the environmental impact categories listed in Table 3.2. (CML, 1992; Finnveden *et al.*, 1993; SETAC, 1993). Further development of this important stage of a lifecycle assessment is vital.

The objective of the present analysis was limited to producing a lifecycle inventory, so a full impact analysis is not attempted. In the following examples, however, global warming categories are calculated as an example of how impact analysis can be applied.

12.4   A case study

A good way to demonstrate comparisons between alternative IWM systems is by considering the different ways of managing the waste of a hypothetical area (Box 12.4). Let us take an area of one million households, producing waste of the amount and typical composition for France (see Chapter 5 for relevant data). As a baseline scenario (Case 1), it is assumed that this waste is all collected mixed (i.e. no home sorting) and transported to a local land-fill site. (Details of all other assumptions used in these examples are given in the LCI Example Boxes at the end of this chapter.) Whilst the LCI spreadsheet will give a full inventory of environmental categories, in this example we have selected energy consumption and the amount of final inert waste remaining in landfills as indicators. Within the air emissions we have also selected, as an example, the gases which contribute to global warming (mainly methane and carbon dioxide), which have been converted into a score for global warming potential.

Box 12.4 A CASE STUDY: THE SYSTEMS COMPARED.

Waste amount and composition   1 000 000 households with French waste generation amount and composition.

Case No.	1	2	3	4	5
Collection method	commingled	commingled	commingled	Biowaste (narrow definition) and Restwaste collected separately	Dry recyclables (glass, newspaper, plastic, metals). and restwaste collected separately
Frequency	3 × per week	3 × per week	3 × per week	4 collections per week (2 per fraction); glass collected in central bring system	4 collections per week (3 for restwaste, 1 for recyclables)
Treatment method	All landfilled	All incinerated Fe recovered	All composted Fe recovered	Biowaste composted Restwaste landfilled	Recyclables sorted and sold Restwaste incinerated
Variants			3. Compost sold as product 3a. Compost landfilled as residue	4a. 3 collections per week 4b. Biowaste bin washed out every 2 weeks 4c. Add paper to biowaste	5a. 3 collections per week 5b. Close to home bring system for recyclables 5c. Central bring system for recyclables

The overall energy consumption, global warming potential and final inert waste due to the baseline system (Case 1) are given in Boxes 12.5–12.7. The effect of altering the way that the waste is treated, and/or collected can then be investigated. Starting first with changing the way that waste is treated, Boxes 12.5–12.7 show the same result categories assuming that the waste is handled by mass burn incineration, prior to landfilling of the residue (Case 2). There is a clear improvement in each of the categories shown.

Using the same commingled collection, it is possible to treat the mixed waste biologically, either by composting or biogasification (Case 3). The LCI results show that the overall impact of this particular scenario will vary, according to whether the treated material ('compost') can be marketed or alternatively requires further disposal. In many parts of Europe there would currently be a limited market for such a product. There is more chance of securing markets if source-separated biowaste is used as feedstock for the biological treatment as examined in Case 4. In addition, glass is collected in a bank system. In this scenario, the use of an additional collection vehicle for the biowaste will add further vehicle air emissions to the lifecycle inventory, but these may be avoided if both the biowaste and restwaste collection can be integrated (either by alternating collections, or by collecting both on the same visit) (Case 4a). If we assume that the biowaste collected is narrowly defined (food and garden waste only), then it is likely that residents will have to wash out the bins due to the odour produced (at least in the warmer months) (Case 4b). An alternative is to use a wider biowaste definition, which includes some of the paper fraction. Since this gives structure to the biowaste and absorbs any leachate, bin washing is less likely to be needed (Case 4c). Boxes 12.5–12.7 show how these options vary in their energy consumption, final solid waste and greenhouse gas emissions.

Any other combinations can also be investigated; such flexibility is the strength of the IWM-1 LCI spreadsheet. Collection of recyclable materials can be added to any of the previous scenarios, but is looked at particularly in Case 5. Here glass, paper, plastic and metal are collected separately in a separate collection and sorted centrally, using an additional collection visit (Case 5) or integrated into the normal collection system (Case 5a). The effect of collecting the same materials with a bring system is also investigated (Cases 5b, 5c).

These integrated systems show significant improvements over the basic waste management system considered here, in terms of energy consumption, final solid waste and greenhouse gas emissions. The overall energy consumption is particularly reduced when the reprocessing savings of the recovered materials relative to virgin are included (Box 12.5).

The above examples model a hypothetical waste management system, but they do demonstrate how the spreadsheet can be used to compare the different options available for managing the municipal solid waste of any area.

Box 12.5. OVERALL ENERGY CONSUMPTION OF VARIOUS SOLID WASTE MANAGEMENT SYSTEMS

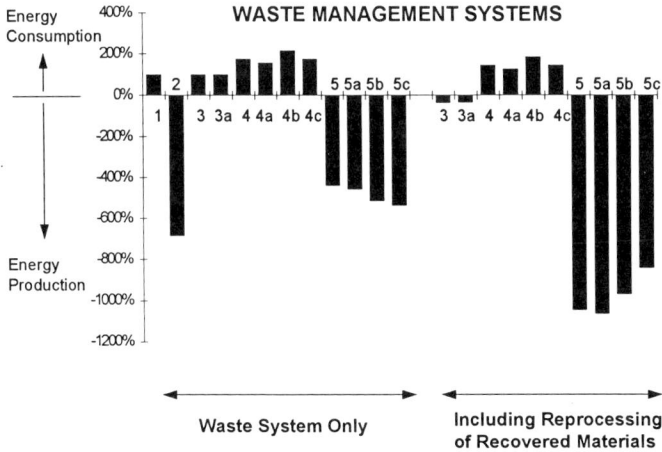

Note: Energy consumption (thermal energy equivalent) indexed relative to basic waste management system (Case 1).

Key to Waste Management Systems Compared.

1. Basic System: commingled collection of household waste followed by landfilling.

2. Commingled collection and mass-burn incineration of household waste.

3. Commingled collection and composting of household waste, with market for the compost.
   3a. As in case 3, but with no market for the compost.

4. IWM system with separate collection and composting of biowaste, landfilling of restwaste.
   4a. As in case 4, with an integrated collection system.
   4b. As in case 4, including washing of the biobin.
   4c. As in case 4, but including paper in the biowaste collection.

5. IWM system with separate kerbside collection of dry recyclables and incineration of the restwaste.
   5a. As in case 5, with an integrated collection system.
   5b. As in case 5, with a close-to-home bring system for dry recyclables.
   5c. As in case 5, with a central bring system for dry recyclables.

For full details of options compared, see LCI Examples 1-5, (this chapter)

Box 12.6. OVERALL FINAL SOLID WASTE VOLUME PRODUCED BY VARIOUS SOLID WASTE MANAGEMENT SYSTEMS

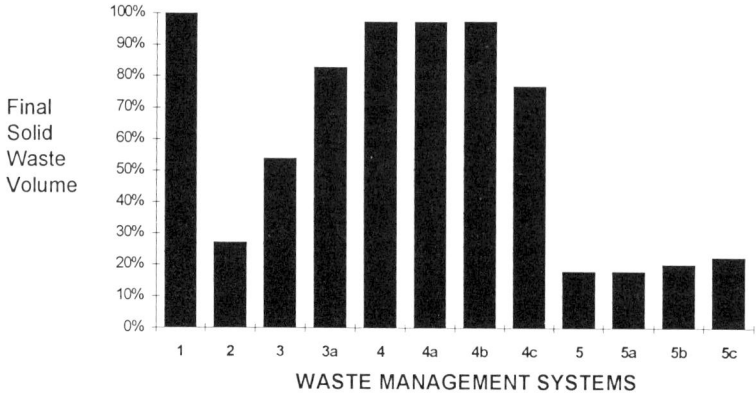

Final Solid Waste Volume

WASTE MANAGEMENT SYSTEMS

Note: Final Solid Waste Volume indexed relative to basic system (Case 1)

Box 12.7. GLOBAL WARMING POTENTIAL CAUSED BY AIR EMISSIONS FROM VARIOUS SOLID WASTE SYSTEMS

Relative Global Warming Potential

WASTE MANAGEMENT SYSTEMS

Notes:
1. Global Warming Potential (GWP) is shown relative to the basic waste management system (Case 1)
2. GWP values used are $CO_2 = 1$, $CH_4 = 35$, $N_2O = 260$. (Values for 20yr timescale, IPCC, 1992)

Given local data on waste generation, waste composition and treatment costs, it would be possible to apply this analysis to any area.

12.5 Identifying improvement opportunities

In discussions of integrated waste management in this book, there has been considerable emphasis on the use of Total Quality management techniques. One of the key concepts of Total Quality is continuous improvement; the process of continually monitoring performance and looking for ways to improve a system. As it provides a way to monitor performance, the LCI of waste can not only be useful in choosing between different options, as demonstrated above, it can also be used within any waste management system, to identify areas for potential improvement.

12.5.1 The importance of operations in the home

To identify where the greatest potential for improvement exists, it is first necessary to determine where the largest impacts occur, and then find ways of reducing them. Essentially, this consists of running a sensitivity analysis on the system, to show which possible alterations will result in significant improvements. When this is done, some unexpected results can occur. Clearly, if the environmental performance of the waste treatment processes is improved, there will be overall system improvements. If, for example, more of the landfill gas could be collected and burned with energy recovery, or more landfill sites were lined with leachate collection and treatment, there would be fewer potentially harmful emissions to air and water. What is perhaps surprising, however, is the significant effect that the behaviour of householders can have on the overall system.

Taking the basic waste management scenario used above (Case 1), where all household waste is collected commingled and landfilled directly, the effect of householders' behaviour can be predicted (Figure 12.2). If each household were to make one special trip of 2 km each way by car each week to a material bank site to drop off recyclable materials, the energy consumption of the whole system would increase by 138%. Similarly, if each household in the system were to use an extra plastic collection bag per week, there would be an 84% increase in overall energy consumption. If every household were to wash out their bin with warm water every month, this would also have an effect on energy consumption, which would increase by 82%.

The importance of the actions of the householders should perhaps not come as a complete surprise. There is a large gearing effect. Further, the household is the source of the waste that needs to be managed, and the place that the initial sorting of waste can occur. Householders can have a very dramatic effect on the impacts of waste management by reducing the

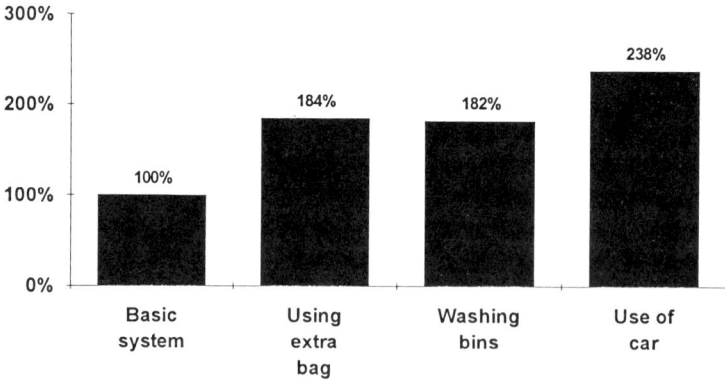

Notes:

1. Basic system involves commingled collection and landfilling of household waste, as in LCI Example 1, (this chapter).
2. Extra bag use scenario involves basic system with use of one extra 20g LDPE collection bag per household per waste collection.
3. Bin washing scenario involves basic system plus washing of collection bin with warm water each week.
4. Use of car scenario involves basic system plus one special trip by car per household per week to deliver recyclables to materials banks (2km each way). Assumes 90% petrol:10% diesel cars.
5. Overall energy consumption on basis of thermal energy equivalent.

Figure 12.2 Effect of householder behaviour on overall energy consumption.

amount of waste that they put into the waste management system in the first place, by choosing to use, for example, concentrated or compact products that contribute less waste themselves, and use less packaging per use. Using refill packs or light-weighted containers will also tend to reduce the overall generation of household waste. If households were to generate less waste, the overall environmental impact of managing the total amount of waste would fall accordingly. Similarly, for the waste that they still generate, households can ensure that as much as possible is valorised by effectively separating the waste into the different categories requested.

One of the most important opportunities for overall environmental improvement, therefore, is likely to occur within each household. The household's behaviour in generating, and separating the waste (if required) and also in its use of car transport, collection bags and bin washing will have major effects on the overall system performance. Any impacts that occur at the household stage will be repeated in every household in the system, so can end up having a highly significant overall effect.

This importance of consumer behaviour in determining the overall environmental impact of both products and services has been highlighted before. The well known LCA practitioner Dr. Ian Boustead has often told the story that the overall energy consumption for producing a bottle of whisky can be doubled simply by the consumer driving a short distance to the shop to purchase this item alone. Hindle *et al.* (1993) have similarly shown that it is by focusing on the usage of a product by consumers that the largest potential environmental improvements can be identified. In waste management too, householders' behaviour has a critical effect on the overall environmental impact. Communicating with these vital players in the system is essential. Investing time and money in educating households about the effect their own actions in generating and handling waste can have on the environment would seem to be time and money well spent.

12.5.2   System improvements

Whilst the case studies described above may not themselves be strictly applicable to a particular area, they do provide some general pointers as to how an integrated waste management system can be optimised.

Optimising collection.   Collection of the waste usually represents a significant part of the economic cost of a waste management system, and as the above examples show, it can also be a significant source of environmental impacts. Optimising the collection system will therefore improve the overall performance of most systems, in both economic and environmental terms. This means collecting all of the waste from households in the form necessary for the chosen waste management options in as few visits as possible. At the same time, the comfort level of the householders must be maintained for full participation and cooperation to be achieved.

Including paper in the biobin.   The example shows some of the advantages of adopting a wider biowaste definition when source-separated organic material is collected for composting (Case 4), and including non-recyclable paper in the biowaste. The major advantage is the diversion of more material from landfill (i.e. less final solid waste produced, along with the emissions that landfilling produces) along with the production of more useful product (compost) (see Chapter 9 for full details).

The importance of waste to energy technologies.   The above examples show that recovering the energy from waste materials has the largest individual effect on overall energy consumption.

Benchmarking waste treatment processes and identifying outliers.   The case studies above used the generic data supplied in this book. When the spreadsheet is used, the most useful results will come from inserting locally sourced data for the region and system in question. In such cases, the generic data can be considered as a benchmark against which the local process performance can be assessed. If the local level of energy consumption or emission generation differs widely from the benchmark figure, this will act as a prompt to re-check the data, and perhaps seek ways of carrying out the relevant process more efficiently.

12.6   Future directions

The application of the emerging science of lifecycle assessment in general, and lifecycle inventory in particular, heralds a new approach to waste management. It offers the possibility of taking an overall view of the waste management system, and allows measurement of progress towards the goals of environmental and economic sustainability. It also helps move the discussion on from the relative merits of individual technologies such as recycling and incineration where it has been concentrated for some time.

The overall value of the LCI model developed in this book will become clear as it is applied to different solid waste management systems in specific parts of Europe, or elsewhere in the world. Whilst improvements to the LCI spreadsheet can and will be made, it represents a workable tool that can be used to both compare future integrated waste management options and optimise existing systems. As with all lifecycle studies, it is likely to come up with a few surprises and offer some interesting new insights.

There are undoubtedly improvements to be made. This study represents no more than a first attempt to define the structure, scope and suitable boundaries for such an LCI approach, and to collate what data are available. We trust that this will not be the final form of the model. We hope that experts in individual waste treatment processes will contribute their knowledge to improve each of the modules of the model. An inevitable, yet valuable, consequence of attempting to build this model has been the demonstration of gaps in the data. As more appropriate data are forthcoming, they can be incorporated in the spreadsheet. We can already point to some areas for improvement: the use of country-specific data for electricity and fuel production rather than European average values, for example. More recent and more reliable waste composition data are also needed, at local, regional and national levels, along with an agreed method for classifying waste materials. These can all be included in future updates. There is also a need for more input from LCA experts. Currently there is much debate about the methodology for impact analysis, and it is to be hoped that scientifically based and generally accepted methods will soon be available. When this happens, impact analysis can be added on to the lifecycle inventory developed here.

The best way to develop the LCI tool will be by using it. We hope that waste managers, regulators, legislators and waste producers will find the approach that it takes both stimulating and helpful. Their experience and feedback will further improve this lifecycle inventory tool as a way to assess

LCI BOX 8

8. RESULTS TABLE

(a) COST	COLLECTION	SORTING	BIOLOGICAL TREATMENT	THERMAL TREATMENT	LANDFILL	IWM MODEL TOTAL	RECYCLING SAVINGS	OVERALL TOTAL
Overall (000 ecu)								
per h.hld (ecu)								

(b) ENVIRONMENTAL IMPACTS								
ENERGY								
Consumption								
Petrol 000 litres								
Diesel 000 litres								
Electricity MW h								
Nat. gas 000 cu m								
TOTAL GJ Thermal								
Recovery								
Electricity MWh								
TOTAL GJ Thermal equiv.								
NET ENERGY USE GJ th								
FINAL SOLID WASTE								
non-hazardous (kt)								
hazardous (kt)								
TOTAL WEIGHT (kt)								
TOTAL VOLUME (000m3)								

EMISSIONS								
AIR EMISSIONS(kg)								
Particulates								
CO								
CO_2								
CH_4								
NO_x								
N_2O								
SO_x								
HCl								
HF								
H_2S								
HC								
Chlor. HC								
Dioxins/Furans								
Ammonia								
Arsenic								
Cadmium								
Chromium								
Copper								
Lead								
Mercury								
Nickel								
Zinc								
WATER EMISSIONS(kg)								
BOD								
COD								
Suspended Solids								
Total Org. Compounds								
AOX								
Chlorinated HCs								
Dioxins/Furans								
Phenol								
Ammonia								
Toatal Metals								
Arsenic								
Cadmium								
Chromium								
Copper								
Iron								
Lead								
Mercury								
Nickel								
Zinc								
Chloride								
Fluoride								
Nitrate								
Sulphide								

Note: Emissions and solid waste include power generation

(c) STATISTICS	AVERAGE	Paper	Glass	Metal-Fe	Metal non-Fe	Plastic-film	Plastic-rigid	Textiles	Organic
Materials Recovery rate									
Organic Recovery Rate									
Overall Recovery Rate									
Diversion from Landfill									
	Total	Paper	Glass	Metal-Fe	Metal non-Fe	Plastic-film	Plastic-rigid	Textiles	Compost
Sec. materials (Kt/yr):									

LCI Example 1.

**Spreadsheet Inputs used for a Basic Waste Management System (Case 1)
Commingled collection of household waste followed by landfilling.**

Box 1. Waste Inputs.	
Country:	*France*
System area population:	*2,500,000*
Average no. of persons per household:	*2.5* (giving 1,000,000 households)
Waste amount and composition:	Default values used.
Detailed composition:	Metals: *90%* Ferrous; *10%* Non-Ferrous
	Plastic: *50%* Film; *50%* Rigid
No Commercial waste included.	

Box 2a. Collection of Household Waste.	
Kerbside Collection Systems	
Bins: Total weight of bins used:	*15kg/household*
Average lifespan of bins:	*10 years*
Collection Vehicles:	
Total no. of collection visits per year :	*156* (3x per week)
Average collection truck fuel consump:	*30* litres/1000 property visits.

Boxes 3,4,5 and 6.
No Inputs (left blank).

Box 7. Landfilling.	
Landfill management:	
Energy consumption of landfill site	*0.6* litre of diesel/tonne input
Landfill gas collected?	*YES,* Collection Efficiency: *40%*
Energy recovered from landfill gas:	*NO*
Leachate site lined, with leachate collection?	*YES* Collection Efficiency *70%*

(Note: All boxes not defined above are left blank.)

LCI Example 2.

Spreadsheet Inputs used for a basic waste management system with mass burn incineration of commingled waste (Case 2)

Inputs as for Case 1 above (basic waste management system), with the following alterations:

Box 6. Thermal Treatment.

6a Mass Burn Incineration of MSW/Restwaste and Sorting Residues.

Does Mass burn incineration of restwaste occur?	*YES*	(Insert "1" in box)

Incineration Process:

Energy Recovery?	*YES*	Electricity Gen. Efficiency:	*20%*
Residues:			
Ferrous Recovery?	*YES*	Recovery Efficiency:	*90%*
Ash/clinker re-used?	*NO*		
Distance to non-hazardous waste landfill:	*10km*		
Distance to Hazardous waste landfill:	*20km*		

(Note: All boxes not defined above are left blank.)

LCI Example 3.

Spreadsheet Inputs used for a basic waste management system including composting of commingled waste (Case 3).

Inputs as for Case 1 above (basic waste management system), with the following alterations:

Box 5. Biological Treatment

Biological Treatment of Restwaste?	*YES*	(Insert "1" in box)

Presort:
    Recovery of Materials: Ferrous metal: *90%* (other materials 0%)

Biological Treatment Process:
    Define process used: Insert "1" for COMPOSTING

Outputs:

Compost - % marketable	**Case 3:**	*100%*
	Case 3a:	*0%*

    Residue treatment:

Sorting residue:	*100% to Landfill*
Compost residue:	*100% to Landfill*
Transport distance:	*10 km to Landfill*

(Note: All boxes not defined above are left blank.)

LCI Example 4.

Spreadsheet inputs used for an IWM system including composting of source-separated biowaste, with landfilling of the restwaste (Case 4).

Inputs as for Case 1 above (basic waste management system), with the following alterations:

Box 2a. Collection of Household Waste
Materials Collection Banks:
    Residents' Transport to Material Bank sites:
    Average no. of special trips/ hhld/yr: *12* (once per month)
    Average car journey length: *2km*

    Amounts collected:
        In single material containers: *30 kg/hhld/yr* (for glass)
    Collection and Transport to Bulking Depot or MRF:
        Average diesel fuel use/tonne   *0.5 litres*

Kerbside Collection Systems:
    Biowaste bin/bag contents: Paper : *0 kg/hhld/yr*   Organic: *200 kg/hhld/yr*

Collection Containers:
    Bins: Total weight used: *25kg /household* (one large and one small bin)
        Average lifespan of bin: *10 years* (as above)
Collection Vehicles:
    No. of collection visits/property /yr: *208* (4x per week)

Box 4. Materials Recycling.
    Transport from Sorting Facility/ Collection Bank to Reprocessing plant.
        Distance (one way)   *100km* (for glass)

Box 5. Biological Treatment
Input: Biological treatment of unsorted waste?   *NO* (leave box blank or insert "0")

Biological Treatment Process:
    Define Process Used: *COMPOSTING* (Insert "1" in box)

Outputs:
    Compost - % marketable: *100%*
    Residue Treatment:
        Sorting Residue: *100%* to Landfill
        Compost Residue: *100%* to Landfill
        Transport distance: *10 km* to Landfill

CASE 4a. As above, but with 3 collections per property per week.

Box 2a. Collection of Household Waste
    Collection Vehicles:
    No. of collection visits/property /yr: *156* (3x per week)

CASE 4b. As Case 4, but with biowaste bin washed out every 2 weeks.

Box 2a. Collection of Household Waste
    Collection containers:
    Bins: # bin washes/hhld/yr: *26*

CASE 4c. As Case 4, but with paper added to the biowaste collection.

Box 2a. Collection of Household Waste
    Kerbside Collection Systems:
    Biowaste bin/bag contents: Paper : *125 kg/hhld/yr*   Organic: *200 kg/hhld/yr*

LCI Example 5.

Spreadsheet Inputs used for an IWM system including materials recycling, with incineration of the restwaste (Case 5).

Inputs as for Case 1 above (basic waste management system), with the following alterations:

Box 2a. Collection of Household Waste

Kerbside Collection System:
    Waste Fractions Collected:
    Dry Recyclables (kg/hhld/yr):   Paper: *125*   Glass: *60*   Metal (Fe):*25* (non Fe):*4*
                               Plastic (Rigid): *20*
    Kerbside sort?   *YES* (Insert "1")

    Collection containers:
        Bins: Total weight used: *17* kg (large bin plus box for recyclables)
                     Average lifespan: *10* yrs

    Collection Vehicles:
    Total No. of collection vehicle visits/property/yr: *208* (4x per week)

Box 3a. Material Recovery Facility (MRF) Sorting.

Processing:
Energy/fuel consumption: Electrical: *22* kWh/tonne   Diesel: *1.0* litre/tonne

Destination for Outputs: *100%* of materials to Recycling

Residue Treatment: *100%* to Incineration
Transport Distance: *10km* to Incinerator.

Box 4. Materials Recycling.
    Transport from Sorting Facility/ Collection Bank to Reprocessing plant.
        Distance (one way)   *100km* (for all materials)

Box 6a. Mass Burn Incineration of MSW/Restwaste and Sorting Residues.

Mass burn incineration of Restwaste occurs?   *YES* (Insert "1" in box)

Incineration Process:
    Energy Recovery?   *YES* (Insert "1" in box)

Residues:
    Ferrous recovery from ash?   *YES* (Insert "1" in box)
    Transport distance to landfill for non haz. residue   *10km*
    Transport distance to landfill for haz. residue   *20km*

CASE 5a. As Case 5 above, but with only 3 collections per property per week (integrated collection).

Box 2a. Collection of Household Waste
Kerbside Collection Vehicles:
Total No. of collection vehicle visits/property/yr: *156* (3x per week)

CASE 5b. As Case 5, but with a close-to-home bring system for mixed recyclables.

Box 2a. Collection of Household Waste
Materials Collection Banks:
Residents' transport to Materials bank sites:
Car journey length: *0* (Close to home allows delivery on foot)
Amounts collected in mixed material banks: Paper: *93.75*   Glass: *45*   Metal (Fe): *18.75*
Metal (non Fe): *3*   Plastic (Rigid): *15*
(assume 75% of kerbside collected amounts due to lower convenience, and lower participation rate)
Collection and transport to MRF:
Average diesel consumption: *3* litres/tonne collected
Kerbside Collection Vehicles:
Total No. of collection vehicle visits/property/yr: *156* (3x per week - restwaste only)

CASE 5c. As Case 5, but with a central bring system for mixed recyclables.

Box 2a. Collection of Household Waste
Materials Collection Banks:
Residents' transport to Materials bank sites:
Average No. of special trips to site: *12* /hhld/yr (once per month)
Average Car journey length: *2 km* (each way)
Amounts collected in mixed material banks: Paper: *62.5*   Glass: *30*   Metal (Fe): *12.5*
Metal (non Fe): *2*   Plastic (Rigid): *10*
(assume 50% of kerbside collected amounts due to lower convenience, and lower participation rate)
Collection and transport to MRF:
Average diesel consumption: *0* litres/tonne collected (collected centrally)
Kerbside Collection Vehicles:
Total No. of collection vehicle visits/property/yr: *156* (3x per week - restwaste only)

the environmental and economic sustainability of integrated solid waste management systems.

12.7 Operating the IWM-1 LCI spreadsheet

The results box for the LCI spreadsheet is shown in LCI Box 8. Details of the data inserted in the LCI spreadsheet for the options compared in the case study are given in the LCI Example Boxes 1–5.

References

CML (1992) *Environmental Life Cycle Assessment of Products. Backgrounds – 1992*, ed. R. Heijungs. Centrum voor Milieukunde, Leiden.

Finnveden, G., Andersson-Sköld, Y., Samuelsson, M.-O., Zetterberg, L. and Lindfors, L.-G. (1993) *Classification (impact analysis) in connection with life cycle assessments – a preliminary study*. In *Product Life Cycle Assessment – Principles and Methodology*. Nordic Council of Ministers, pp. 172–231.

Hindle, P., White, P.R. and Minion, K. (1993) *Achieving real environmental improvement using value: impact assessment. Long Range Planning*, **26**(3), 36–48.

IPCC (1992) Intergovernmental Panel on Climate Change: 1992 IPCC Supplement. IPCC Secretariat, World Meteorological Organisation, Geneva, Switzerland.

SETAC (1993) *Impact Assessment*. Report of the SETAC workshop in Sandestin, Florida, 1992. Society of Environmental Toxicology and Chemistry – Europe, Brussels.

Postscript

Journey's end! No doubt you have found it difficult at times. We did. We make no apology, but hope you have found the journey as enlightening and rewarding as we have.

During the journey we have looked at many new things. Sometimes we have looked at things we already knew but from a different perspective. As we look back, we would note the following as being especially important.

There will always be solid waste to manage. No matter how creative we are in reducing the amount of solid waste, waste there will still be, and this must be handled in an environmentally and economically sustainable manner. We set out ways in which environmental sustainability can be worked towards and how progress can be measured. We also set out how the costs of waste management can be measured and controlled.

Integrated waste management and lifecycle inventory are tools to help us move towards affordable environmental sustainability (they also allow us to measure our progress). We demonstrate that there are environmental and economic benefits from dealing with all aspects of the waste stream within a single management system. We also demonstrate the value to waste treatment of combining different treatment methods: recycling, biological treatment, thermal treatment and landfilling. Each of these techniques has a valid and valuable role to play in the management system. The choice of which ones to employ will be determined by local considerations of the specific waste stream, technical factors, costs and the will of the public. Theory argues for the use of each of the four basic treatment systems, but practicalities may often preclude this complete approach. The inventory stage of a lifecycle assessment should be used to help ensure that the best environmental option is implemented within the practical constraints faced by a particular locality. The inventory can also be used to highlight and help prioritise operational areas for environmental improvement effort.

Using lifecycle inventory is better than arbitrary approaches. Lifecycle inventory results allow for more objective and transparent discussions, based on facts and data rather than on feeling and beliefs. Arbitrary approaches can be misleading. The hierarchy of waste management is an arbitrary approach, although it has been a useful 'default' in the absence of a true lifecycle approach. We hope that regulators will take care to avoid being prescriptive in arbitrary ways. We welcome the recognition of lifecycle analysis in the European Union common position reached in December 1993 on the proposed Packaging Waste Directive. We fervently hope that this will be maintained as the text is further debated and trust that this book will enable more informed decisions to be taken in the future.

A total quality approach can deliver real environmental and economic improvements. Total Quality Management encourages people constantly to seek improvements by measuring the effectiveness of the current system and designing possible improvements. These can be implemented and checked to ensure that real improvements have indeed been made. From this new point the 'virtuous circle' of improvement starts again. 'Sustainable development' is a societal goal expressed in Total Quality language; i.e. it is a state that may never be fully attained but towards which we constantly strive by continuously improving. It is appropriate to use the tools of total quality management as we journey towards the goal. Continuous improvement must be exercised in both the environmental and economic aspects of solid waste management.

The lifecycle inventory model provided with the book provides a novel, practical tool for studying both environmental and economic aspects of waste management systems. The model deals with a variety of waste materials from the cradle of the dustbin to the grave of ultimate disposal as inert material. Achieving the status of 'inert material' may come after many decades (or even centuries) in a landfill. To date, lifecycle inventories have studied individual materials or products through their lifecycle including the inventory effects of disposal. Using this model, it is possible to study both the environmental and economic aspects of waste management systems themselves.

The environmental lifecycle model only provides the inventory stage of a full lifecycle assessment. It is recognised that there is a need for the development of the lifecycle assessment tool to enable impact analysis and valuation to be carried out. The work of the Society of Environmental Toxicology and Chemistry (SETAC) and the International Organisation for Standardisation (ISO) on the development and standardisation of lifecycle assessment methods will be important in this area. Until such time as this can be done with a broad measure of scientific agreement, we must be content to make decisions based on the inventory alone. The means for carrying out an LCI are broadly agreed within the scientific community and are followed in the model.

Partly because the lifecycle assessment is incomplete when only the inventory is done, but mainly because of the quality of available data, the conclusions drawn through use of the model must be interpreted with reasonable caution. Nevertheless, it is far better to take a decision with an imperfect understanding of the environmental effects than to simply make a best guess.

The environmental benefit (or cost) of recycling can be estimated using the model. It is clearly shown that the processing of recovered materials in order to recycle them is outside the boundary of the waste management LCI of this book, but it is included as an optional add-on module. It is possible, therefore, to compare the overall impacts of a waste management system that incorporates recycling, with the alternative where all waste material is incinerated and/or landfilled, and all new products are made from virgin materials. This will give the overall benefit, or cost, of recycling in general. Using this approach, it is possible to choose between the use of virgin or recycled material for a particular product, on an environmental basis. Only if the analysis supports the use of recycled material is there an environmental reason for developing markets for recycled material. Such 'end-use' markets should then be the reason for starting collection schemes for recyclables.

Similarly, for individual products or packaging, the decision as to whether to design them for recycling or disposal will need to be made on the basis of the full lifecycle, from the sourcing of the raw materials, not just on the part of the lifecycle that it spends as solid waste. Such LCIs of individual packages or products are complementary to the LCI of solid waste developed in this book. Individual product LCIs need data on waste management; the model developed here can be used as a source of such information.

More and higher quality data are needed, not for their own sake but in order to make better decisions. This is the cry of everyone who attempts an LCI analysis. Throughout the book we critically review the available data and choose what we consider to be the most representative as the default data in the model. There is no area in which the available data can be considered to be sufficient. The obvious incompatibility in the available statistics on solid waste amounts in different Member States of the European Union is just one among many cases that can be cited.

The importance of getting the basic data right is underlined by the marked differences in the composition of municipal solid waste (MSW) in Northern Europe compared with Southern Europe. MSW consists of a diverse range of materials. Two major fractions in all countries are paper/board and food/garden waste. Plastics, glass and metals occur at a lower level. Southern European countries have higher levels of food/garden waste than northern countries.

In looking at the waste management options for a particular locality, it may well be that more precise data are available. The model allows such data to be used and we encourage such practice.

For a particular system, or geographical area, there may be much more precise environmental data about some or all aspects of the operation. Using such data within the model can be helpful in identifying areas for improvement. The opportunity to reduce energy usage by moving from a two truck to a one truck collection system for household waste collection involving recyclables is one obvious example. The LCI model confirms the intuitive view that this is likely to be an important improvement within the complete LCI of the waste. Important? Yes. But how many schemes have acted on intuition and developed single vehicle operations? Have the few who have done so been able to ensure that they have not introduced deleterious effects in another part of the overall system? Such considerations can be now be handled, based on data. This is a pre-requisite for a Total Quality improvement process.

Material-specific inventory data needs to be developed. This is very important when considering materials recycling, thermal and biological treatment, and landfilling. We do not expect full precision, but it would be very helpful to have a clearer understanding of the contribution of individual materials to the emissions of a particular process. This would help determine the best treatment options for individual materials. With plastics, for example, which recycling targets make sense from an environmental point of view? Where is the break-even point at which thermal treatment with energy recovery becomes more environmentally sustainable?

It is not only in the environmental area that more and higher quality data are needed. Comparing costs between systems is extremely difficult. Nowhere is this more marked than in collection and sorting. We welcome the work being done by ERRA to implement standard accounting procedures in their various demonstration projects so that like can be compared with like.

A variety of waste management systems are required to meet local needs. We are convinced that there is no single, 'best' system that every locality should have. The varieties of waste compositions, housing and street structures alone are sufficient to rule out any chance of finding a practical, single approach. The hopelessness of such a search is underlined by considerations such as existing infrastructure, local availability of markets and differing public expectations and habits. Variety can be found in all aspects of integrated waste management:

1. *Collection*: The work of ERRA clearly shows the variety of methods that can be used to collect materials for recycling. The blue box collection scheme of Adur, England simply would not work in Barcelona or Prato. Similarly, the English suburban householder is unlikely

to accept a variety of rubbish and recycling bins every couple of hundred metres along the street. Similarly, the two-weekly collection of putrescible waste as practised in Lemsterland would be totally unacceptable in Barcelona where waste has to be collected daily, especially in summer.

2. *Sorting*: The design of a materials recovery facility (MRF) will be determined in large measure by the design of the scheme and the treatment methods chosen for the waste. With a degree of sorting and, hence, control at the point of collection (as in Adur), the MRF does not need to be designed to deal with as much unacceptable material as does the MRF in Barcelona. Similarly the sorting prior to producing refuse derived fuel is very different from the sorting prior to composting which is very different from the sorting in a MRF. Nevertheless, it is possible to generate recyclable materials in all three operations.

3. *Recycling*: The choice of materials collected and sorted for recycling should be guided (indeed controlled) by considerations both of available markets and likely overall environmental benefit. The Packaging Producer Responsibility Plan produced by industry for government in the UK (PRG, 1994) makes this point well. To take just one example: the available capacity for paper packaging recycling in the UK can be filled by material collected from commercial premises. This is economically and environmentally the best way to fill the capacity. It is cheaper to collect the material from such sources where it is already sorted and where the energy expenditure per tonne collected will be lower than from households. Further, the material is cleaner which leads to lower environmental effects when re-processing, as well as to lower costs.

4. *Biological treatment*: There are many ways in which this can be accomplished. The two basic processes (aerobic and anaerobic) can each be broken down into a variety of different approaches involving differences in plant technologies and feed-stock. At its very simplest, kitchen food waste and garden rubbish can be composted in the garden. (Some local authorities regard this activity as source reduction and are prepared to provide supporting funding.) At its most complex, a biogasification plant is a sophisticated piece of engineering operated as a factory producing both compost and energy.

5. *Thermal treatment*: Some thermal treatment is done simply to render the waste inert and reduce its volume. Other methods recover energy with the production of inert, volume-reduced waste as a welcome, additional benefit. Producing energy is eminently environmentally sensible. However, if a locality already has a thermal treatment facility that does not produce energy, it would be a nonsense not to use it, all other factors (especially emissions) being up to standard, simply

because it did not yield captured energy. Energy production does not necessarily mean mass burn thermal treatment. There are opportunities to increase energy yield by burning high calorific solid waste that has been sorted. There are opportunities to use high calorific solid waste as an alternate fuel to coal, oil or natural gas.

6. *Energy from waste*: There is a temptation to use this phrase simply as a euphemism for 'mass burn incineration'. As the IWM model shows, energy can be derived from waste in a variety of ways: anaerobic biological treatment; thermal treatment using different techniques; from recovered landfill gas.

People are important to the effective running of solid waste management systems. This aspect is only lightly touched upon in the book, but it is central to the concepts of Total Quality and of a 'management system'. An integrated waste management system must be set up to enable each person involved in the system to meet the needs of their customers and to have their needs as customers met by their suppliers. Thus, for example householders need to know clearly and exactly how they should present their waste materials to the collector. What they are asked to do should be simple and understandable. The Barcelona project of ERRA clearly demonstrates the vital importance of this principle. Similarly operators in thermal treatment or aerobic biological treatment plants should know exactly what to do in order to ensure that the emissions are properly monitored and controlled. Further, they should be encouraged to find and test out new ways of operating that will lead to lower emissions.

To obtain real environmental improvements in the management of solid waste, it is essential to get the economics right. Responsible economic management will seek to ensure that economic costs are minimised within a waste management system that seeks to deliver a real environmental improvement. Reducing costs will often lead directly to environmental improvements; energy savings is a clear example. It will also lead to the further development of markets for recycled materials and for compost. It is therefore vital that environmental effects and economics be considered together. The model in this book does this.

Clearly it is not responsible management when costs are reduced by taking short cuts that are clearly and significantly detrimental to the environment. Legislation is required to ensure that such criminal practice can be punished. It is important, however, that legislation in the waste management area does not become so detailed that it risks inhibiting innovation nor that it requires certain practices or targets to be met that cannot be justified on lifecycle environmental considerations.

Partnerships are required to make real progress. Effective partnerships will have people with different skills and responsibilities working together against the shared goals of environmental and economic sustainability. Manufacturers must find ways to design waste out of their products and to use the quality products produced from waste; householders must act responsibly by dealing with their waste in ways that aid its collection, sorting and subsequent treatment; local authorities must implement householder-friendly systems that yield affordable environmental improvements; waste management companies must ensure that they operate with environmental and economic efficiency; regulators must ensure that their regulations support on-going development and do not dictate ineffective approaches; opinion leaders must ensure that they fully consider the facts and the creative possibilities before they start to lead public opinion.

A lifecycle inventory of solid waste provides a means to consider the environmental and economic consequences of change before change is initiated. Integrated waste management provides a holistic approach to managing household and commercial waste against the goals of environmental and economic sustainability. A Total Quality approach to management provides the framework within which people can work together towards making the environmental and economic sustainability of solid waste management a reality.

As always, when one journey ends another begins. . . .

Reference

PRG (1994) *Real value from packaging waste – a way forward*. Report of the Producer Responsibility Industry Group, Villiers House, 41–47 The Strand, London, 38 pp.

Appendix 1 – Secondary materials nomenclature

Household waste is disposed of in a variety of ways. The four options are:

- Recycle   specified materials
- Compost   specified paper (no laminates); kitchen and garden waste
- Incinerate   unspecified refuse; derived fuel
- Landfill   unspecified

The ERRA Secondary Materials Nomenclature attempts to define the possible recycling and composting streams (Figures A1.1–A1.4) in a household waste management programme, breaking them down into the various materials for which end-markets or uses exist, or are anticipated in the near future.

A standardised terminology is essential if recycling and composting programmes are to be analysed and compared. Many programmes, for example, may sort and market 'plastics', but the exact nature of the materials included may vary from PET soft drink bottles only to a range of rigid bottles, tubs and films made from different resins.

The structure of the ERRA Secondary Materials Nomenclature is hierarchical with the first level referring to a material group such as plastics, glass or paper and the lowest level referring to a specific type of material. As far as possible, these reflect existing end-markets, known trends in technology etc. The hierarchical structure means that new categories can be added as they are identified. It is thus end-market driven. To ensure accuracy each material hierarchy has also been reviewed with assistance from several of the European Material Associations.

The structure is driven by end-markets and needs flexibility to expand as end-markets grow. However new end-markets do not necessarily mean modification to the structure.

Reproduced with permission of the European Recovery and Recycling Association (ERRA).

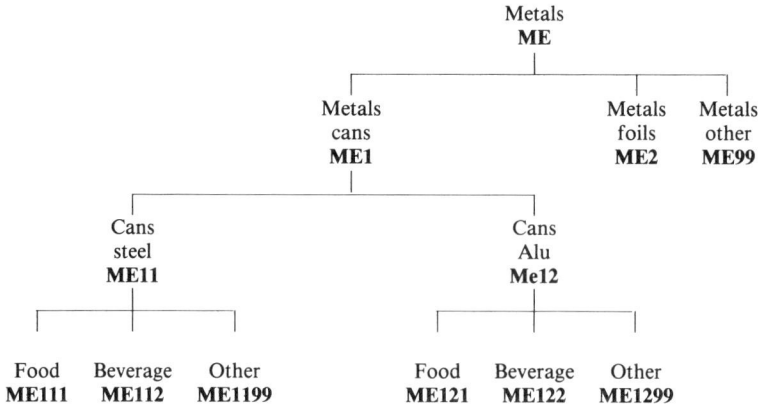

Figure A1.1 Metals.

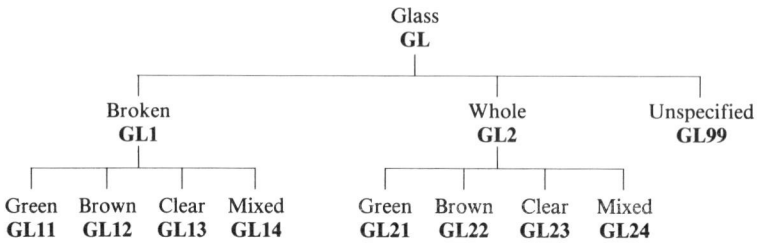

Figure A1.2 Glass (excluding flat glass, light bulbs, Pyrex, etc.).

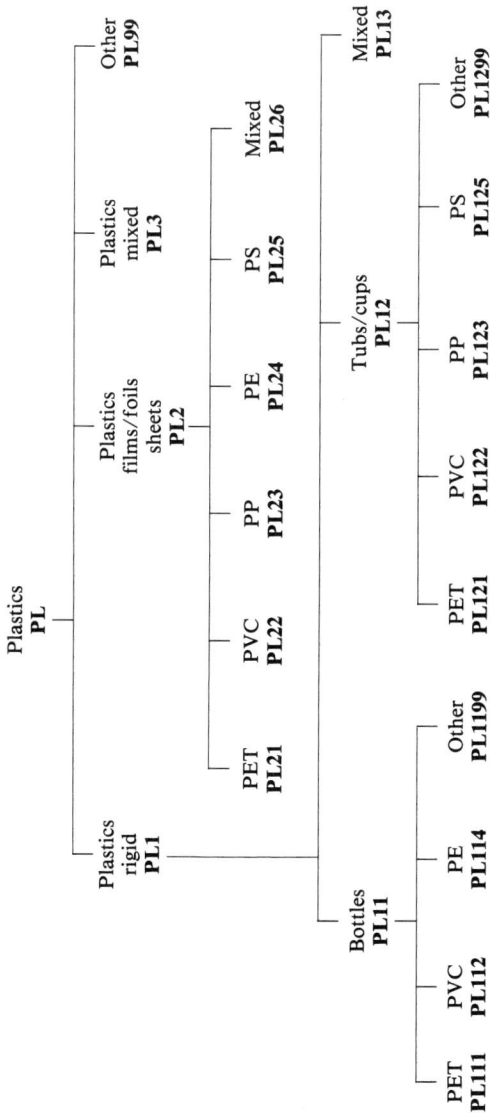

Figure A1.3 Plastics.

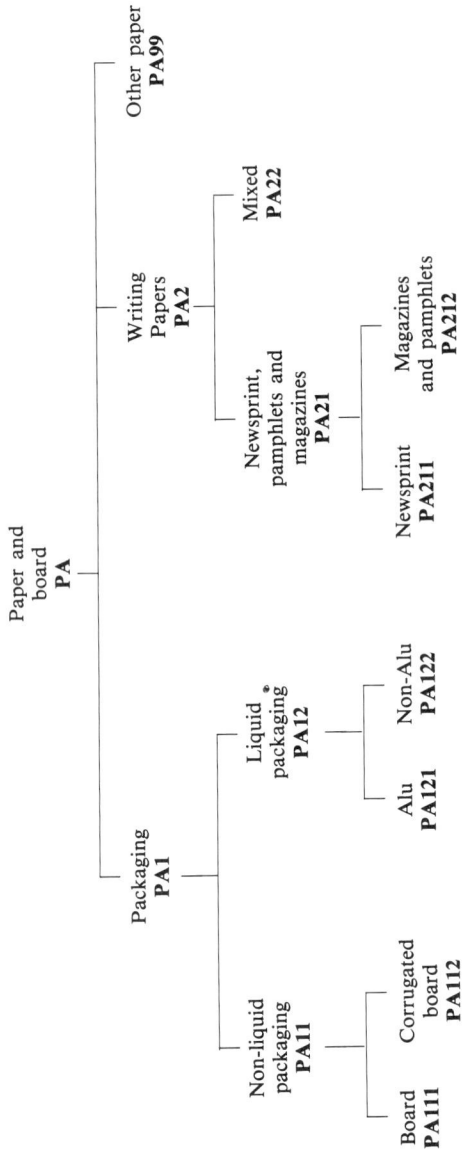

Figure A1.4 Paper.

Appendix 2 – Waste analysis procedure

The waste stream generated by any household may be disposed of in a variety of different ways, depending on local circumstances and customs. In most predominantly residential areas, collection from the property is the most common practice, but in many places this is supplemented by the provision of a central depository to which excessive or bulky items can be taken. Many households will also have access to Glass Bottle Banks and other similar facilities through which recyclable materials can be recovered and kept away from final disposal.

Whilst household waste is usually a very significant part of a community's waste stream, it is not usually the largest contributor to society's waste problem. The waste generated by agriculture, industry and commerce is many times greater although its composition is normally different to that of the domestic waste stream.

The focus of this analysis procedure is the domestic waste stream and in particular on that portion which can usefully be separated for recovery and recycling. The aim is to offer a method for the collection of quantitative and qualitative data on household waste.

Household waste recovery programmes

In order that the results from a waste analysis may be applied to the whole of the community to which it pertains, it is essential that the sample taken and analysed is truly representative of the community. The selection of the ideal sample is therefore a local issue and will depend on:

- number of households in the area
- type of housing
- social background
- type of refuse collection system

In many cases, this demographic information is available from proprietary databases such as the ACORN programme from CACI. By using such data to give an overall view of the community, a group of properties, streets, districts or refuse collection routes can be selected to represent the wider area. For example, if a community comprises 25% of 'type A' households, 30% of 'type B' and 45% of 'type C', then the waste collected from a single refuse collection route which also comprises 25% 'type A', 30% 'type B'

Table A Size of sample

Number of households involved	Minimum sample size (households)
Less than 1000	10% or 50*
1000–9999	5.0% or 100*
10 000–49 999	2.5% or 500*
More than 50 000	1.0% or 1250*

* whichever is greater

Table B Weight of sample

Number of households involved	Minimum sample weight (kg)
Less than 1000	500
1000–9999	1000
10 000–49 999	5000
More than 50 000	12 500

and 45% of 'type C' households may be considered as representative of the community.

The ideal period in which to conduct a waste analysis programme is during the 12 months prior to the start up of a recovery project. As waste composition may be subject to seasonal variation, an analysis should ideally be carried out at three-monthly intervals to give the annual picture.

Where budgetary or operational factors do not allow this, a minimum of two analyses, six months apart, should be undertaken. e.g. mid-summer and mid-winter.

With projects already in operation, or where the projects will be operational in less than 12 months, the latter programme should be adopted. This will be the case with the majority of ERRA projects as many have been running for some time.

The size of the sample to be taken is dependent on the number of properties involved and the degree of confidence required of the results of the analysis. Further guidance can be obtained from statistical calculations but for the most part, a simplified and less costly sample can be taken as shown in Table A.

Assuming that the average household produces at least 10 kg of refuse a week and that it is collected on a weekly basis, following the above sampling procedure will produce the amounts of material shown in Table B.

These then are the samples on which the analysis will be performed. However, as the actual classification of materials will be carried out manually, even the smallest of these samples represents a mammoth task.

It is therefore necessary to reduce the size of the sample which will be classified to a more manageable level of between 100 and 200 kg. This must nevertheless be done with due regard to maintaining the accuracy of the sample.

There are several ways of effecting this – some very simple and others requiring a high degree of potentially expensive mechanisation. The method set out in *Coning and quartering* below requires only the minimum of equipment but may take some time to perform, especially with the larger samples, due to the number of times the procedure will require repeating.

Once the sample has been reduced to a manageable 100–200 kg, the analysis can begin.

Firstly, the sample should be screened through a 20 mm mesh sieve so that any pieces of less than 20 mm are removed and weighed separately. These are classified as 'fines'. The balance of the material should then be sorted manually using a classsification system. Any materials which cannot be allocated to one of these categories should be designated as one of:

- miscellaneous combustibles (MC)
- miscellaneous non-combustibles (MNC)

The weight of each category of material should be recorded along with the other information.

The moisture content and bulk density evaluation should be carried out on the surplus 'two quarters' retained from the final reduction operation referred to above. Details of the test methods are given in the last section.

'Coning and quartering' sampling technique

1. Sample dropped onto floor area and mixed by mechanical shovel.
2. Sample collated into uniform pile of approximately 0.8 m high.
3. Pile divided into two by a straight line through the centre of the stack.
4. Pile further divided by a second line (roughly) perpendicular to the first.
5. Either pair of opposite quarters removed to leave half of original sample.

Process is repeated until the required sample size is achieved.

Determination of moisture content and bulk density

These optional evaluations may be carried out on the surplus 'two quarters' from the final sample size reduction operation. These should be combined and a further reduction performed. Each pair of opposite quarters should then be combined and used as two separate samples, A and B.

Sample A: Moisture content

1. Weigh the sample to an accuracy of not less than 0.1 kg. Record this weight as W1.
2. Spread the sample over a number of trays so that the height of the sample is not more than 250 mm.
3. Place in a fan assisted oven at $90° \pm 2°C$ for 24 hours.
4. Allow the sample to cool to room temperature and re-weigh. Record this weight as W2.
5. The moisture content can be calculated as:

$$\% H_2O = \frac{(W1 - W2) \times 100}{W1}$$

Sample B: Bulk density

1. Weigh a container of known volume, not less than 200 litres or more than 300 litres. Record this weight as W1.
2. Pour the sample into the container until it is overflowing.
3. Settle the contents of the container by dropping it three times from a height of 100 mm.
4. Top up the container volume with the remainder of the sample. If insufficient material is available, the unoccupied container volume (V2) should be measured.
5. Weight the container and its contents. Record this weight as W2.

For a filled container, the bulk density is calculated as:

$$\text{Bulk density} = \frac{W2 - W1}{V1} \text{ in kg/litre}$$

For a partially filled container, the bulk density is calculated as:

$$\text{Bulk density} = \frac{W2 - W1}{V1 - V2} \text{ in kg/litre}$$

Reproduced with permission of the European Recovery and Recycling Association (ERRA).

Appendix 3 – Programme ratios

A number of ratios have been developed for use in analysing recycling programmes; terms such as diversion rate and recovery rate are widely used but their meaning has often been unclear and different programmes and jurisdictions have applied the measurements in different ways.

A number of key measurements have been identified to monitor progress against legislative targets and to support the management of specific programmes. These have been defined precisely so that programmes can be compared.

The ratios have been divided into two groups:

Group 1: Area specific. Applied to a defined geographical area or jurisdiction and reflecting legislative requirements. For example, measurements for progress towards national targets.

Each ratio is to be applied to a specific waste stream and at a specified point in the processing system (see Table 1). Most of the ratios are to be applied to the household waste stream. However, the concept can be applied to any other defined waste stream.

Table 1 Area specific

Ratio: Waste stream	Definition	Point of application
Diversion rate Household waste	$\dfrac{\text{Amount of material recovered from generators served}}{\text{Total amount of available waste from generators served}}$	End of final process prior to sale to end-market (e.g. after MRF)
Packaging recovery rate Packaging waste	$\dfrac{\text{Amount of packaging material recovered from generators served}}{\text{Total amount of packaging material available in the waste stream of the generators served}}$	End of final process prior to sale to end-market
Packaging recycling rate Packaging waste	$\dfrac{\text{Amount of packaging material recycled from generators served}}{\text{Total amount of packaging material available in the waste stream of the generators served}}$	End of final process prior to sale to end-market

Table 2.1 Process specific – Consumer participation process

Ratio: Waste stream	Definition	Point of application
Participation rate Household waste	Number of generators participating at least once in a four-week period — Total number of generators served by the programme in a four-week period	Before collection
Set-out rate Household waste	Number of generators putting out on collection day — Number of generators served by the programme on that day	Before collection
Capture rate Household waste	Amount of a specific material collected per participating generator — Total amount of the specific material available in the waste stream of the participating generator served	After collection/ pre MRF
Collection recovery rate Household waste	Amount of a specific material collected from generators served — Total amount of the specific material available in the waste stream of generators served	After collection/ pre MRF

Table 2.2 Process specific – Sorting process

Ratio: Waste stream	Definition	Point of application
Recovery rate Household waste	Amount of a specific material recovered from generators served — Total amount of the specific material available in the waste stream of the generators served	End of final process prior to sale to end-market
Recycling rate Household waste	Amount of a specific material recycled from generators served — Total amount of the specific material available in the waste stream of the generators served	End of final process prior to sale to end-market
End-market ratio Household waste	Amount of a specific material sold from MRF — Amount of the specific material recovered from MRF	End of final process prior to sale to end-market
Residue ratio Household waste	Amount of material sent to final disposal from MRF — Amount of material received at MRF	End of final process prior to sale to end-market

For each ratio a point of application has also been defined. For example, a diversion rate is to be measured at the end of the final sorting process (end of Material Recovery Facility processing) prior to sale to the end-market.

Group 2: Process specific. Management tools designed to assess the efficiency of the programme. These ratios are further subdivided into ratios that apply to: the consumer participation process (see Table 2.1); and the sorting process (see Table 2.2).

Reproduced with permission of the European Recovery and Recycling Association (ERRA).

Appendix 4 – Terms and definitions

Bring collection system
Related term:

Householder required to take recoverables to one of a number of collection points.

Drop-off
Related term:

Synonym for bring collection system.

Close to home

Applies to any bring collection system that locates containers within an intended walking distance of the households served.

Contamination

Misplaced materials:
1. non-targeted or dirty materials set out by the householder
2. failure of the collection and processing systems to
   a) maintain separation or
   b) effectively sort materials to required specifications.

Final disposal

Landfill or incineration without energy recovery.

Generator

Household, commercial outlet, institution etc.

Kerbside collection system

Householder places recoverables in container/bag which s/he positions, on a specific collection day, in the immediate vicinity of the property.

Related term:

Door to door

Synonym for kerbside collection system.

MRF

Materials Recovery Facility – a place where collection recyclables are delivered for processing before being sold.

Multi-material

Programmes collecting more than one specified material using a coherent collection system.

Packaging material

All products made of any materials of any nature to be used for the containment, protection, handling, delivery and presentation of goods, from raw materials to processed goods, from the producer to the user or the consumer. Disposables used for the same purpose are to be considered as packaging too. (EC proposed Packaging Directive.)

Participating	Contributing some or all requested materials to recovery programme.
Passbys	The number of households/generators a vehicle is required to drive by in a day – i.e. the total number of potential pick-up points in a kerbside collection programme. NB: Applies to kerbside collection only.
Processing	The upgrading of materials at a MRF or other facility to meet a market specification. Upgrading operations include, for example, sorting, densification, shredding etc.
Recovered	Materials recycled, composted, regenerated or whose energy value can be used as an energy source (as defined in Annex IIB to 75/442/EEC). Equivalent to 'valorisation'.
Recyclables	Post-use materials that can be recycled 'for the original purpose or for other purposes, but excluding energy recovery' (EC Directive – recycling).
Recycled	Recovered for the original purpose or for other purposes, but excluding energy recovery.
Separation	A physical and conscious holding apart of individual material groups from each other.
Served	Offered facilities of a recovery programme, e.g. kerbside collection, bring system.
Sold	Positive or negative value possible.
Sorting	Handling of the material in order to separate it from other materials. Sorting may occur at the household (by the resident), at the point of collection (by the resident or by collection staff) and/or at the MRF.
Targeted materials	Items/materials that a recovery programme has requested the householder to separate from household waste.
Valorisation	Transformation of material which extracts value from processed recoverables.
Waste stream	Household waste or packaging waste as indicated on each ratio.

Reproduced with permission of the European Recovery and Recycling Association (ERRA).

Appendix 5 – Currency conversion values

Values as at 8th March 1994.

Country/currency	Amount for 1 ecu
BELGIUM/LUX franc	39.85
DENMARK krone	7.56
GERMANY mark	1.94
GREECE drachma	280.4
SPAIN peseta	159.3
FRANCE franc	6.58
IRELAND pound	0.79
ITALY lira	1907.1
NETHERLANDS guilder	2.17
PORTUGAL escudo	199.1
UK pound sterling	0.76
USA dollar	1.13
CANADA dollar	1.53
JAPAN yen	118.8
SWITZERLAND franc	1.62
NORWAY krone	8.39
SWEDEN krona	9.03
FINLAND markka	6.25
AUSTRIA schilling	13.6
ICELAND krona	82.1

Source: Official Journal of the European Communities, 8/3/94.

Appendix 6 – How to use the IWM LCI spreadsheet

1. Load IWM-1 LCI spreadsheet from the diskette, using either Excel 4.0 or Lotus 123 v.2.01 software. (Later versions can also be used)

2. The spreadsheet will appear as a series of vertical boxes. These boxes are shown at the end of Chapters 5–11 as the LCI boxes.

3. The user needs to go through boxes 1–7, filling in the required data in the shaded cells (in the Excel version) or the coloured cells (in the Lotus 123 version)

Instructions for inserting the data are given in the spreadsheet. More detailed instructions can be found in the section at the end of the relevant book chapter (Chapters 5–11). Note that only shaded or coloured cells need to be completed. Other values will be calculated and inserted automatically.

4. Data can only be inserted in the shaded/coloured cells. The other cells are locked to prevent them being altered accidentally.

5. Results from the spreadsheet are shown in the last of the boxes, LCI Box 8, which is found at the bottom left-hand corner of the spreadsheet.

6. The fixed data used in the spreadsheet and the method of calculation can be found in the columns to the right of the input boxes. Users of the Excel version can find the relevant data tables by using the 'Go To' command.

The authors would welcome any comments about the spreadsheet, in writing, to:

Dr. P. White, Procter & Gamble, Newcastle Technical Centre, Whitley Rd, Longbenton, Newcastle upon Tyne, NE12 9TS, UK.

Index—Figures

Index—Tables

Index—Boxes

Index—LCI Boxes

Index—LCI Data Boxes

Index—Subjects

Figures in **bold type**, Tables in *italic* and Boxes in ***bold italics***

Index—Authors cited

Note where there are more than two co-authors, the first author alone is listed.